Computational Atomic Structure

Computational Atomic Structure

An MCHF Approach

Charlotte Froese Fischer

Vanderbilt University

Tomas Brage

Lund University

Per Jönsson

Lund University

CRC Press
Taylor & Francis Group
Boca Raton London New York

CRC Press is an imprint of the
Taylor & Francis Group, an **informa** business

CRC Press
Taylor & Francis Group
6000 Broken Sound Parkway NW, Suite 300
Boca Raton, FL 33487-2742

© 1997 by Taylor & Francis Group, LLC
CRC Press is an imprint of Taylor & Francis Group, an Informa business

Visit the Taylor & Francis Web site at
http://www.taylorandfrancis.com

and the CRC Press Web site at
http://www.crcpress.com

Contents

Preface

The last decade has seen tremendous advancements in computer technology, not only in computational speed but also in memory and storage resources. Calculations which twenty years ago could only be done on a main-frame computer, can now be performed easily on many personal computers. This book, *Computational Atomic Structure*, was written partly in response to this change.

Our book differs from other books on atomic theory or quantum mechanics in that it deals not only with theory but also the computational aspects of actually determining an atomic property. An important goal of this book is to review the atomic structure theory needed for the computation of atomic properties using programs from the MCHF ATOMIC STRUCTURE PACKAGE that are available from a number of ftp sites. Such calculations can be at a simple, Hartree–Fock level or at a more advanced, multiconfiguration Hartree–Fock level. Another goal is to explain the computational procedures that are used. Included in the book are sample calculations, performed in an interactive mode, to show what input is expected from the user, and sections of screen output. The latter serves two purposes—it gives the reader an idea of the information that can be obtained and allows the person installing the system to confirm proper execution of the code. In the interest of space, only selected output is provided. Complete output will be provided with the code where examples too complex for a book will be presented. Finally, it was anticipated that some users may wish to modify the programs. To encourage such changes, we deal briefly with the computational aspects and outline some of the numerical procedures.

In the MCHF approach, the first step is always to find an approximate wave function for bound states of an atomic system. An energy criterion is used for this purpose and several chapters are devoted to the study of factors affecting energy level structures. Once a wave function has been determined, a number of atomic properties can be computed. The chapter on transition probabilities is the 'grand finale' that ties everything together. However, the book does not stop there. It includes a brief excursion into the theory of continuum processes, just enough to establish a bridge between these two areas.

With the book and the code it should be possible to explore many atomic properties. An experimental approach is recommended—some of these experiments will be more accurate than others. The goal should always be an understanding of the important effects. However, the present code should not be viewed as a 'state-of-the-art' code. Its development started in the mid 1960s and much could be done to improve it. Indeed, for large-scale calculations, the authors have already modified this version for sparse-matrix methods and dynamic memory allocation. The resulting code, however, is not standard whereas the present version, with at most minor modifications, will compile on both PCs and workstations. It is, in a sense, a small-case version where more intermediate data can be printed, making it more suitable as a learning tool. Missing entirely is a graphical interface, partly because of the lack of a standard. Such an interface could be a valuable contribution.

The code will be available from several sites:

(i)   From the Internet URL `http://kurslab.fysik.lu.se/casbook.html` which will contain further instructions.

(ii) From the ftp site `fusion.vuse.vanderbilt.edu` through anonymous ftp. After login, the user should `cd pub/cas`

(iii) From IOP at `ftp://ftp.iop.org/pub/books/software/cas`

At all of these sites we will make the source code available; this needs a FORTRAN77 compiler to generate executables. The README file provides further instructions and should be downloaded first. Linux executables also are available for sites without a compiler.

These programs have been tested on many platforms and applied to many atomic systems. But, as research software, they cannot be guaranteed for any particular purpose. Neither the authors nor the publisher offer any warranties, nor do they accept any liabilities with respect to the programs and their application.

Your comments and suggestions are welcome. Please send electronic mail to `cff@vuse.vanderbilt.edu`, `brage@kurslab.fysik.lu.se` or `pj@speedy.te.mah.se`.

Charlotte Froese Fischer
Tomas Brage
Per Jönsson

Acknowledgments

This book owes a debt of gratitude to many for their support and assistance. The code itself relies heavily on contributions from both Alan Hibbert and Michel Godefroid. Others too have contributed, and are mentioned in the documentation of the programs. We are grateful to all these people.

Chapter 1

Introduction

1.1 Introduction

According to quantum mechanics (Messiah 1965) a stationary state of an N-electron atom is described by a wave function $\psi(q_1, \ldots, q_N)$, where $q_i = (r_i, \sigma_i)$ represents the space and spin co-ordinates† of the electron labelled i. The wave function is assumed to be continuous with respect to the space variables and is a solution to the wave equation

$$\mathcal{H}\psi(q_1, \ldots, q_N) = E\psi(q_1, \ldots, q_N), \tag{1.1}$$

where $\mathcal{H}$ is the Hamiltonian operator for the atomic system. The wave equation is an eigenvalue problem, and solutions exist only for certain values of E. These values are known as the eigenvalues of the operator, and they represent the possible values of the total energy of the system. Following the mathematical terminology, the set of all eigenvalues is known as the eigenvalue spectrum of the operator.

The operator $\mathcal{H}$ depends on the atomic system as well as on the underlying quantum mechanical formalism. For non-relativistic calculations, the normal starting point is Schrödinger's equation where the Hamiltonian, in atomic units (see appendix C) is given as

$$\mathcal{H} = \sum_{i=1}^{N} \left(-\frac{1}{2}\nabla_i^2 - \frac{Z}{r_i} \right) + \sum_{i>j}^{N} \frac{1}{r_{ij}}. \tag{1.2}$$

Here Z is the nuclear charge of the atom, r_i is the distance of electron i from the nucleus and r_{ij} is the distance between electron i and electron j. The above Hamiltonian is valid under the assumptions that relativistic effects can be neglected and that the atomic nucleus can be treated as a point charge of infinite mass. Later, we shall include the relativistic and higher-order nuclear effects as small corrections.

† For a brief discussion about the electron spin, see appendix A.1.2.

1

1.2 Properties of the wave function

Some important properties of the wave function can be obtained from general principles and symmetry considerations.

1.2.1 Normalization

The Hamiltonian operator (1.2) has both a discrete and a continuous spectrum. Wave functions, or eigenfunctions, belonging to the discrete spectrum are square integrable and represent bound states. Often these eigenfunctions are assumed to be normalized, i.e.

$$\int_q |\psi(q_1, \ldots, q_N)|^2 \, dq_1 \ldots dq_N \equiv \langle \psi | \psi \rangle = 1, \qquad (1.3)$$

where the integral sign means integration over all space co-ordinates and summation over all spin co-ordinates. For normalized wave functions,

$$|\psi(q_1, \ldots, q_N)|^2 dq_1 \ldots dq_N \qquad (1.4)$$

can be interpreted as the probability of finding the electrons in the generalized volume element $dq_1 \ldots dq_N$ centred at $q_1, \ldots, q_N$ (Messiah 1965, chapter 4).

Eigenfunctions belonging to the continuous spectrum have infinite square integrals and describe systems with one or more free electrons. These eigenfunctions, often known as continuum functions, may be labelled by one or more continuous variables k in addition to a set of discrete quantum numbers. Although the continuum functions have infinite square integrals they can be normalized in the Dirac sense with respect to k (see Messiah 1965, chapter 5). We will come back to this in section 1.3.2 in connection with continuum functions for hydrogen.

1.2.2 Antisymmetry

The electrons in an atom are all identical, i.e. there is no property that can be used to distinguish or label them, and thus the Hamiltonian operator must be invariant with respect to any permutation of the electron co-ordinates. As a consequence, if $\psi(q_1, \ldots, q_N)$ is an eigenfunction of $\mathcal{H}$ with eigenvalue E, so also is

$$\wp_{ij} \psi(q_1, \ldots, q_i, \ldots, q_j, \ldots, q_N) = \psi(q_1, \ldots, q_j, \ldots, q_i, \ldots, q_N), \qquad (1.5)$$

where $\wp_{ij}$ is the operator that permutes the electron co-ordinates q_i and q_j. Thus, the general eigenfunction of $\mathcal{H}$ with eigenvalue E can be obtained as a linear combination of functions for which the co-ordinates have been permuted. It is, however, an experimental fact that only linear combinations that result in completely antisymmetric eigenfunctions give a correct description of the atomic system (Messiah 1965, chapter 14). One possbile representation for such an antisymmetric eigenfunction is given by the linear combination

$$\frac{1}{\sqrt{N!}} \sum_{\wp} (-1)^p \wp \psi(q_1, \ldots, q_N) \tag{1.6}$$

where p is the parity of the permutation and the sum $\wp$ is over all $N!$ permutations. For future use, we introduce the antisymmetrization operator $\mathcal{A}$ defined by

$$\mathcal{A} = \frac{1}{\sqrt{N!}} \sum_{\wp} (-1)^p \wp. \tag{1.7}$$

As an example of the action of this operator, we consider a three-electron system for which we have

$$\mathcal{A}\psi(q_1, q_2, q_3) = \frac{1}{\sqrt{3!}} [\psi(q_1, q_2, q_3) - \psi(q_1, q_3, q_2) \\ + \psi(q_2, q_3, q_1) - \psi(q_2, q_1, q_3) + \psi(q_3, q_1, q_2) - \psi(q_3, q_2, q_1)]. \tag{1.8}$$

An antisymmetric wave function is, as is easily verified, identically zero when two electrons with the same spin occupy the same position in space. Due to the continuity of the wave function with respect to the spatial variables, it follows that the absolute value of the wave function is small whenever two electrons with the same spin are close together.

1.2.3 Angular properties

The non-relativistic Hamiltonian commutes with the total orbital angular momentum operator $L = \sum_{i=1}^{N} l_i$ as well as with the total spin angular momentum operator $S = \sum_{i=1}^{N} s_i$, i.e.

$$[\mathcal{H}, L] = [\mathcal{H}, S] = 0 \tag{1.9}$$

and thus $\mathcal{H}, L^2, L_z, S^2$ and S_z define a set of mutually commuting operators. This implies that simultaneous eigenfunctions of all these operators exist (Messiah 1965, chapter 5), and we have

$$\begin{aligned}
\mathcal{H}\psi(q_1, \ldots, q_N) &= E\psi(q_1, \ldots, q_N) \\
L^2\psi(q_1, \ldots, q_N) &= L(L+1)\psi(q_1, \ldots, q_N) \\
L_z\psi(q_1, \ldots, q_N) &= M_L\psi(q_1, \ldots, q_N) \\
S^2\psi(q_1, \ldots, q_N) &= S(S+1)\psi(q_1, \ldots, q_N) \\
S_z\psi(q_1, \ldots, q_N) &= M_S\psi(q_1, \ldots, q_N).
\end{aligned} \tag{1.10}$$

These simultaneous eigenfunctions will be denoted $\psi(\gamma L M_L S M_S; q_1, \ldots, q_N)$ where γ represents additional quantum numbers needed to completely specify the state.

1.2.4   Parity

In addition to the angular momentum quantum numbers LM_LSM_S, the eigenfunctions of the Hamiltonian can be denoted by their parity. The parity operator Π is defined by the relation

$$\Pi\psi(r_1, \sigma_1, \ldots, r_N, \sigma_N) = \psi(-r_1, \sigma_1, \ldots, -r_N, \sigma_N). \qquad (1.11)$$

From the definition it is clear that $\Pi^2 = 1$ and so the eigenvalues of Π are $\pi = \pm 1$. The parity operator commutes with the Hamiltonian and the angular momentum operator, and hence the atomic eigenfunctions may be taken also as eigenfunctions of Π. Eigenfunctions belonging to the eigenvalues $+1$ and -1 of the parity operator are said, respectively, to be *even* and *odd*.

1.3   One-electron systems

One-electron systems are fundamental in that they are the only atomic systems for which Schrödinger's equation can be solved exactly. Moreover, the results concerning these systems form the basis of most of the approximate methods for the more complex many-electron systems. We shall therefore treat the one-electron systems in some detail.

Introducing spherical co-ordinates (r, θ, φ), the Schrödinger equation for a single electron in a general spherical potential $U(r)$ can be written (Messiah 1965, chapter 9)

$$\mathcal{H}\phi(r, \theta, \varphi, \sigma) = E\phi(r, \theta, \varphi, \sigma), \qquad (1.12)$$

where†

$$\mathcal{H} = -\frac{1}{2}\nabla^2 + U(r) \equiv -\frac{1}{2}\left(\frac{1}{r}\frac{\partial^2(r\bullet)}{\partial r^2} - \frac{l^2}{r^2}\right) + U(r). \qquad (1.13)$$

For an electron moving in the Coulomb field of a nucleus with charge Z the spherical potential is given by‡

$$U(r) = -\frac{Z}{r}. \qquad (1.14)$$

The Hamiltonian (1.13) commutes with the angular momentum operators l and s and thus $\mathcal{H}, l^2, l_z, s^2$ and s_z constitute a set of mutually commuting operators. The simultaneous eigenfunctions of these operators can be written

$$\phi(q) = R(r)Y_{lm_l}(\theta, \varphi)\chi_{m_s}(\sigma) = \frac{1}{r}P(r)Y_{lm_l}(\theta, \varphi)\chi_{m_s}(\sigma) \qquad (1.15)$$

---

† The action of the operator $\frac{1}{r}\frac{\partial^2(r\bullet)}{\partial r^2}$ on a function $R(r)$ is defined by $\frac{1}{r}\frac{\partial^2(rR(r))}{\partial r^2}$.
‡ The treatment presented here can be extended to more general potentials, the only requirement being that $\lim_{r\to 0} rU(r) = c$, a constant, and that as $r \to \infty$, $U(r)$ is negative and approaches zero at least as rapidly as $-1/r$.

where the spherical harmonics $Y_{lm_l}(\theta, \varphi)$ and spin functions $\chi_{m_s}(\sigma)$ are eigenfunctions of the orbital and spin angular momentum operators, respectively:

$$l^2 Y_{lm_l}(\theta, \varphi) = l(l+1) Y_{lm_l}(\theta, \varphi), \qquad l_z Y_{lm_l}(\theta, \varphi) = m_l Y_{lm_l}(\theta, \varphi) \qquad (1.16)$$

$$s^2 \chi_{m_s}(\sigma) = \tfrac{1}{2}\left(\tfrac{1}{2} + 1\right) \chi_{m_s}(\sigma), \qquad s_z \chi_{m_s}(\sigma) = m_s \chi_{m_s}(\sigma). \qquad (1.17)$$

In spherical co-ordinates the inversion $r \to -r$ is given by

$$(r, \theta, \varphi) \to (r, \pi - \theta, \varphi + \pi). \qquad (1.18)$$

Noting that the spherical harmonics satisfy

$$Y_{lm_l}(\pi - \theta, \varphi + \pi) = (-1)^l Y_{lm_l}(\theta, \varphi) \qquad (1.19)$$

it is seen that the one-electron eigenfunctions have even parity for even l and odd parity for odd l.

To determine the radial functions we insert

$$\phi(q) = \frac{1}{r} P(r) Y_{lm_l}(\theta, \varphi) \chi_{m_s}(\sigma)$$

into the Schrödinger equation. Using equation (1.16), it is seen that $P(r)$ is a solution to the radial equation

$$\left(\frac{d^2}{dr^2} - 2U(r) - \frac{l(l+1)}{r^2} + 2E \right) P(r) = 0. \qquad (1.20)$$

Since $\phi(q)$ must be everywhere finite, $P(r)$ should satisfy the boundary condition $P(0) = 0$.

1.3.1 Bound state solutions

According to the general theory, (Messiah 1965, chapter 9), square integrable solutions to the radial equation (1.20) satisfying the boundary condition at the origin exist only for certain discrete energy values $E < 0$. The different solutions $P(r)$ belonging to these energy values may be distinguished by the value l to which they correspond and by the number of nodes away from the origin. It is customary to denote the orbital angular momentum with the spectroscopic notation

$$\begin{aligned} l = \;& 0\;1\;2\;3\;4\;5\;6\;7\;8\ldots \\ & s\;p\;d\;f\;g\;h\;i\;k\;l\ldots \end{aligned} \qquad (1.21)$$

and to use a principal quantum number n defined by

$$n = l + \nu + 1 \qquad (1.22)$$

where ν is the number of nodes of the radial function not counting the zero at the origin. As an example, the radial function $P(2s; r)$ corresponds to an orbital angular momentum $l = 0$ and has one node away from the origin.

The total one-electron wave function

$$\phi(q) = \frac{1}{r} P(nl; r) Y_{lm_l}(\theta, \varphi) \chi_{m_s}(\sigma) \tag{1.23}$$

can thus be specified entirely by four quantum numbers nlm_lm_s (the spin quantum number s is always $1/2$ and need not be specified). Such functions are called *spin-orbitals*, and we will denote them either as $\phi(nlm_lm_s; q)$, or $(q|nlm_lm_s)$, or, if the co-ordinates are unimportant, as $|nlm_lm_s\rangle$. For the spin-orbitals to be normalized, $P(nl; r)$ should satisfy

$$\int_0^\infty P^2(nl; r)\, \mathrm{d}r = 1. \tag{1.24}$$

Even for normalized spin-orbitals there is a trivial arbitrariness between $P(nl; r)$ and $-P(nl; r)$. To resolve this we use the convention that the radial functions should be positive near the origin.

For small r, the radial function can be expanded in a power series

$$P(nl; r) = a_0 r^s + a_1 r^{s+1} + \dots \tag{1.25}$$

where, since $\phi(q)$ must be finite everywhere, $s \geqslant 1$. Inserting this expansion in the radial equation, collecting terms of equal order in r, it is seen that

$$\frac{1}{r} P(nl; r) \propto r^l \tag{1.26}$$

for small r. Thus, the probability for the electron to be close to the nucleus rapidly decreases for increasing l. Only for $l = 0$ does the electron have a finite probability of being at the site of the nucleus. For $r \to \infty$ it is easily seen that the radial function decreases exponentially

$$P(nl; r) \propto \mathrm{e}^{-\sqrt{2|E|}\, r}. \tag{1.27}$$

In table 1.1 some of the normalized radial functions are shown for hydrogen where $U(r) = -1/r$.

1.3.2 Continuum state solutions

For $E > 0$ the electron is no longer bound to the nucleus and it behaves, for large r, essentially like a free particle. It can be shown that there is a solution to the radial equation (1.20) for every positive E, and a continuous spectrum of positive eigenvalues adjoins the discrete levels of negative energy

Table 1.1. Normalized radial functions for $U(r) = -1/r$.

$$P(1s; r) = 2re^{-r}$$

$$P(2s; r) = \frac{1}{\sqrt{2}} re^{-r/2} \left(1 - \frac{1}{2}r\right)$$

$$P(2p; r) = \frac{1}{2\sqrt{6}} r^2 e^{-r/2}$$

$$P(3s; r) = \frac{2}{3\sqrt{3}} re^{-r/3} \left(1 - \frac{2}{3}r + \frac{2}{27}r^2\right)$$

$$P(3p; r) = \frac{8}{27\sqrt{6}} r^2 e^{-r/3} \left(1 - \frac{1}{6}r\right)$$

$$P(3d; r) = \frac{4}{81\sqrt{30}} r^3 e^{-r/3}$$

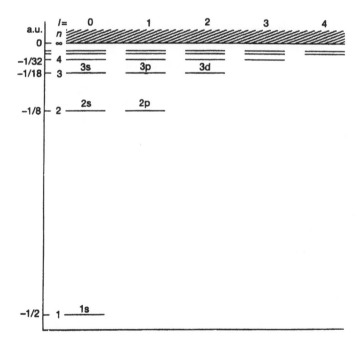

Figure 1.1. The discrete and continuous part of the eigenvalue spectrum for hydrogen.

(see figure 1.1). Introducing k so that $E = k^2/2$ (in atomic units), the radial equation can be written

$$\left(\frac{d^2}{dr^2} - 2U(r) - \frac{l(l+1)}{r^2} + k^2\right) P(kl; r) = 0 \qquad (1.28)$$

where k together with the angular momentum quantum number l are now used to label the different radial functions. For an electron moving in the Coulomb field of the nucleus with charge Z the asymptotic form of the radial function is

$$P(kl; r) \propto \sin\left(kr - \frac{1}{2}l\pi + \frac{Z}{k}\log 2kr + \sigma_l\right) \qquad (1.29)$$

(Abramowitz and Stegun 1965) where σ_l, the Coulomb phase shift, is

$$\sigma_l = \arg \Gamma\left(l + 1 - i\frac{Z}{k}\right) \qquad (1.30)$$

with Γ being the gamma function. If the continuum solutions are normalized in the Dirac sense so that

$$\int_0^\infty P(kl; r) P(k'l; r)\, dr = \delta(E - E') \qquad (1.31)$$

then

$$|\phi(klm_l m_s; q)|^2 \, dE \, dq \qquad (1.32)$$

can be interpreted as the probability that the electron is in the generalized volume element dq centred at q and has an energy in the interval dE centred at $E = k^2/2$. We shall be dealing mainly with bound states in the first part of this book, and the discussion about continuum states is postponed until chapter 10.

1.4   Many-electron systems

For many-electron systems the exact forms of the eigenfunctions are not known, and instead approximate eigenfunctions must be found. One possibility that provides a lot of insight into the nature of approximate wave functions is to replace the non-relativistic Hamiltonian $\mathcal{H}$ by one for which the Schrödinger equation is solvable.

1.4.1   Central-field approximation

In the central-field approximation the full Hamiltonian is replaced with the separable Hamiltonian $\mathcal{H}_0$

$$\mathcal{H} \approx \mathcal{H}_0 = \sum_{i=1}^{N}\left(-\frac{1}{2}\nabla_i^2 - \frac{Z}{r_i} + V(r_i)\right) \qquad (1.33)$$

where the central potential $V(r)$ approximates the effects of the Coulomb repulsion among the electrons.

The approximate Hamiltonian $\mathcal{H}_0$, as well as the full Hamiltonian, commutes with the total angular momentum operators L^2, L_z, S^2 and S_z and we may choose the eigenfunctions of $\mathcal{H}_0$ to be eigenfunctions also of these operators. If

$$\mathcal{H}_0\psi_0(q_1, \ldots, q_N) = E_0\psi_0(q_1, \ldots, q_N) \tag{1.34}$$

then, since $\mathcal{H}_0$ is separable, the eigenvalues and eigenfunctions can be written

$$E_0 = \sum_{i=1}^{N} E_i \tag{1.35}$$

and

$$\psi_0(q_1, \ldots, q_N) = \prod_{i=1}^{N} \phi(\alpha_i; q_i), \tag{1.36}$$

where the individual spin-orbitals are solutions to the one-electron equation

$$\left[-\frac{1}{2}\nabla^2 + U(r)\right]\phi(\alpha; q) = E\phi(\alpha; q) \tag{1.37}$$

with $U(r) = -(Z/r) + V(r)$. As shown previously, the one-electron function, or spin-orbital, can be written

$$\phi(\alpha; q) = \frac{1}{r}P(nl; r)Y_{lm_l}(\theta, \varphi)\chi_{m_s}(\sigma), \tag{1.38}$$

where the solutions are characterized by the quantum numbers $\alpha = nlm_lm_s$. Note that for a general potential $U(r)$ the one-electron energy E, contrary to the Coulomb case, depends both on n and l.

The Hamiltonian $\mathcal{H}_0$ is invariant with respect to permutations of the electron co-ordinates and so any permutation of the co-ordinates in the product function (1.36) also leads to an eigenfunction. By combining the permuted product functions we can form an antisymmetric function

$$\Phi(q_1, \ldots, q_N) = \mathcal{A}\prod_{i=1}^{N} \phi(\alpha_i; q_i). \tag{1.39}$$

This function can also be represented as a so-called *Slater determinant*

$$\Phi(q_1, \ldots, q_N) = \frac{1}{\sqrt{N!}}\begin{vmatrix} \phi(\alpha_1; q_1) & \phi(\alpha_1; q_2) & \ldots & \phi(\alpha_1; q_N) \\ \phi(\alpha_2; q_1) & \phi(\alpha_2; q_2) & \ldots & \phi(\alpha_2; q_N) \\ \vdots & & & \\ \phi(\alpha_N; q_1) & \phi(\alpha_N; q_2) & \ldots & \phi(\alpha_N; q_N) \end{vmatrix}. \tag{1.40}$$

In this representation it is seen that the total wave function $\Phi(q_1, \ldots, q_N)$ vanishes identically if two electrons have the same value of the four quantum numbers $\alpha = nlm_lm_s$. Thus, for allowed states of the atom no two electrons can have the same value of the four quantum numbers. This is the exclusion principle in the form originally discovered by Pauli (1925). Note also that the determinant vanishes whenever $q_i = q_j$; that is, whenever two electrons with the same spin have the same space co-ordinates.

To determine the parity of the Slater determinant we note that each of the spin-orbitals building the Slater determinant has parity $(-1)^l$. The Slater determinant itself must therefore have the well defined parity

$$\pi = (-1)^{l_1}(-1)^{l_2}\ldots(-1)^{l_N} = (-1)^{\sum_i l_i} \tag{1.41}$$

which is even or odd dependent on whether the sum of the orbital angular momentum quantum numbers is even or odd.

1.4.2   Electron configuration

As seen above, an eigenfunction of the central-field Hamiltonian $\mathcal{H}_0$ can be written as a Slater determinant. The corresponding energy E_0 is then given by the sum of the energies of the spin-orbitals appearing in the determinant

$$E_0 = \sum_{i=1}^{N} E_i. \tag{1.42}$$

Spin-orbitals with the same values of the n and l quantum numbers are said to belong to the same *subshell* (or shell, for short) and are called equivalent spin-orbitals. Correspondingly, we may loosely speak of electrons belonging to the same subshell and refer to them as equivalent electrons. Since the energy of a spin-orbital depends only on the n and l quantum numbers, the energy E_0 is entirely determined by the *electron configuration*, i.e. the distribution of spin-orbitals with respect to the subshells.

A general electron configuration is given by

$$(n_1l_1)^{w_1}(n_2l_2)^{w_2}\ldots(n_ml_m)^{w_m}, \qquad N = \sum_{a=1}^{m} w_a \tag{1.43}$$

where $w_1, w_2\ldots$, etc are the occupation numbers of the spin-orbitals in the different subshells. The corresponding energy can then be written

$$E_0 = \sum_{a=1}^{m} w_a E_{n_a l_a} \tag{1.44}$$

where E_{nl} denotes the energy of the spin-orbitals in an nl subshell. Often the electron configurations are denoted by means of the spectroscopic symbols given in section 1.3.1. For instance $1s^2 2s^2 2p^2$ means that there are two electrons in each of the $1s$, $2s$ and $2p$ subshells.

According to the Pauli exclusion principle there can be only one electron in each spin-orbital, and thus there can be at most $2(2l + 1)$ electrons in a subshell nl. A subshell which is fully occupied is said to be closed, in contrast to a partially occupied shell which is said to be open. The energy of a configuration is, according to (1.44), given by the occupation numbers of each subshell. Therefore, the ground (lowest-energy) configuration for a particular atom should be obtained by successively filling the electron subshells with the lowest energies E_{nl} leading to a number of closed subshells and at most one open shell. This is the so-called 'auf Bau' principle formulated by Bohr (1922) to explain the periodic table of the elements.

The concept of a configuration has a simple interpretation. For light

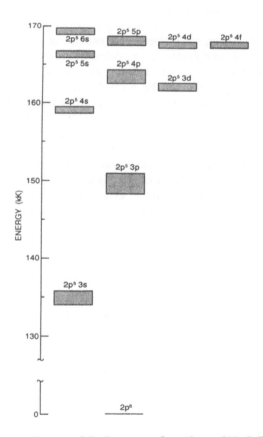

Figure 1.2. Schematic diagram of the lowest configurations of Ne I. The levels of each configuration lie within the limited energy range shown by the corresponding shaded block.

atoms, the experimental energy levels often appear in closely spaced groups (see figure 1.2). When a central-field calculation is performed using a suitable potential $V(r)$, it is found that the average energy of each of these groups corresponds reasonably well with the energy of a certain configuration, and thus it is possible to assign configuration labels to the groups. Furthermore, if the assignment of the configuration has been done correctly, it is seen that the number of states† in a group is equal to the number of determinants of the corresponding configuration.

1.4.3 Configuration state functions (CSFs)

From the central-field approximation we can obtain approximate energy levels and eigenfunctions of the full non-relativistic Hamiltonian (1.2). In general, these approximate eigenfunctions in the form of Slater determinants are not eigenfunctions of the total angular momentum operators (1.10), as are the exact eigenfunctions. However, by a linear combination of determinants belonging to the same electron configuration, i.e. determinants with the same set of n and l quantum numbers, but with different m_l and m_s, eigenfunctions of the angular momentum operators can be formed. The functions obtained in this way are better approximations to the exact eigenfunctions of the non-relativistic Hamiltonian than the Slater determinants themselves and they are referred to as *configuration state functions* or CSFs. The CSFs are denoted $\Phi(\gamma L M_L S M_S)$ or $|\gamma L M_L S M_S\rangle$ where γ represents the electron configuration and all additional quantum numbers needed to completely specify the state. As an example, we consider the $1s^2 2s^2 2p^2$ configuration. To this configuration there correspond 15 Slater determinants characterized by the different combinations of the magnetic quantum numbers m_l and m_s. From these Slater determinants the following 15 CSFs can be formed:

$$\Phi(1s^2 2s^2 2p^2 \ L M_L S M_S), \qquad \begin{cases} L = 0, \quad M_L = 0 \\ S = 0, \quad M_S = 0 \end{cases}$$

$$\Phi(1s^2 2s^2 2p^2 \ L M_L S M_S), \qquad \begin{cases} L = 2, \quad M_L = -2, -1, 0, 1, 2 \\ S = 0, \quad M_S = 0 \end{cases}$$

$$\Phi(1s^2 2s^2 2p^2 \ L M_L S M_S), \qquad \begin{cases} L = 1, \quad M_L = -1, 0, 1 \\ S = 1, \quad M_S = -1, 0, 1. \end{cases}$$

$$(1.45)$$

† The number of states belonging to a degenerate level can be inferred experimentally by placing the atom in an external magnetic field where the degeneracy with respect to the M quantum numbers is removed.

1.4.4 *LS* terms

In the central-field approximation all the Slater determinants belonging to a certain configuration, and thus also the CSFs formed from these determinants, correspond to the same energy level. When the non-central part of the electron interaction

$$-\sum_{i=1}^{N} V(r_i) + \sum_{i<j}^{N} \frac{1}{r_{ij}}$$
(1.46)

is taken into account, the different CSFs will, depending on their total angular momentum quantum numbers, correspond to different energies. These energy levels are called the *LS terms* of the configuration and are given by the expectation values of the different CSFs,

$$E = \langle \Phi(\gamma L M_L S M_S) | \mathcal{H} | \Phi(\gamma L M_L S M_S) \rangle.$$
(1.47)

The above expectation value is independent of M_L and M_S (see appendix A.4.1), and each LS term is $(2L + 1)(2S + 1)$-fold degenerate.

Since the LS terms are independent of the M_L and M_S quantum numbers the latter are often omitted. In those cases where the M_L and M_S quantum numbers are of no significance the CSFs will simply be denoted $\Phi(\gamma LS)$ or $\Phi(\gamma^{(2S+1)}L)$ where L refers to the spectroscopic notation,

$$\begin{aligned} L = \quad 0 \;\; 1 \;\; 2 \;\; 3 \;\; 4 \;\; 5 \;\; 6 \;\; 7 \;\; 8 \ldots \\ S \;\; P \;\; D \;\; F \;\; G \;\; H \;\; I \;\; K \;\; L \ldots \end{aligned}$$
(1.48)

and $2S + 1$ is the so-called multiplicity of the term. Thus the CSFs of (1.45) are written in spectroscopic notation as $\Phi(1s^2 2s^2 2p^2 \, {}^1S)$, $\Phi(1s^2 2s^2 2p^2 \, {}^1D)$, and $\Phi(1s^2 2s^2 2p^2 \, {}^3P)$, respectively. The latter notation will generally be used throughout the remainder of the book. For odd parity states a superscript 'o' is sometimes added after the letter denoting L, as in $2s2p \, {}^3P^o$.

1.4.5 Multiconfiguration expansions

In many cases, the CSFs are surprisingly good approximations to the exact eigenfunctions ψ of the full Hamiltonian. Better approximations, however, can be obtained as a linear combination of CSFs,

$$\Psi(\gamma LS) = \sum_{i=1}^{M} c_i \Phi(\gamma_i LS).$$
(1.49)

The exact eigenfunction is then usually labelled in the same way as the dominant CSF in the expansion. The difficulty with this *multiconfigurational* approach for obtaining approximate eigenfunctions lies in the selection of the appropriate central-field potential $U(r)$ that enters equation (1.37) for the spin-orbitals. This problem, however, can be avoided entirely if the variational method is applied instead to determine the spin-orbitals, or more precisely, the radial parts of the spin-orbitals. The variational method will be discussed in the remaining sections of this chapter.

1.5   The variational method

Variational methods for solving the Schrödinger equation rely on a reformulation of the eigenvalue problem. Solving the Schrödinger equation for bound states is equivalent to finding functions ψ that leave the energy functional

$$\mathcal{E}(\psi) = \langle \psi | \mathcal{H} | \psi \rangle / \langle \psi | \psi \rangle \tag{1.50}$$

stationary to first order with respect to variations $\delta\psi$ in ψ satisfying the boundary conditions. In addition to the boundary conditions, the variation must also have the expected integrability, continuity and differentiability properties. Without stating it explicitly we will assume that this is always the case. To show the equivalence of the two problems, consider the variation $\delta\mathcal{E}$ of $\mathcal{E}$, which by definition is given by

$$\mathcal{E}(\psi + \delta\psi) - \mathcal{E}(\psi) = \delta\mathcal{E} + \mathcal{O}((\delta\psi)^2). \tag{1.51}$$

Using (1.50), keeping terms only to first order in $\delta\psi$, and multiplying by $\langle \psi | \psi \rangle$ we obtain

$$\begin{aligned} \delta\mathcal{E} \, \langle \psi | \psi \rangle &= \langle \delta\psi | \mathcal{H} - \mathcal{E}(\psi) | \psi \rangle + \langle \psi | \mathcal{H} - \mathcal{E}(\psi) | \delta\psi \rangle \\ &= 2\langle \delta\psi | \mathcal{H} - \mathcal{E}(\psi) | \psi \rangle, \end{aligned} \tag{1.52}$$

where the latter equality follows from the fact that $\mathcal{H}$ is Hermitian for bound states. Now, if $\mathcal{E}(\psi)$ is stationary the variation $\delta\mathcal{E}$ vanishes and thus

$$\langle \delta\psi | \mathcal{H} - \mathcal{E}(\psi) | \psi \rangle = 0 \tag{1.53}$$

from which it follows that

$$(\mathcal{H} - \mathcal{E}(\psi)) | \psi \rangle = 0. \tag{1.54}$$

Conversely, if ψ is an eigenfunction to $\mathcal{H}$ then $\delta\mathcal{E} = 0$ and $\mathcal{E}(\psi)$ is stationary.

The energy functional (1.50) is defined in terms of unnormalized functions ψ. In many situations it is convenient to restrict variations to the space of normalized functions so that

$$\langle \psi | \psi \rangle = \langle \psi + \delta\psi | \psi + \delta\psi \rangle = 1. \tag{1.55}$$

The solution of this variational problem, which is referred to as an optimization problem under constraint, relies on the following theorem. If ψ is a solution to the optimization problem under the normalization constraint, then there exists a so called *Lagrange multiplier* λ such that the functional

$$\mathcal{F}(\psi) = \mathcal{E}(\psi) + \lambda\langle\psi|\psi\rangle \tag{1.56}$$

is stationary to first order with respect to *all* variations $\delta\psi$ in ψ satisfying the boundary conditions. (see Fletcher 1987, chapter 9).

1.5.1 Approximate variational solutions

The variational problem (1.56) can of course never be solved exactly, and instead we have to find approximate solutions. One way of doing so is to choose a variational function ψ_v that depends on a number of parameters $\alpha = (\alpha_1, \ldots, \alpha_n)$

$$\psi_v = \psi_v(\alpha; q_1, \ldots, q_N). \tag{1.57}$$

Then these parameters are determined from the stationary condition of the functional

$$\mathcal{F}(\alpha) = \mathcal{E}(\psi_v) + \lambda\langle\psi_v|\psi_v\rangle \tag{1.58}$$

with respect to variations of the parameters, leading to the conditions

$$\frac{\partial\mathcal{F}(\alpha)}{\partial\alpha_i} = 0, \quad i = 1, \ldots, n \tag{1.59}$$

with λ such that $\langle\psi_v|\psi_v\rangle = 1$. This is a finite, though possibly non-linear, problem that can be solved on a computer, and the resulting ψ_v and $\mathcal{E}(\psi_v)$ represent the best estimates, from a purely energetic point of view, of the exact eigenfunction and eigenvalue within the function space spanned by the variational function. Clearly, the variational function should have the same parity and angular symmetry as the exact eigenfunction. Furthermore, the variational function should be flexible and incorporate the correct qualitative features of the exact eigenfunction.

In addition to the normalization constraint (1.55), the variational parameters are often subject to a number of other constraints. These constraints can generally be written

$$\mathcal{C}_i(\alpha) = 0, \quad i = 1, \ldots, m \tag{1.60}$$

where $\mathcal{C}_i$ are so-called constraint functions. In this case Lagrange multipliers have to be introduced for each of the constraints, and the problem is to find the parameters that leave the functional

$$\mathcal{F}(\alpha) = \mathcal{E}(\psi_v) + \sum_{i=1}^{m}\lambda_i\mathcal{C}_i(\alpha) \tag{1.61}$$

stationary with respect to allowed variations. In addition, the Lagrange multipliers must be such that the stationary solution satisfies all constraints.

1.5.2 The matrix eigenvalue problem

A simple, but very important, variational function is given by the expansion

$$\Psi = \sum_{i=1}^{M} c_i \Phi(\gamma_i LS). \tag{1.62}$$

Here the CSFs $\Phi(\gamma_i LS)$ are assumed to be known and only the coefficients c_i need to be determined. Most frequently, the CSFs are orthonormal so the normalization condition becomes simply

$$\langle\Psi|\Psi\rangle = \sum_{i=1}^{M} c_i^2 = 1. \tag{1.63}$$

Inserting the above expansion into (1.58) and demanding that the functional be stationary with respect to variations in the coefficients yields the equations,

$$\mathbf{Hc} = -\lambda\mathbf{c}, \tag{1.64}$$

where $\mathbf{H}$ is the Hamilton matrix with elements

$$H_{ij} = \langle\Phi(\gamma_i LS)|\mathcal{H}|\Phi(\gamma_j LS)\rangle \tag{1.65}$$

and $\mathbf{c} = (c_1, \ldots, c_M)^t$ the column vector of the expansion coefficients. Only when $-\lambda$ is an eigenvalue of $\mathbf{H}$ does a normalized solution exist. Hence the constrained variational problem leads to a matrix eigenvalue problem. Since the Hamiltonian matrix is Hermitian, the eigenvalue equation has exactly M orthonormal solutions

$$\mathbf{c}_k = (c_{1k}, \ldots, c_{Mk})^t, \qquad \mathbf{c}_k^t\mathbf{c}_l = \delta_{kl} \tag{1.66}$$

with corresponding real eigenvalues

$$-\lambda_1 \leqslant \ldots \leqslant -\lambda_k \ldots \leqslant -\lambda_M. \tag{1.67}$$

Out of these M solutions, one or several, depending on the expansion, are good approximations to the corresponding exact wave functions. The variational energies $\mathcal{E}(\Psi)$ for the different solutions are, as is easily verified, equal to the associated matrix eigenvalues, $-\lambda$. For this reason the Lagrange multiplier associated with the normalization constraint (1.63) is often denoted by E so that $\mathcal{E}(\Psi) = E$. The above method for obtaining approximate wave functions is called the *configuration interaction* method and will be discussed further in chapter 7 in connection with relativistic corrections.

1.5.3 The Hartree–Fock problem

In the central-field approximation as defined in (1.33), each electron moves in the same potential, $-(Z/r)+V(r)$. As mentioned earlier, the choice of $V(r)$ is non-trivial. Hartree (1927) argued intuitively that each electron had its own potential and that the Coulomb replusion potential, $V(nl; r)$, for an nl electron could be determined from the spherically averaged charge distribution (or electron cloud) of the other electrons in the system. From this assumption he derived what are now known as the *Hartree equations*. These are coupled radial equations in that the charge distribution of one electron depends on the other and vice versa. Hartree proposed that these equations be solved by an iterative procedure he called the *self-consistent field* method.

The Hartree wave function is a spherically symmetric wave function that is simply a product of radial functions. Fock (1930) noted that these solutions did not satisfy the Pauli exclusion principle. By considering simple systems which could be represented by a single determinant such as the lithium or sodium ground state and by applying the variational principle, he derived equations remarkably like the Hartree equations, except for the presence of some extra terms arising from antisymmetry called the *exchange terms*.

In atomic physics today, the variational wave function for a many-electron system is chosen in the form of a configuration state function $\Psi = \Phi(\gamma LS)$ where the radial functions $[P(n_1l_1; r), P(n_2l_2; r), \ldots, P(n_ml_m; r)]$ are undetermined, and the stationary condition with respect to variations in the latter leads to the so-called *Hartree–Fock* equations. The derivation and solution of the Hartree–Fock equations will be discussed in detail in chapter 3. If instead the variational function is chosen in the form of a multiconfiguration expansion,

$$\Psi = \sum_{i=1}^{M} c_i \Phi(\gamma_i LS), \qquad (1.68)$$

the stationary condition with respect to variations in the radial functions leads to a set of coupled differential equations similar to the Hartree–Fock equations. These differential equations themselves are coupled to the matrix eigenvalue equation resulting from the variation of the expansion coefficients and these two problems must be solved simultaneously. The method based on the latter variational function is known as *multiconfiguration Hartree–Fock* and will be discussed in chapter 4.

1.5.4 The Hylleraas–Undheim–MacDonald theorem

For states that are the lowest of their symmetry the variational problem is a minimization problem, and the approximate eigenvalue approaches the exact eigenvalue from above. That is

$$E \leqslant \mathcal{E}(\Psi), \qquad (1.69)$$

where E is the exact eigenvalue (see exercise (vi)). For excited states this is not generally true, and variational solutions may give energies that are too low. For multiconfiguration expansions, however, we have a stronger result that relies on a theorem derived independently by Hylleraas and Undheim (1930) and MacDonald (1933). Suppose the variational function is expanded in an orthonormal basis set

$$\Psi = \sum_{i=1}^{M} c_i \Phi(\gamma_i L S) \tag{1.70}$$

and that the eigenvalues of the $M \times M$ Hamiltonian matrix $H_{ij} = \langle \Phi(\gamma_i L S) | \mathcal{H} | \Phi(\gamma_j L S) \rangle$ are arranged according to

$$E_1^M < E_2^M < \ldots < E_M^M. \tag{1.71}$$

If the basis set is augmented by $\Phi(\gamma_{M+1} L S)$ and the eigenvalues of the $(M + 1) \times (M + 1)$ Hamiltonian matrix are

$$E_1^{M+1} < E_2^{M+1} < \ldots < E_{M+1}^{M+1}, \tag{1.72}$$

we have an interleaving of the eigenvalues,

$$E_{k-1}^M < E_k^{M+1} < E_k^M. \tag{1.73}$$

If the basis $\Phi(\gamma_1 L S), \ldots, \Phi(\gamma_M L S)$ is extended to span the whole function space of the correct $L S$ symmetry, then

$$E_k^{\text{exact}} < E_k^M; \tag{1.74}$$

that is, the kth lowest eigenvalue of the $M \times M$ Hamiltonian matrix is an upper bound to the exact energy of the kth lowest excited state of the given angular symmetry and parity.

In summary, in any calculation for a state lowest in its symmetry, the computed energy will be an upper bound to the exact energy. The variational principle may be applied to excited states, in which case the energy is stationary at a solution, but the computed energy may not be an upper bound. Requiring orthogonality to all lower states assures that the energy is an upper bound. This theorem makes it clear that the lower states need not be exact states.

1.6 Summary

This chapter has been a general review of atomic theory as it applies to the chapters which follow. Each chapter will deal with different aspects of an atomic structure computation, and the decisions that need to be made in the course of a sucessful calculation.

At this point it is useful to summarize the basic assumptions of the remaining chapters.

Schrödinger's equation

As a first step, we solve the non-relativistic Schrödinger equation

$$(\mathcal{H} - E)\psi = 0. \tag{1.75}$$

We will refer to an approximate solution for a state labelled γLS as an *atomic state function* (ASF).

Atomic state functions

We will assume that the ASF, $\Psi(\gamma LS)$, is a linear combination of CSFs

$$\Psi(\gamma LS) = \sum_{i=1}^{M} c_i \Phi(\gamma_i LS). \tag{1.76}$$

This approach is a *multiconfiguration* approximation and the expansion coefficients are referred to as *mixing coefficients*.

Configuration state functions

Each configuration state function (CSF) is an eigenfunction of the total orbital momentum and spin-angular momentum operators L^2, L_z, S^2 and S_z. In addition it is an eigenfunction of the parity operator Π. The CSF is an antisymmetric sum of products of one-electron *spin-orbitals*, one for each electron.

Spin-orbitals

Each spin-orbital is a one-electron function of the form

$$\phi(nlm_lm_s; q) = \frac{1}{r} P(nl; r) Y_{lm_l}(\theta, \varphi) \chi_{m_s}(\sigma) \tag{1.77}$$

where the *radial function* $P(nl; r)$ depends only on nl quantum numbers.

Mixing coefficients

The CSFs form a basis for the ASF . Requiring the energy functional to be stationary with respect to variations in the mixing coefficients, leads to the matrix eigenvalue problem

$$(\mathbf{H} - E\mathbf{I})\mathbf{c} = 0 \tag{1.78}$$

where the Hamiltonian matrix, $\mathbf{H}$, has the elements

$$H_{ij} = \langle \Phi(\gamma_i LS) | \mathcal{H} | \Phi(\gamma_j LS) \rangle \tag{1.79}$$

The eigenvector c associated with the corresponding eigenvalue E, determines the mixing coefficients and the total energy respectively. This problem is also called a *configuration interaction* problem.

Additional effects

Relativistic and finite nuclear mass (isotope) effects are treated by methods appropriate for small effects that may be included in the Hamiltonian of a configuration interaction problem.

Much of the book will be concerned with determining the optimum set of radial functions for spin-orbitals used in defining the CSFs and with efficient expansions that result in reliable estimates of atomic state functions. Once these have been determined, other atomic properties may be computed.

1.7   Exercises

(i)   Show that the bound state solutions $P(nl; r)$ of equation (1.20) have the asymptotic behaviour

$$\frac{1}{r} P(nl; r) \propto r^l, \qquad \text{for } r \ll 1$$

$$P(nl; r) \propto e^{-\sqrt{2|E|}\, r}, \qquad \text{for } r \gg 1.$$

(ii)   When $U(r) = -Z/r$ in (1.20), the normalized solution $P(Z; r)$ is said to be *hydrogenic*. By introducing the transformation $\rho = Zr$, show that $P(Z; r) = Z^{1/2} P(1; \rho)$. Use table 1.1 and derive expressions for $P(Z; r)$ for the $1s$, $2s$, and $2p$ states.

(iii)   Consider a variational function for the ground state of hydrogen, namely

$$\psi_v = \frac{1}{r} P(r) Y_{00}(\theta, \varphi)$$

where $P(r)$ is of the form (1) Are^{-Br}, (2) $Ar/(r^2 + B^2)$ and (3) $Ar^2 e^{-Br}$

(a)   For each case, determine parameters A and B and the associated Lagrange multiplier λ such that the energy functional is stationary under the normalization constraints. Verify that the variational function of form (1) gives the exact solution. From a purely energetic point of view which of the forms, (2) or (3), yields the best approximation to the exact wave function? Hint: You may wish to use a symbol manipulation program to evaluate some of the integrals.

(b)   Plot the normalized radial wave functions $P(r)$ for the cases (1)–(3).

(c)   Explain why the functional form (2) gives a better energy than (3).

(d)   A variational function that gives a good value of the total energy does not necessarily give good values for other expectation values. To illustrate this, calculate the expectation value $\langle \psi_v | r | \psi_v \rangle$ for the three variational functions. Which of the two functions, (2) or (3), gives the best approximation of the exact expectation value?

(e) The total energy is only one measure of the quality of a variational function. Other measures are possible. If the exact wave function is known, as is the case for hydrogen, an alternative measure is given by $1 - |\langle \psi | \psi_v \rangle|^2$ where ψ is the exact wave function. According to this measure which of the functions (2) and (3) is the best approximation of the exact wave function?

(iv) Consider a variational function for the ground state of helium of the form

$$\psi_v = A \frac{P(Z_{\text{eff}}; r_1)}{r_1} \frac{P(Z_{\text{eff}}; r_2)}{r_2} Y_{00}(\theta_1, \varphi_1) Y_{00}(\theta_2, \varphi_2) \chi(1, 2)$$

where $P(Z_{\text{eff}}; r) = re^{-Z_{\text{eff}}r}$ is an unnormalized hydrogenic $1s$ orbital corresponding to an effective charge Z_{eff} and $\chi(1, 2) = \frac{1}{\sqrt{2}} \left[\chi_{1/2}(\sigma_1)\chi_{-1/2}(\sigma_2) - \chi_{-1/2}(\sigma_1)\chi_{1/2}(\sigma_2) \right]$ is a coupled spin-function.

(a) Show that the variational function is antisymmetric and has the correct orbital and spin angular symmetry.

(b) Determine the A, Z_{eff} and the associated Lagrange multiplier λ so that the energy functional is stationary under the normalization constraint. Compare the variational energy with the exact, non-relativistic energy $E = -2.90372$ Hartree. Hint: $\langle \psi_v | \frac{1}{r_{12}} | \psi_v \rangle = A^2 \frac{5}{32} \frac{1}{Z_{\text{eff}}^2}$.

(c) Let $Z_{\text{eff}} = Z - S$ where $Z = 2$. Which effective screening S of the nuclear charge is obtained for the variational solution?

(v) Consider the function $\Psi = \sum_{i=1}^{M} c_i \Phi(\gamma_i LS)$ where $\Phi(\gamma_i LS)$ are known CSFs that are assumed to be orthonormal. Show that the stationary condition of the functional $\mathcal{F}(\mathbf{c})$ with respect to variations in the expansion coefficients c_i leads to the matrix eigenvalue problem $\mathbf{Hc} = -\lambda \mathbf{c}$.

(vi) If E is the lowest eigenvalue of the Hamiltonian $\mathcal{H}$, show that $E \leqslant \mathcal{E}(\Psi)$ for all normalized variational functions Ψ.

Chapter 2

Configuration State Functions and Matrix Elements of the Hamiltonian

2.1 Configuration state functions

In the configuration model of a many-electron atomic system, CSFs form the basis for an atomic state function.

CSFs can be constructed in two equivalent ways, either as a linear combination of Slater determinants belonging to the same configuration, i.e. determinants with the same set of n and l quantum numbers, but with different m_l and m_s quantum numbers, or using expressions based on coefficients of fractional parentage and explicit antisymmetrization. The codes featured in this book follow the latter approach which is derived from angular momentum theory. An outline of the underlying theory is provided in appendix A. A comprehensive treatment can be found in the book by Cowan (1981, chapter 9).

2.1.1 Coupling of two equivalent electrons

Let us consider a group of two equivalent electrons $(nl)^2$. The products of the spin-orbitals

$$(q_1|n_1l_1m_{l_1}m_{s_1})(q_2|n_2l_2m_{l_2}m_{s_2}),\qquad(2.1)$$

where $n_1l_1 = n_2l_2 = nl$, are simultaneous eigenfunctions of the operators l_1^2, l_{1z}, l_2^2, l_{2z}, s_1^2, s_{1z}, s_2^2 and s_{2z}. With the multiple use of the Clebsch–Gordan expansion (A.18), once for coupling the orbital angular momenta and once for coupling the spin angular momenta, we obtain an eigenfunction of L^2, L_z, S^2 and S_z, namely

$$
\begin{aligned}
(q_1,q_2|n_1l_1n_2l_2LM_LSM_S) = \sum_{m_{l_1}m_{l_2}}\sum_{m_{s_1}m_{s_2}} & \langle l_1l_2m_{l_1}m_{l_2}|l_1l_2LM_L\rangle \\
\times & \langle s_1s_2m_{s_1}m_{s_2}|s_1s_2SM_S\rangle \\
\times & (q_1|n_1l_1m_{l_1}m_{s_1})(q_2|n_2l_2m_{l_2}m_{s_2}). \quad (2.2)
\end{aligned}
$$

Using the symmetry relation (A.21),

$$|\gamma j_1 j_2 J M\rangle = (-1)^{j_1+j_2-J}|\gamma j_2 j_1 J M\rangle, \tag{2.3}$$

for interchanging the order of the coupling we obtain

$$(q_1, q_2|n_1 l_1 n_2 l_2 L M_L S M_S\rangle = (-1)^{l_1+l_2-L+s_1+s_2-S}(q_2, q_1|n_2 l_2 n_1 l_1 L M_L S M_S\rangle. \tag{2.4}$$

But now $n_1 l_1 = n_2 l_2 = nl$ and so

$$(q_1, q_2|(nl)^2 L M_L S M_S\rangle = (-1)^{L+S+1}(q_2, q_1|(nl)^2 L M_L S M_S\rangle. \tag{2.5}$$

Thus, when $L + S$ is even, the eigenfunction of the total angular momentum operators is automatically antisymmetric. When $L + S$ is odd, the eigenfunction is symmetric and does not represent a physical state. This can also be viewed somewhat differently. Applying the Clebsch–Gordan expansion to a product of equivalent electrons, a number of possible values for L and S can be obtained. However, when the restrictions on the permissible values of m_{l_1}, m_{s_1}, m_{l_2} and m_{s_2} imposed by the Pauli exclusion principle are taken into account, only L and S such that $L + S$ is even are allowed.

2.1.2 Coefficients of fractional parentage

To form antisymmetric eigenfunctions of the total angular momenta belonging to a group of three equivalent electrons $(nl)^3$, we start by coupling a spin-orbital $(q_3|nlm_l m_s)$ onto the antisymmetric eigenfunction $(q_1, q_2|(nl)^2\overline{LM_L SM}_S\rangle$ to form an eigenfunction

$$(q_1, q_2, q_3|((nl)^2\overline{LS}nl)L M_L S M_S\rangle. \tag{2.6}$$

Using the recoupling formula (A.27) twice, once for the orbital angular momenta and once for the spin-angular momenta, we see that (2.6) can be expressed as a sum over coupled states

$$(q_1, q_2, q_3|(nl(nl)^2 L'S')L M_L S M_S\rangle. \tag{2.7}$$

The summation, in general, involves terms with $L' + S'$ odd, and hence the eigenfunction (2.6) is not antisymmetric with respect to the interchange of the co-ordinates q_2 and q_3. However, it may be possible to find a linear combination,

$$(q_1, q_2, q_3|(nl)^3\alpha L M_L S M_S\rangle$$
$$= \sum_{\overline{LS}}(l^2\overline{LS}|\}l^3\alpha LS)(q_1, q_2, q_3|((nl)^2\overline{LS}nl)L M_L S M_S\rangle$$

$$\tag{2.8}$$

that, using the recoupling formula (A.27) for all terms in the linear combination, results in a zero coefficient for those recouplings with $L' + S'$ odd. The function formed in this way is a completely antisymmetric eigenfunction of the total angular momentum operators. In general, there may be more than one such linear combination and to distinguish these, an additional quantity α has to be introduced. In the cases considered in this book, α is the *seniority number* (Racah 1943, Cowan 1981). For partially filled f-shells, seniority alone is not sufficient to distinguish all possible states.

The above procedure of forming antisymmetric eigenfunctions can be generalized to configurations of the form $(nl)^w$. If $(q_1, \ldots, q_{w-1} | (nl)^{w-1} \overline{\alpha L S})$ is an antisymmetric eigenfunction for the $(nl)^{w-1}$ configuration, then a completely antisymmetric eigenfunction for $(nl)^w$ can be written

$$(q_1, \ldots, q_w | (nl)^w \alpha L M_L S M_S)$$
$$= \sum_{\overline{\alpha L S}} (l^{w-1} \overline{\alpha L S} | \} l^w \alpha L S)(q_1, \ldots, q_w | ((nl)^{w-1} \overline{\alpha L S} nl) L M_L S M_S),$$

(2.9)

where the summation is now over all allowed terms $\overline{\alpha L S}$ of $(nl)^{w-1}$. The expansion coefficients $(l^{w-1} \overline{\alpha L S} | \} l^w \alpha L S)$, known as *coefficients of fractional parentage* (cfp), are normally evaluated using computer programs, but tables also are available. The tables published by Nielson and Koster (1963) contain all coefficients for the p^w, d^w and f^w configurations.

Thus, starting from the allowed terms of the subshell with occupation number two and using the fractional parentage expansions, the allowed terms of a subshell with any occupation can be derived. Table 2.1 lists the various terms and their seniority as a function of the occupation number of the subshells we will be concerned with. The notation is that used in the angular momentum programs. The terms for more than half-filled subshells $l^{2(2l+1)-w}$ are the same as those for l^w.

2.1.3 Construction of CSFs for general configurations

A general configuration consists of groups of equivalent electrons

$$(n_1 l_1)^{w_1} (n_2 l_2)^{w_2} \ldots (n_m l_m)^{w_m}, \qquad N = \sum_{a=1}^{m} w_a.$$

(2.10)

To construct the CSFs we start with the products of the antisymmetric eigenfunctions for the different groups of equivalent electrons

$$(Q_1 | (n_1 l_1)^{w_1} \alpha_1 L_1 M_{L_1} S_1 M_{S_1})(Q_2 | (n_2 l_2)^{w_2} \alpha_2 L_2 M_{L_2} S_2 M_{S_2})$$
$$\times \ldots \times (Q_m | (n_m l_m)^{w_m} \alpha_m L_m M_{L_m} S_m M_{S_m})$$

(2.11)

Table 2.1. List of possible terms and their seniority for commonly occurring subshells as produced by the TERMS program.

```
Subshell TERMS (2S+1, L, seniority)

 s(1)    2S1
 s(2)    1S0

 p(1)    2P1
 p(2)    1S0 1D2 3P2
 p(3)    2P1 2D3 4S3

 d(1)    2D1
 d(2)    1S0 1D2 1G2 3P2 3F2
 d(3)    2D1 2P3 2D3 2F3 2G3 2H3 4P3 4F3
 d(4)    1S0 1D2 1G2 3P2 3F2 1S4 1D4 1F4 1G4 1I4 3P4 3D4 3F4 3G4 3H4 5D4
 d(5)    2D1 2P3 2D3 2F3 2G3 2H3 4P3 4F3 2S5 2D5 2F5 2G5 2I5 4D5 4G5 6S5

 f(1)    2F1
 f(2)    1S0 1D2 1G2 1I2 3P2 3F2 3H2
```

where $\mathcal{Q}_1$ represents the co-ordinates $q_1, \ldots, q_{w_1}$, $\mathcal{Q}_2$ the co-ordinates $q_{w_1+1}, \ldots, q_{w_1+w_2}$, etc, with $\mathcal{Q}_m$ of the final factor representing $q_{N-w_m+1}, \ldots, q_N$. With the repeated use of the Clebsch–Gordan expansion (A.18) we can couple the product functions to total angular momenta $LM_L SM_S$ according to some specified coupling scheme. In this book the coupling is from left to right, for which the notation

$$\gamma LM_L SM_S = (n_1 l_1)^{w_1} \alpha_1 L_1 S_1 \ (n_2 l_2)^{w_2} \alpha_2 L_2 S_2 L_{12} S_{12}$$
$$\times (n_3 l_3)^{w_3} \alpha_3 L_3 S_3 \ L_{123} S_{123} \ldots (n_m l_m)^{w_m} \alpha_m L_m S_m \ LM_L SM_S$$
$$(2.12)$$

can be used. This coupling is shown graphically in figure 2.1. The function coupled in this way may be written

$$(q_1, \ldots, q_N | \gamma LM_L SM_S)^u, \qquad (2.13)$$

where the index u indicates that this function is antisymmetric with respect to co-ordinate permutations within each subshell, but not antisymmetric with respect to co-ordinate permutations between different subshells. The additional antisymmetrization can, however, be accomplished through the restricted permutations

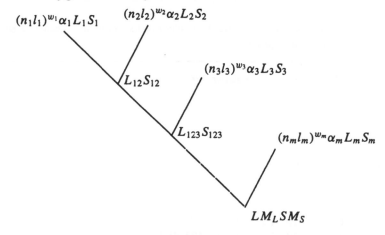

Figure 2.1. Coupling of subshells for a CSF.

$$(q_1, \ldots, q_N | \gamma L M_L S M_S)$$

$$= \left(\frac{\prod_{a=1}^{m} w_a!}{N!} \right)^{1/2} \sum_{\wp} (-1)^P \wp (q_1, \ldots, q_N | \gamma L M_L S M_S)^u,$$

$$(2.14)$$

where the sum is over all permutations involving co-ordinate exchange only between two different subshells such that the co-ordinate number within each subshell remains in an increasing order.

2.1.4 Properties of the CSFs

When coupling groups of equivalent electrons, none of the coupling formulae are concerned with the principal quantum numbers. Consequently, for a configuration γ, as given by (2.12), the function $(q_1, \ldots, q_N | \gamma L M_L S M_S)^u$ may be separated into a product of a radial factor, $\prod_{i=1}^{N} R(n_i l_i; r_i)$, and a spin-angular factor $(q'_1, \ldots, q'_N | \gamma' L M_L S M_S)^u$ where $q' = (\theta, \varphi, \sigma)$ and γ' defines angular and spin coupling.

As an example, let us consider the $1s2p\ {}^3P$ configuration state function, using the spectroscopic notation where M_L and M_S quantum numbers are omitted. Then, assuming orbitals are associated with co-ordinates by their position, we have

$$(q_1, q_2 | 1s2p\ {}^3P)^u = R(1s; r_1) R(2p; r_2)(q'_1, q'_2 | sp\ {}^3P)^u. \qquad (2.15)$$

The antisymmetrization operator for this case is

$$\mathcal{A} = \frac{1}{\sqrt{2}} (1 - \wp_{12}), \qquad (2.16)$$

where $\wp_{12}$ represents the permutation of co-ordinates 1 and 2. The effect of this permutation on the radial factor is obvious. For the spin-angular factor we have

$$\wp_{12}(q_1', q_2'|sp\ {}^3P)^u = (q_2', q_1'|sp\ {}^3P)^u \qquad \text{(by definition)}$$
$$= (q_1', q_2'|ps\ {}^3P)^u \qquad \text{(by (2.4))}.$$

Apply the antisymmetrization operator to (2.15) and using the above property, we get

$$(q_1, q_2|1s2p\ {}^3P) = \frac{1}{\sqrt{2}}R(1s;r_1)R(2p;r_2)(q_1', q_2'|sp\ {}^3P)^u$$

$$-\frac{1}{\sqrt{2}}R(2p;r_1)R(1s;r_2)(q_1', q_2'|ps\ {}^3P)^u.$$

$$(2.17)$$

2.1.5 Computational aspects

Several restrictions apply to the possible CSFs. The MCHF atomic structure package limits the orbital quantum numbers l and L to a maximum value of 10, but the occupation of *unfilled* subshells with $l \geqslant 3$ is restricted to at most two and their terms are restricted to those of d^2. Normally, the program GENCL is used to generate configuration states to be included in wave function expansions. After a complete run, GENCL writes information to a file cfg.inp which later can be read and interpreted by the other programs in the atomic structure package. In the cfg.inp file the configuration states are specified by first listing the subshells and their occupation in a linear form where, for example, $2p^3$ becomes 2p(3), followed on a separate line, by multiplicity $(2S + 1)$, the orbital symmetry (L) and the seniority, first for each subshell, and then for the resultant when coupling these subshells from left to right. For the latter, seniority is no longer relevant and the GENCL program simply assigns it a zero value. Table 2.2 displays the use of GENCL for producing an expansion for the configuration $1s^2 2s\ 2p^2 3s$ belonging to the ${}^3P$ term. Notice that for this configuration there are *two* possible CSFs. In this example, the $1s$ subshell is fully occupied, or closed, and can therefore only couple to ${}^1S$. For computational reasons, closed subshells are best specified separately. Occasionally, n values higher than $n = 9$ are desired. Though the programs in the package are not designed for extremely high n, higher n values in the ASCII collating sequence are ':', ';', '<', '=', '>', '?' for $n = 10, 11, 12, 13, 14, 15$, respectively. A number of programs assume that there will be at most five subshells (filled or unfilled) outside the closed subshells common to all CSFs.

2.2 Matrix elements of the Hamiltonian

When performing calculations using variational functions in the form of CSF expansions the matrix elements of the Hamiltonian need to be evaluated. The

Table 2.2. Example showing the use of GENCL for the production of the `cfg.inp` file.

```
             Header  ?
> Example
      Closed Shells  ?
> 1s
      Reference Set  ?
>2s(1)2p(2)3s(1)
                2  ?
>

          Active Set  ?
>

        Replacements  ?
>

        Final Terms  ?
>3P
                2  ?
>
```

Listing of the `cfg.inp` file

```
   Example
   1s
   2s( 1)  2p( 2)  3s( 1)
      2S1     3P2     2S1     2P0     3P0
   2s( 1)  2p( 2)  3s( 1)
      2S1     3P2     2S1     4P0     3P0
```

Hamiltonian can be written as the sum of a one-electron operator

$$F = \sum_{i=1}^{N} f_i = \sum_{i=1}^{N} \left(-\frac{1}{2} \nabla_i^2 - \frac{Z}{r_i} \right) \tag{2.18}$$

and a two-electron operator

$$G = \sum_{i<j}^{N} g_{ij} = \sum_{i<j}^{N} \frac{1}{r_{ij}}. \tag{2.19}$$

It is advantageous to treat these operators separately. The one-electron operator gives non-zero matrix elements only between CSFs that differ by at most one orbital. For the two-electron operator the CSFs may differ by at most two orbitals. Both the one- and two-electron operators are scalars (commute with $\mathbf{L}$ and $\mathbf{S}$), and thus the matrix elements are diagonal with respect to the quantum numbers, L, S, M_L and M_S. In addition, as was discussed in section 1.4.4, the matrix elements are independent of M_L and M_S. For this reason we will frequently leave out the magnetic quantum numbers in the notation for the matrix elements.

Due to the antisymmetry of the CSFs the matrix elements of the one- and two-electron operators do not depend on the indices i and j, and thus the matrix elements can be evaluated as $N\langle\gamma LS|f_N|\gamma'LS\rangle$ and $\frac{N(N-1)}{2}\langle\gamma LS|g_{N-1N}|\gamma'LS\rangle$, respectively. The operators f_N and g_{N-1N} act only on the variables $\mathbf{r}_N$ and $\mathbf{r}_{N-1}, \mathbf{r}_N$, respectively, but the evaluation of the matrix elements involves integration over *all* co-ordinates. Consequently, some spin-orbitals participate *actively* in the integration over the operators whereas others are *spectators*. Integration over the co-ordinates of the spectators often is zero or unity because of the orthonormality of the spherical harmonics and spin-functions. If we assume orthonormality of the radial functions so that

$$\int_0^\infty P(nl; r)P(n'l; r)dr = \delta_{nn'} \tag{2.20}$$

the integrals are zero or unity in all cases. In the present chapter we assume that this condition is satisfied, but in section 4.5 we will see how this condition can be relaxed resulting in additional terms and overlap integrals.

The evaluation of matrix elements is complicated by the coupling and antisymmetrization of the CSFs. Algebraic techniques pioneered by Racah, however, provide analytical expressions for the angular part, and the general matrix elements can be written in terms of radial integrals only (for a compact treatment see Fano (1965)). Let us consider some simple cases before introducing the codes which perform integrations for the general case.

2.2.1 One-electron matrix elements

Introducing the radial integral†

$$I(a, b) = -\frac{1}{2}\int_0^\infty P(a; r)\mathcal{L}P(b; r)\, dr, \tag{2.21}$$

where $l_a = l_b = l$, say, and the operator $\mathcal{L}$ is defined as

$$\mathcal{L} = \frac{d^2}{dr^2} + \frac{2Z}{r} - \frac{l(l+1)}{r^2}, \tag{2.22}$$

the general one-electron matrix element can be written,

$$\langle\gamma LS|\sum_{i=1}^{N}\left(-\frac{1}{2}\nabla_i^2 - \frac{Z}{r_i}\right)|\gamma'LS\rangle = \sum_{ab}w_{ab}I(a, b), \tag{2.23}$$

where w_{ab} are angular coefficients and the sum ab is on the orbitals, a being occupied in γ and b in γ'.

† Here we use the abbreviated notation a for $n_a l_a$.

As an example we consider the diagonal matrix element

$$\langle 1s2p \,^3P| \sum_{i=1}^{2} \left(-\frac{1}{2}\nabla_i^2 - \frac{Z}{r_i} \right) |1s2p \,^3P\rangle, \tag{2.24}$$

and the off-diagonal† matrix element

$$\langle 1s2p \,^3P| \sum_{i=1}^{2} \left(-\frac{1}{2}\nabla_i^2 - \frac{Z}{r_i} \right) |1s3p \,^3P\rangle. \tag{2.25}$$

Performing all algebra we obtain the rather simple expressions

$$\langle 1s2p \,^3P| \sum_{i=1}^{2} \left(-\frac{1}{2}\nabla_i^2 - \frac{Z}{r_i} \right) |1s2p \,^3P\rangle = I(1s, 1s) + I(2p, 2p),$$

$$\tag{2.26}$$

$$\langle 1s2p \,^3P| \sum_{i=1}^{2} \left(-\frac{1}{2}\nabla_i^2 - \frac{Z}{r_i} \right) |1s3p \,^3P\rangle = I(2p, 3p).$$

2.2.2   Two-electron matrix elements

To evaluate the two-electron Coulomb matrix elements, the operator must be expressed in a more suitable tensorial form. This can be done by first expanding the operator in terms of Legendre polynomials,

$$\frac{1}{r_{12}} = \sum_{k=0}^{\infty} \frac{r_<^k}{r_>^{k+1}} P_k(\cos\omega), \tag{2.27}$$

where

$$r_> = \max(r_1, r_2),$$
$$r_< = \min(r_1, r_2), \tag{2.28}$$

and ω is the angle between r_1 and r_2, and then using the spherical harmonic addition theorem

$$P_k(\cos\omega) = \sum_q (-1)^q C_q^{(k)}(\theta_1, \varphi_1) C_{-q}^{(k)}(\theta_2, \varphi_2) = \mathbf{C}^{(k)}(\theta_1, \varphi_1) \cdot \mathbf{C}^{(k)}(\theta_2, \varphi_2)$$

$$\tag{2.29}$$

to obtain the familiar multipole expansion,

$$\frac{1}{r_{12}} = \sum_{k=0}^{\infty} \frac{r_<^k}{r_>^{k+1}} \mathbf{C}^{(k)}(\theta_1, \varphi_1) \cdot \mathbf{C}^{(k)}(\theta_2, \varphi_2), \tag{2.30}$$

---

† Off-diagonal with respect to the configuration.

for the operator. Using the multipole expansion it can be shown that the general expression for the matrix element can be written

$$\langle \gamma LS | \sum_{i<j} \frac{1}{r_{ij}} | \gamma' LS \rangle = \sum_{abcd;k} v_{abcd;k} R^k(ab, cd), \qquad (2.31)$$

where $v_{abcd;k}$ are angular coefficients and the sum on $abcd$ is over orbitals ab occupied in γ and orbitals cd occupied in γ'. $R^k(ab, cd)$ is a so-called Slater integral given by

$$R^k(ab, cd) = \int_0^\infty \int_0^\infty P(a; r_1) P(b; r_2) \frac{r_<^k}{r_>^{k+1}} P(c; r_1) P(d; r_2) \, dr_1 \, dr_2. \quad (2.32)$$

The Slater integral has many symmetries. In fact,

$$R^k(ad, cb) \equiv R^k(cb, ad) \equiv R^k(cd, ab) \equiv R^k(ab, cd). \qquad (2.33)$$

A special notation is used for Slater integrals that depend only on two sets of quantum numbers:

$$F^k(a, b) \equiv R^k(ab, ab) \quad \text{and} \quad G^k(a, b) \equiv R^k(ab, ba). \qquad (2.34)$$

As an example of Coulomb matrix elements we consider

$$\langle 1s2p \,\, ^3P | \frac{1}{r_{12}} | 1s2p \,\, ^3P \rangle \qquad (2.35)$$

and

$$\langle 1s2p \,\, ^3P | \frac{1}{r_{12}} | 2p3d \,\, ^3P \rangle, \qquad (2.36)$$

which can be written (using the orbital notation of the code)

$$\langle 1s2p \,\, ^3P | \frac{1}{r_{12}} | 1s2p \,\, ^3P \rangle = F^0(1s, 2p) - \frac{1}{3} G^1(1s, 2p),$$
$$(2.37)$$
$$\langle 1s2p \,\, ^3P | \frac{1}{r_{12}} | 2p3d \,\, ^3P \rangle = -\frac{\sqrt{2}}{3} R^1(2p, 3d; 1s, 2p) + \frac{\sqrt{2}}{5} R^2(2p, 3d; 2p, 1s).$$

A diagonal Coulomb matrix element for an eigenfunction that has not been fully antisymmetrized through the restricted permutation (2.14) would only contain F^k integrals and for this reason these integrals are referred to as representing the 'direct' interaction whereas the G^k integrals represent the 'exchange' interaction. For equivalent electrons no explicit antisymmetrization through restricted permutations are needed. Thus, only F^k integrals appear in the diagonal Coulomb matrix element for such CSFs. As an example we consider

$$\langle 2p^2 \,\, ^3P | \frac{1}{r_{12}} | 2p^2 \,\, ^3P \rangle, \qquad (2.38)$$

for which we have

$$\langle 2p^2 \,\, ^3P | \frac{1}{r_{12}} | 2p^2 \,\, ^3P \rangle = F^0(2p, 2p) - \frac{1}{5} F^2(2p, 2p). \qquad (2.39)$$

2.2.3 Average energy of a configuration

For an approximate atomic state function (ASF) consisting of a single CSF, the energy functional is a single, diagonal matrix element. The expressions (2.23) and (2.31) for the one- and two-electron parts, respectively, can be simplified. Consider the general configuration state function defined in secton 2.1 where w_a electrons are assigned to the $n_a l_a$ subshell and m subshells are occupied. Performing the angular integration, the Hamiltonian or energy matrix element can be written

$$\mathcal{E}(\gamma LS) = \sum_{a=1}^{m} w_a \left[I(a,a) + \left(\frac{w_a - 1}{2} \right) \sum_{k=0}^{2l_a} f_k(l_a) F^k(a,a) \right]$$
$$+ \sum_{a=2}^{m} \left\{ \sum_{b=1}^{a-1} w_a w_b \left[F^0(a,b) + \sum_{k=|l_a-l_b|}^{(l_a+l_b)} g_k(l_a,l_b) G^k(a,b) \right] \right\}.$$

$$(2.40)$$

In general, the coefficients $f_k(l_a)$ and $g_k(l_a, l_b)$ depend not only on the configuration but also on the coupling. A useful concept introduced by Slater (1960) is the *average energy of a configuration*, denoted as $\mathcal{E}(av)$. This is a weighted average of all possible LS terms where the weighting factor is $(2L+1)(2S+1)$. In this case the coefficients have a simple formula, independent of the coupling, namely

$$f_k(l_a) = 1, \quad k = 0$$
$$= -\frac{2l_a + 1}{4l_a + 1} \begin{pmatrix} l_a & k & l_a \\ 0 & 0 & 0 \end{pmatrix}^2, \quad k > 0 \qquad (2.41)$$
$$g_k(l_a, l_b) = -\frac{1}{2} \begin{pmatrix} l_a & k & l_b \\ 0 & 0 & 0 \end{pmatrix}^2.$$

Furthermore, LS term energies can be expressed as

$$\mathcal{E}(\gamma LS) = \mathcal{E}(av) + \Delta\mathcal{E}(\gamma LS) \qquad (2.42)$$

where $\Delta\mathcal{E}(\gamma LS)$ has some useful properties:

(i)   Whenever a configuration has only one term, $\Delta\mathcal{E}(\gamma LS) = 0$. Consequently, all configurations consisting of only filled subshells or filled subshells and one electron have $\Delta\mathcal{E}(\gamma LS) = 0$.

(ii) $\Delta\mathcal{E}(l^w LS) = \Delta\mathcal{E}(l^{4l+2-w} LS)$; that is, the deviation for electrons and 'holes' is the same.

(iii) For a general configuration, the deviation is a sum of deviations for the unfilled subshells and deviations between subshells.

Because of these properties, tables have been prepared by Slater (1960) tabulating the deviations from the average energy for a number of cases.

Table 2.3. Example of the use of the program NONH for determining the energy expression for the wave function expansion shown in the cfg.inp file.

Listing of the cfg.inp file.

```
  1s( 1)   2p( 1)
     2S1      2P1      3P0
  2p( 1)   3d( 1)
     2P1      2D1      3P0
```

NONH program output

```
>Nonh
 FULL PRINT-OUT ? (Y/N)
>y
 STATE  (WITH  2 CONFIGURATIONS):
 -------------------------------
 CONFIGURATION  1 ( OCCUPIED ORBITALS= 2 ):  1s( 1)  2p( 1)
                         COUPLING SCHEME:      2S1     2P1
                                                           3P0
 CONFIGURATION  2 ( OCCUPIED ORBITALS= 2 ):  2p( 1)  3d( 1)
                         COUPLING SCHEME:      2P1     2D1
                                                           3P0
 ALL INTERACTIONS ? (Y/N)
>y
 <   1 |H|   1 > = < 1s( 1) 2p( 1) |H|  1s( 1) 2p( 1) >
                 COEFFICIENTS OF VARIOUS INTEGRALS
 E-TOTAL(ET) E-AVERAGE(EAV) (ET - EAV) INTEGRALS WITH OVERLAPS
    1.000000     1.000000      0.000000  I ( 1s, 1s)
    1.000000     1.000000      0.000000  I ( 2p, 2p)
    1.000000     1.000000      0.000000  F 0( 1s, 2p)
   -0.333333    -0.166667     -0.166667  G 1( 1s, 2p)

 <   2 |H|   1 > = < 2p( 1) 3d( 1) |H|  1s( 1) 2p( 1) >
                 COEFFICIENTS OF VARIOUS INTEGRALS
 E-TOTAL(ET) E-AVERAGE(EAV) (ET - EAV) INTEGRALS WITH OVERLAPS
   -0.471405     0.000000     -0.471405  R 1( 2p, 3d; 1s, 2p)
    0.282843     0.000000      0.282843  R 2( 2p, 3d; 2p, 1s)
 <   2 |H|   2 > = < 2p( 1) 3d( 1) |H|  2p( 1) 3d( 1) >
                 COEFFICIENTS OF VARIOUS INTEGRALS
 E-TOTAL(ET) E-AVERAGE(EAV) (ET - EAV) INTEGRALS WITH OVERLAPS
    1.000000     1.000000      0.000000  I ( 2p, 2p)
    1.000000     1.000000      0.000000  I ( 3d, 3d)
    1.000000     1.000000      0.000000  F 0( 2p, 3d)
    0.200000     0.000000      0.200000  F 2( 2p, 3d)
   -0.066667    -0.066667      0.000000  G 1( 2p, 3d)
   -0.257143    -0.042857     -0.214286  G 3( 2p, 3d)
    3 matrix elements        3 non-zero matrix elements
    100.0000 % dense          NF= 3 NG= 3 NR= 2 NL= 4
    Total number of terms = 12 The total number of integrals = 11
```

2.2.4   Computational aspects

In the MCHF atomic structure package the program NONH derives the expressions for the Hamiltonian matrix elements between CSFs as listed in the file cfg.inp. This file could be produced by GENCL, specifying 1s(1)2p(1) as the first member of the reference configuration, and 2p(1)3d(1) as the second, the final term being 3P. In simple cases like these, a text editor can also be used to produce the file. Table 2.3 shows the use of NONH for displaying these energy expressions. Because of symmetry, only the elements on or below the diagonal need to be considered. In showing the execution of the NONH program, user input follows a > symbol, whereas computer output is indented. The calculation computed three matrix elements of which all are non-zero, resulting in an interaction matrix that is 100% dense. The number of different F^k, G^k, R^k and I integrals is reported, the total number of terms in the energy expression and the total number of integrals. The program also produces an int.lst file which will be explained later, in chapter 4. Here we simply observe that NONH may be used to obtain energy matrix elements.

2.3   Exercises

(i)   Show that the F^k and G^k Slater integrals are always positive.

(ii)   Using GENCL determine all possible LS terms of the configuration 1s(2)2s(2)2p(2). Then, using NONH and the results of exercise (i), determine which of these terms has the lowest energy.

(iii)   By Hund's rule, the term with the greatest value of S for the given electron configuration, and of these the term with the greatest value of L, has the lowest energy. Using NONH, confirm this for the 1s(2)2s(2)2p(3) configuration.

(iv)   Using NONH, evaluate the matrix elements:

  (a) $\langle 1s^2\, 2s^2\, 2p^2\; ^3P | \mathcal{H} | 1s^2\, 2s^2\, 2p^2\; ^3P \rangle$

  (b) $\langle 1s^2\, 2s^2\, 2p^2\; ^3P | \mathcal{H} | 1s^2\, 2s^2 2p3p\; ^3P \rangle$

  (c) $\langle 1s^2\, 2s^2\, 2p^2\; ^3P | \mathcal{H} | 1s^2\, 2s^2\, 3p^2\; ^3P \rangle$.

  Which of these matrix elements is associated with the most angular data?

(v)   Using NONH, confirm that only F^k integrals appear in the expression for $\langle 2p^3\; ^2P | \mathcal{H} | 2p^3\; ^2P \rangle$. Show that for $\langle 2p^2(^3P)3p\; ^2P | \mathcal{H} | 2p^2(^3P)3p\; ^2P \rangle$ also G^k integrals appear.

Chapter 3

Hartree–Fock Calculations

3.1 The Hartree–Fock approximation

When Hartree–Fock theory was first under development, Hartree–Fock approximations to many-electron wave functions often were assumed to be single Slater determinants. As explained in chapter 1, such wave functions, in general, are not eigenfunctions of the total angular momentum operators. In current atomic physics it is customary to define the Hartree–Fock approximation as a stationary solution of the variational problem (1.60) for a single configuration state function and the Hartree–Fock energy, the stationary energy associated with this solution. A special case is the fixed- or frozen-core Hartree–Fock (FCHF) approximation where orbitals defining the core are not allowed to vary.

As shown in the previous chapter, the energy functional associated with a configuration state function can be written as a linear combination of one- and two-electron radial integrals. Thus, we need to consider variations of these integrals with respect to a variation in the radial functions from $P(nl; r)$ to $P(nl; r) + \delta P(nl; r)$.

Starting with the one-electron integral

$$I(nl, nl) = -\frac{1}{2} \int_0^\infty P(nl; r) \mathcal{L} P(nl; r) \, dr, \tag{3.1}$$

we have

$$\begin{aligned} \delta I(nl, nl) &= -\frac{1}{2} \int_0^\infty \delta P(nl; r) \mathcal{L} P(nl; r) \, dr - \frac{1}{2} \int_0^\infty P(nl; r) \mathcal{L} \delta P(nl; r) \, dr \\ &= -\int_0^\infty \delta P(nl; r) \mathcal{L} P(nl; r) \, dr \end{aligned} \tag{3.2}$$

where the latter follows from integration by parts and the zero boundary conditions at the origin and at infinity.

Next we consider the radial integrals arising from the two-electron Coulomb operator, the F^k and G^k Slater integrals. In order to derive the first-order variation of these integrals (and R^k in general) it is convenient to first replace variables (r_1, r_2) by (r, s) and introduce the functions

$$Y^k(ab; r) = r \int_0^\infty \frac{r_<^k}{r_>^{k+1}} P(a; s)P(b; s)\, ds$$

$$= \int_0^r \left(\frac{s}{r}\right)^k P(a; s)P(b; s)\, ds$$

$$+ \int_r^\infty \left(\frac{r}{s}\right)^{k+1} P(a; s)P(b; s)\, ds. \qquad (3.3)$$

Then

$$F^k(a, b) = \int_0^\infty P^2(a; r) \left(\frac{1}{r}\right) Y^k(bb; r)\, dr$$

$$= \int_0^\infty P^2(b; s) \left(\frac{1}{s}\right) Y^k(aa; s)\, ds, \qquad (3.4)$$

since the order of integration in the definition of F^k may be reversed. Similarly

$$G^k(a, b) = \int_0^\infty P(a; r)P(b; r) \left(\frac{1}{r}\right) Y^k(ab; r)\, dr. \qquad (3.5)$$

It is then straightforward to show that

$$\delta F^k(a, b) = 2(1 + \delta_{ab}) \int_0^\infty \delta P(a; r)P(a; r) \left(\frac{1}{r}\right) Y^k(bb; r)\, dr \qquad (3.6)$$

and

$$\delta G^k(a, b) = 2 \int_0^\infty \delta P(a; r)P(b; r) \left(\frac{1}{r}\right) Y^k(ab; r)\, dr. \qquad (3.7)$$

Instead of proceeding directly to the derivation of the general Hartree–Fock equations, let us start with some simple cases, each illustrating certain points, and concluding with the general case.

3.2 The Hartree–Fock equation for $1s\, 2p\ ^3P$

Our first example is that treated in detail in chapter 2. In this case, the only unknown functions are $P(1s; r)$ and $P(2p; r)$. According to section 1.5, the variational principle requires that the energy functional be stationary with respect to small variations in these functions. Combining the results of (2.26) and (2.37) we get

$$\mathcal{E}(1s2p\ ^3P) = I(1s, 1s) + I(2p, 2p) + F^0(1s, 2p) - (1/3)G^1(1s, 2p). \qquad (3.8)$$

However, this expression was derived on the assumption that all radial functions were normalized. For such constrained variation, Lagrange multipliers need to be introduced and the variational principle applied to

$$\mathcal{F}(P(1s; r), P(2p; r)) = \mathcal{E}(1s2p\ ^3P) + \lambda_{1s,1s}\langle 1s|1s\rangle + \lambda_{2p,2p}\langle 2p|2p\rangle \quad (3.9)$$

where $\langle nl|n'l\rangle = \int_0^\infty P(nl; r)P(n'l; r)\,dr$. Let us consider the variation of $\mathcal{F}$ with respect to a variation in $P(1s; r)$. Clearly, the variation of terms independent of $P(1s; r)$ are zero and the variation of the rest can be expressed in the form

$$\int_0^\infty \delta P(1s; r)Q(1s; r)\,dr = 0. \quad (3.10)$$

The contributions to $Q(1s; r)$ are as follows:

From $I(1s, 1s)$: $-\mathcal{L}P(1s; r)$

From $F^0(1s, 2p)$: $(2/r)Y^0(2p2p; r)P(1s; r)$

From $-(1/3)G^1(1s, 2p)$: $-(2/3r)Y^1(1s2p; r)P(2p; r)$ (3.11)

From $\lambda_{1s,1s}\langle 1s|1s\rangle$: $2\lambda_{1s,1s}P(1s; r)$.

Equation (3.10) will be satisfied for all variations $\delta P(1s; r)$ only if $Q(1s; r) = 0$. Collecting terms and changing sign we get

$$\left(\mathcal{L} - \frac{2}{r}Y^0(2p2p; r) - \varepsilon_{1s,1s}\right)P(1s; r) + \frac{2}{3r}Y^1(1s2p; r)P(2p; r) = 0 \quad (3.12)$$

where $\varepsilon_{1s,1s} = 2\lambda_{1s,1s}$. Applying the stationary condition to variations in $P(2p; r)$ leads to the condition

$$\left(\mathcal{L} - \frac{2}{r}Y^0(1s1s; r) - \varepsilon_{2p,2p}\right)P(2p; r) + \frac{2}{3r}Y^1(1s2p; r)P(1s; r) = 0. \quad (3.13)$$

Each of these equations may be written in the form

$$\left(\frac{d^2}{dr^2} + \frac{2}{r}[Z - Y(nl; r)] - \frac{l(l+1)}{r^2} - \varepsilon_{nl,nl}\right)P(nl; r) = \frac{2}{r}X(nl; r) \quad (3.14)$$

where $Y(nl; r)$ accounts for the screening of the nucleus by the presence of the other electrons and $X(nl; r)$ arises from antisymmetry and is called the exchange term. The parameter $\varepsilon_{nl,nl}$ is called the diagonal energy parameter. The boundary conditions are $P(nl; 0) = 0$ and $P(nl; r) \to 0$ as $r \to \infty$. Written in this form, the equation is a second-order differential equation of boundary value type, but analysing $X(nl; r)$ more carefully, it becomes clear that the equation is an integro-differential equation. For example, suppose we define the Y^k function of (3.3) in terms of an operator

$$Y^k(a\bullet; r) = r \int_0^\infty \frac{r_<^k}{r_>^{k+1}} P(a; s) \bullet ds \quad (3.15)$$

so that $Y^k(ab; r) = Y^k(a\bullet; r)P(b; r)$. In other words, the integral operator operating on a function includes the function in the integrand. Then (3.12) for the radial function $P(1s; r)$ becomes

$$\left(\mathcal{L} - \frac{2}{r}Y^0(2p2p; r) + \frac{2}{3r}P(2p; r)Y^1(2p\bullet; r) - \varepsilon_{1s,1s}\right)P(1s; r) = 0. \quad (3.16)$$

In terms of this operator, which combines the properties of a differential operator with that of an integral operator, the equation is an eigenvalue problem. Such equations are called 'integro-differential'. In the present case, the equation for $P(1s; r)$ is a linear, homogeneous equation; in particular, if $P(nl; r)$ is a solution, $cP(nl; r)$ is also a solution. We will see later that, in some cases, the integro-differential eigenvalue problem is non-linear. The nuclear potential together with the the Y^k functions that contribute to the screening is sometimes referred to as the *local* or *direct* potential whereas Y^k functions that contribute to the exchange function and are of the integral equation type define what is referred to as the *non-local* potential.

3.3 The self-consistent field procedure

From current estimates of $P(1s; r)$ and $P(2p; r)$, functions $Y(nl; r)$ and $X(nl; r)$ can be computed. The diagonal energy parameter can be estimated using the Rayleigh quotient. The latter is obtained by multiplying (3.14) by $P(nl; r)$, integrating, dividing by $\langle nl|nl\rangle$ (which is unity in our case) to yield

$$\varepsilon_{nl,nl} = \langle nl|\mathcal{L}|nl\rangle - \int_0^\infty P(nl; r)\frac{2}{r}[Y(nl; r)P(nl; r) + X(nl; r)]\ \mathrm{d}r. \quad (3.17)$$

Then the differential equation has the form

$$\left(\frac{\mathrm{d}^2}{\mathrm{d}r^2} + f(r) - \varepsilon_{nl,nl}\right)P(nl; r) = g(r) \quad (3.18)$$

with boundary conditions $P(nl; 0) = 0$ and $P(nl; r) \to 0$ as $r \to \infty$. Solving the equations for $P(1s; r)$ and $P(2p; r)$ and normalizing them, we get new, and hopefully better, estimates of the radial functions. This process is repeated until the current estimates and the computed estimates are 'self-consistent'. For the Hartree equations, Hartree described the process in terms of 'fields' from which the term 'self-consistent field' (SCF) derives, but for the Hartree–Fock equations, the emphasis is on the radial charge distribution for each orbital. Then the SCF process can be depicted, in a manner similar to that used by Hartree (1927),

> Initial radial functions
> ⇓
> For each radial function
>   Compute direct and exchange potential
>   Determine diagonal energy parameter
>   Solve the differential equation
> ⇓
> Final radial functions

Solving the Hartree–Fock equations is not entirely straightforward. As long as $g(r)$ is not identically zero, solutions satisfying the boundary conditions exist for all values of $\varepsilon_{nl,nl}$. However, it should be remembered that the Hartree–Fock equations for the $1s3p\ ^3P$ state, for example, are exactly the same as those for the lowest $1s2p\ ^3P$, that the radial equations are eigenvalue problems with many solutions. In solving the coupled systems of equations, it is then important to control convergence to the desired solution. The Hartree–Fock program achieves this by node counting.

Specifically, the number of changes of sign in the solution $P(nl; r)$ should be $n - l - 1$. In practice, node counting is somewhat of an art since, particularly initially when the potential and exchange may be quite 'non-physical', spurious oscillations may be present and we will see later that even the exact solutions may have oscillations in the large-r or 'tail' region of a function. Not as much is known about f orbitals: because of their r^4 behaviour near the origin (see equation (1.26)), small oscillations have been observed in this region but they usually disappear as the iterations converge.

Table 3.1 shows a Hartree–Fock calculation for $1s2p\ ^3P$ using the interactive Hartree–Fock program HF. The program starts with a few prompts for describing the nature of the problem to be solved. ATOM is a convenient label for the case (up to six characters) and does not affect the computation. TERM designates the spectroscopic term and hence the energy expression defining the stationary conditions, and Z is the nuclear charge. In many cases there will be filled subshells and these can be specified separately from the 1s(1)2p(1) open subshells. It is possible to perform calculations where some orbitals are fixed, so the answer to the next prompt specifies the functions to be varied. The program sets a number of default parameters which can be overridden by user input, whenever needed. At this point, the program has all the information needed to attempt to perform the calculation.

Initial estimates are formed for the two radial functions as screened hydrogenic functions with effective nuclear charge

$$Z_{\text{eff}} = Z - \sigma(nl), \tag{3.19}$$

where $\sigma(nl)$ is a screening constant. The hf.log file shown in table 3.2 informs

Table 3.1. Hartree-Fock calculation for $1s2p\ ^3P$.

```
>HF
 Enter ATOM,TERM,Z
 Examples: O,3P,8. or Oxygen,AV,8.
>He,3P,2.
 List the CLOSED shells in the fields indicated (blank line if none)
 ... ... ... ... ... ... ... ... etc.
>
 Enter electrons outside CLOSED shells (blank line if none)
 Example: 2s(1)2p(3)
>1s(1)2p(1)
 There are  2 orbitals as follows:
   1s  2p
 Orbitals to be varied: ALL/NONE/=i (last i)/comma delimited list/H
>all
 Default electron parameters ? (Y/N/H)
>y
 Default values for remaining parameters? (Y/N/H)
>y
         WEAK ORTHOGONALIZATION DURING THE SCF CYCLE=   T
         SCF CONVERGENCE TOLERANCE (FUNCTIONS)     = 1.00D-08
         NUMBER OF POINTS IN THE MAXIMUM RANGE     = 220

         ITERATION NUMBER  1
         ----------------
         SCF CONVERGENCE CRITERIA (SCFTOL*SQRT(Z*NWF)) =   2.0D-08
             EL        ED          AZ          NORM        DPM
             1s     3.5097389   5.6405312   1.0018151   4.18D-03
             2p     0.2599702   0.3599178   1.0946569   3.61D-02
             2p     0.2628234   0.3932430   0.9662618   9.78D-03
             2p     0.2629724   0.3956921   0.9973008   5.58D-04
 LEAST SELF-CONSISTENT FUNCTION IS 1s: WEIGHTED MAXIMUM DPM=4.18D-03
         ITERATION NUMBER  2
         ----------------
         SCF CONVERGENCE CRITERIA (SCFTOL*SQRT(Z*NWF))=  4.0D-08
             EL        ED          AZ          NORM        DPM
             1s     3.4678763   5.6314990   0.9972158   1.85D-03
             2p     0.2631160   0.3966045   0.9984069   3.98D-04
             1s     3.4675011   5.6313578   0.9998921   4.08D-05
             2p     0.2631197   0.3966438   0.9999510   1.04D-05
 LEAST SELF-CONSISTENT FUNCTION IS 1s: WEIGHTED MAXIMUM DPM=4.08D-05
```

Table 3.1. (continued)

```
ITERATION NUMBER  3
-----------------
SCF CONVERGENCE CRITERIA (SCFTOL*SQRT(Z*NWF))=  8.0D-08
     EL        ED            AZ          NORM        DPM
     1s     3.4674906     5.6313545    0.9999972   1.02D-06
     2p     0.2631197     0.3966453    0.9999986   2.71D-07
     1s     3.4674904     5.6313545    0.9999999   2.74D-08
     2p     0.2631197     0.3966453    1.0000000   8.06D-09
LEAST SELF-CONSISTENT FUNCTION IS 1s: WEIGHTED MAXIMUM DPM=2.74D-08

ITERATION NUMBER  4
-----------------
SCF CONVERGENCE CRITERIA (SCFTOL*SQRT(Z*NWF)) =   1.6D-07
     EL        ED            AZ          NORM        DPM
     1s     3.4674904     5.6313545    1.0000000   7.68D-10
     2p     0.2631197     0.3966453    1.0000000   2.49D-10

 Additional parameters ? (Y/N/H)
>n
```

us that the initial screening constants were zero for $1s$ and unity for $2p$, in agreement with an intuitive concept of $1s$ being an inner orbital and $2p$ an outer orbital. The calculation then proceeds to compute the potential, exchange and diagonal energy parameter ED for each radial function in turn. Tabulated is the value of ED, $AZ = P(nl; r)/r^{l+1}$ as $r \to 0$ (1.26), the normalization integral of the computed solution, NORM, and a quantity related to the maximum change in the new solution over the previous estimate, DPM (Froese Fischer 1986). Since the radial functions are assumed to be normalized to unity, as the solutions converge, NORM should approach unity and clearly the DPM should approach zero. After solving the equations in turn, the program solves the least self-consistent orbital a few times before starting a new iteration.

The SCFTOL is a measure of the degree to which the system is expected to be self-consistent. Rather than failing some high expectation, the program relaxes this criterion so that lower self-consistency is met. If the desired degree of self-consistency is not met, the calculation can be 'recycled'. A file wfn.out was formed by HF: by moving (or renaming) this file to wfn.inp and running HF again, higher accuracy may be achieved.

After the iterations have converged, the program can be requested to compute additional information. We will not describe this option here. Further details are found in appendix D.

The hf.log file provides more information about the nature of the solution. Some of the orbital parameters are printed: the diagonal energy parameter,

Table 3.2. hf.log file for $1s2p\ ^3P$

```
        HARTREE-FOCK WAVE'FUNCTIONS FOR  He    3P    Z =  2.0
                 Core =
    Configuration =  1s(   1)  2p(   1)


        INPUT DATA
        ----- ----
                NL  SIGMA METH ACC OPT
        1   1s  1  0    0.0   1 0.0   0
        2   2p  2  1    1.0   1 0.0   0

        INITIAL ESTIMATES
           NL     SIGMA      E(NL)    AZ(NL)    FUNCTIONS
           1s     0.00      0.000     5.657    SCREENED HYDROGENIC
           2p     1.00      0.000     0.459    SCREENED HYDROGENIC

           NUMBER OF FUNCTIONS ITERATED         =     2
           MAXIMUM WEIGHTED CHANGE IN FUNCTIONS =  0.77D-09

   nl      E(nl)        I(nl)      KE(nl)    Rel(nl)  S(nl)     Az(nl)
   1s   3.4674904   -1.999877   1.979761  -0.000105  0.016   5.631354
   2p   0.2631197   -0.397692   0.151676  -0.000002  0.926   0.396645

   nl     Delta(R)     1/R**3      1/R        R       R**2
   1s       2.524      0.0000    1.98982   0.75621   0.76572
   2p       0.000      0.0652    0.27468   4.65644  26.35495

        TOTAL ENERGY (a.u.)
        ----- ------
        Non-Relativistic      -2.13143707   Kinetic      2.13143707
        Relativistic Shift    -0.00010671   Potential   -4.26287415
        Relativistic          -2.13154378   Ratio       -1.999999998
```

$E(nl) = \varepsilon_{nl,nl}$; $I(nl) \equiv I(nl, nl)$; the kinetic energy, $KE(nl)$, defined as

$$E^{\text{kinetic}}(nl) = -\frac{1}{2}\int_0^\infty P(nl; r)\left[\frac{d^2}{dr^2} - \frac{l(l+1)}{r^2}\right] P(nl; r)\, dr; \qquad (3.20)$$

a relativistic shift correction (described later in chapter 7); the effective screening constant $\sigma(nl)$ for the final solution denoted by $S(nl)$ in the computer output and $AZ(nl)$.

Hartree–Fock radial functions differ from screened hydrogenic functions. Even so, it is useful to define an effective screening constant for a Hartree–Fock function in that it provides some global information about the function. Screening constants can be defined in many ways, but Hartree (1957) suggested that the screening constant be such that

$$\langle r \rangle_{HF} = \langle r \rangle_{Z_{\text{eff}}} \tag{3.21}$$

where $\langle r \rangle_{Z_{\text{eff}}}$ is the mean radius of a hydrogenic atom with $Z = Z_{\text{eff}}$. In other words, the screening parameter is defined such that the mean radius of the Hartree–Fock orbital and that of a screened hydrogenic orbital is the same. Table 3.2 shows that the screening of the $1s$ is indeed small but that the $1s$ does not totally screen the nucleus for the $2p$.

The next set of parameters is the expectation values of the moments as indicated, with Delta(R) the expectation of $\frac{1}{4\pi}\delta(r)r^{-2}$. Following this is total energy information. The total energy, the relativistic shift and the resulting relativistic energy is printed. In addition, the non-relativistic energy is divided into the kinetic energy and the potential energy. For an exact calculation, the virial theorem (see, for example, Cowan 1981) requires that the ratio of the potential energy to kinetic energy be exactly -2, and therefore Total Energy $= -$ Kinetic Energy. Deviations from this relationship are indications of numerical inaccuracy, lack of convergence or that some of the orbitals were fixed.

3.4 Hartree–Fock solutions for the ground state of lithium

Angular momentum theory assumes spin-orbitals are orthonormal. Often orthogonality follows from spin and orbital quantum numbers being different, but in the $1s^2 2s\ {}^2S$ ground state of lithium, for example, this assumption requires orthogonality of the $1s$ and $2s$ radial functions. The stationary conditions now apply to the functional

$$
\begin{aligned}
\mathcal{F}(P(1s;r), &P(2s;r)) \\
&= \mathcal{E}(1s^2 2s\ {}^2S) + \lambda_{1s,1s}\langle 1s|1s\rangle + \lambda_{2s,2s}\langle 2s|2s\rangle + \lambda_{1s,2s}\langle 1s|2s\rangle
\end{aligned}
\tag{3.22}
$$

where

$$\mathcal{E}(1s^2 2s\ {}^2S) = 2I(1s, 1s) + I(2s, 2s) + F^0(1s, 1s) + 2F^0(1s, 2s) - G^0(1s, 2s). \tag{3.23}$$

Applying the variational condition we get equations of the form

$$
\left(\frac{d^2}{dr^2} + \frac{2}{r}[Z - Y(nl; r)] - \frac{l(l+1)}{r^2} - \varepsilon_{nl,nl} \right) P(nl; r)
$$

$$
= \frac{2}{r}X(nl; r) + \varepsilon_{nl,n'l}P(n'l; r)
\tag{3.24}
$$

where $\varepsilon_{1s,2s} = \lambda_{1s,2s}/2$ and $\varepsilon_{2s,1s} = \lambda_{1s,2s}$ or

$$\varepsilon_{1s,2s} = (1/2)\varepsilon_{2s,1s}. \qquad (3.25)$$

Multiplying (3.24) by $P(n'l; r)$, integrating and solving for $\varepsilon_{nl,n'l}$, we get

$$\varepsilon_{nl,n'l} = \langle n'l|\mathcal{L}|nl\rangle - \int_0^\infty P(n'l; r)\frac{2}{r}\left[Y(nl; r)P(nl; r) + X(nl; r)\right]\,dr. \quad (3.26)$$

Table 3.3. Hartree–Fock calculation for $1s^2 2s\ ^2S$ of Li.

```
>HF
 Enter ATOM,TERM,Z
 Examples: O,3P,8. or Oxygen,AV,8.
>Li,2S,3.
 List the CLOSED shells in the fields indicated (blank line if none)
 ... ... ... ... ... ... ... ... etc.
> 1s
 Enter electrons outside CLOSED shells (blank line if none)
 Example: 2s(1)2p(3)
>2s(1)
 There are   2 orbitals as follows:
    1s  2s
 Orbitals to be varied: ALL/NONE/=i (last i)/comma delimited list/H
>all
 Default electron parameters ? (Y/N/H)
>y
 Default values for remaining parameters? (Y/N/H)
>y
     WEAK ORTHOGONALIZATION DURING THE SCF CYCLE=   T
     SCF CONVERGENCE TOLERANCE (FUNCTIONS)      = 1.00D-08
     NUMBER OF POINTS IN THE MAXIMUM RANGE      = 220

     ITERATION NUMBER   1
     ----------------
 SCF CONVERGENCE CRITERIA (SCFTOL*SQRT(Z*NWF)) =   2.4D-08
 C( 1s 2s) =   -0.02936   V( 1s 2s) =    -2.40705   EPS = 0.012196
 E( 2s 1s) =   -0.00105   E( 1s 2s) =    -0.00053
             EL       ED         AZ        NORM       DPM
             1s    5.7027840   9.6820916  0.7852403  3.12D-01
             2s    0.2453924   6.6345765  0.0134856  7.91D-01
             2s    0.4477557   0.4456170  3.5165530  7.64D-01
             2s    0.3267466   1.7139117  1.2279486  3.21D-01
< 1s| 2s>= 9.6D-02
LEAST SELF-CONSISTENT FUNCTION IS 1s :WEIGHTED MAXIMUM DPM=4.42D-01
```

Table 3.3. (continued)

```
ITERATION NUMBER  2
----------------

SCF CONVERGENCE CRITERIA (SCFTOL*SQRT(Z*NWF)) =   4.9D-08
C( 1s 2s) =    0.02168   V( 1s 2s) =    -2.22040   EPS =-0.009765
E( 2s 1s) =    0.00300   E( 1s 2s) =     0.00150
            EL        ED           AZ          NORM        DPM
            1s     4.9689427    9.1974739   1.0550911   4.54D-02
            2s     0.3801757    1.3855788   0.9826618   7.41D-02
            2s     0.3928971    1.3776565   1.1131150   4.16D-02
            1s     4.9720462    9.2707885   1.0225031   7.04D-03
< 1s| 2s>=-2.7D-03
LEAST SELF-CONSISTENT FUNCTION IS 2s: WEIGHTED MAXIMUM DPM=4.16D-02


     . . . .


ITERATION NUMBER  7
----------------
SCF CONVERGENCE CRITERIA (SCFTOL*SQRT(Z*NWF)) =   1.6D-06
C( 1s 2s) =    0.00000   V( 1s 2s) =    -2.19991   EPS = 0.000000
E( 2s 1s) =    0.00573   E( 1s 2s) =     0.00286
            EL        ED           AZ          NORM        DPM
            1s     4.9554830    9.2603703   1.0000000   1.43D-08
            2s     0.3926457    1.4468006   0.9999999   1.29D-08
< 1s| 2s>= 8.7D-09

Additional parameters ? (Y/N/H)
>n
```

A similar equation holds for $\varepsilon_{n'l',nl}$. For exact solutions, these two values would also satisfy (3.25) but for approximate estimates the latter will not hold. Thus a computational procedure must ensure that, as the iterations converge, equation (3.25) will also be satisfied.

Table 3.3 shows the interactive execution of the HF program. This time, the input shows 1s(2) being part of a filled subshell and the only other electron in the configuration 2s(1) is an open subshell. The initial estimates are formed and the 2s orbital made orthogonal to the 1s orbital. The output also shows that *weak orthogonalization* is the default, which means that orbitals will be orthogonalized only at the end of an iteration, or cycle.

The calculation now proceeds to a *rotation analysis* phase which, in the present case is not essential, but in most cases, assists the rate of convergence. The stationary conditions which we have applied sequentially imply that the final, converged radial functions are stationary with respect to rotations, i.e. the simul-

Table 3.4. hf.log file for $1s^22s\ ^2S$ of Li.

```
HARTREE-FOCK WAVE FUNCTIONS FOR  Li    2S    Z = 3.0
   Core            = 1s(   2)
   Configuration = 2s(   1)

   INPUT DATA
   ----- ----
         NL  SIGMA METH ACC OPT
1   1s  1  0    1.0   1 0.0   0
2   2s  2  0    2.0   1 0.0   0

   INITIAL ESTIMATES
   NL     SIGMA    E(NL)    AZ(NL)    FUNCTIONS
   1s     1.00     0.000    7.071     SCREENED HYDROGENIC
   2s     2.00     0.000    1.414     SCREENED HYDROGENIC

   NUMBER OF FUNCTIONS ITERATED          =     2
   MAXIMUM WEIGHTED CHANGE IN FUNCTIONS  =  0.20D-07

                 ATOM Li        TERM 2S
   nl    E(nl)      I(nl)     KE(nl)      Rel(nl)    S(nl)    Az(nl)
   1s  4.9554830 -4.443145  3.611956  -0.000329  0.383   9.260370
   2s  0.3926457 -0.827358  0.208815  -0.000016  1.451   1.446801

   nl    Delta(R)    1/R**3     1/R        R       R**2
   1s      6.824     0.0000   2.68503  0.57312   0.44680
   2s      0.167     0.0000   0.34539  3.87366  17.73846

   TOTAL ENERGY (a.u.)
   ----- ------
   Non-Relativistic      -7.43272693   Kinetic        7.43272693
   Relativistic Shift    -0.00054376   Potential    -14.86545385
   Relativistic          -7.43327069   Ratio         -2.000000000
```

taneous perturbation of $P(1s; r)$ and $P(2s; r)$, maintaining orthogonality, so that

$$P(1s; r) \rightarrow P(1s; r) + \epsilon P(2s; r)$$
$$P(2s; r) \rightarrow P(2s; r) - \epsilon P(1s; r), \tag{3.27}$$

where ϵ is assumed to be small. Let us consider the total energy as a function of the rotation parameter ϵ, say $E(1s^22s; \epsilon)$, expanded in a series as†

† In fact, the expansion is a polynomial in ϵ of degree $\leqslant 4$.

$$E(1s^2 2s; \epsilon) = E_0 + \epsilon \left(\frac{\partial E}{\partial \epsilon} \right)_{\epsilon=0} + \epsilon^2 \left(\frac{\partial^2 E}{\partial \epsilon^2} \right)_{\epsilon=0} + \dots . \tag{3.28}$$

This energy should be stationary with respect to rotations, leading to the condition

$$\partial E(1s^2 2s; \epsilon)/\partial \epsilon = 0. \tag{3.29}$$

For converged solutions this condition will be satisfied for $\epsilon = 0$. At intermediate stages,

$$c(\text{ 1s 2s}) = \left(\partial E(1s^2 2s; \epsilon)/\partial \epsilon \right)_{\epsilon=0},$$

will be non-zero. Differentiating the series expansion for $E(1s^2 2s; \epsilon)$ we get

$$\frac{\partial E}{\partial \epsilon} = \left(\frac{\partial E}{\partial \epsilon} \right)_{\epsilon=0} + 2\epsilon \left(\frac{\partial^2 E}{\partial \epsilon^2} \right)_{\epsilon=0} + \dots . \tag{3.30}$$

Assuming the higher-order terms are negligible, a rotation parameter EPS can be computed so that new orbitals satisfy the stationary condition to first order. This requires the computation of both the first- and second-order variation of the energy with respect to rotation. This can be done exactly through an analysis of the variation of each radial integral (Froese Fischer 1986). In the HF (and also the MCHF to be described later),

$$v(\text{1s 2s}) = -2 \left(\frac{\partial^2 E}{\partial \epsilon^2} \right)_{\epsilon=0}$$

so that EPS = c(1s 2s) / v(1s 2s) in the present case.

Once the radial functions have been rotated, the Lagrange multipliers can be computed. Substituting the computed values using (3.24) a value of $\lambda_{1s,2s}$ can be obtained from which the off-diagonal Lagrange multipliers are derived satisfying (3.25). Then the orbitals are updated in turn, solving the differential equations for the stationary conditions, followed by two solutions of the least self-consistent orbital. Notice the relatively large oscillations in ED for the first three iterations of 2s. The solution of a differential equation produces a function that we call P^{out}, where $P^{in} = P^{(m)}$, the current estimate. Normally, $P^{(m+1)} = P^{out}$, but when oscillations occur, convergence can be improved by the introduction of an *accelerating* parameter α. Then $P^{(m+1)} = (1 - \alpha)P^{out} + \alpha P^{(m)}$, normalized. The HF program allows the user to specify α as the accelerating ACC parameter for each orbital, but it also applies its own strategy. Usually only one orbital is subject to substantial oscillations. The orbital 'in trouble' is the one that is corrected successively after each is updated in turn. When this occurs, the program monitors successive values of ED and increases the ACC parameter until oscillations cease, at which point it begins to decrease the parameter. Notice that in iteration number 2, the ED parameter for 2s is now quite stable. Not all the iterations are shown since the calculation converges rapidly thereafter. At the end of each iteration the current estimates are orthogonalized.

The hf.log file is shown in table 3.4. The default screening parameters for the initial estimates were 1.0 and 2.0 for $1s$ and $2s$, respectively. The screening parameters predicted from the solution are quite different, namely 0.383 and 1.451, respectively. The $1s$ subshell exhibits some *self-screening* since it is doubly occupied. At the same time, the $2s$ electron is able to penetrate the $1s$ subshell to some extent.

3.5   The Hartree–Fock solutions for $1s2s$ ^3S and ^1S states in He

The $1s2s$ ^3S and ^1S states in helium are similar in that both have the same configuration and only the coupling of the spin momenta differs. This has an unexpectedly large effect on the Hartree–Fock approximation and the ease with which the HF equations can be solved. Whereas the ^3S state is the lowest of its symmetry the ^1S has the same symmetry as $1s^2$ ^1S and so the Hartree–Fock solution for $1s2s$ ^1S is a stationary state and not an energy minimum. It is an approximation to the second ^1S eigenvalue of the Hamiltonian. The two approximations differ in the way the energy varies with respect to a rotation of the orbital basis.

The energy expression for this system is

$$\mathcal{E}(1s2s) = I(1s, 1s) + I(2s, 2s) + F^0(1s, 2s) \pm G^0(1s, 2s) \qquad (3.31)$$

where the $+$ sign applies to the ^1S state, and the $-$ to the ^3S. The space part of the configuration states is proportional to (see section 2.1.4)

$$P(1s; r_1)P(2s; r_2) \pm P(2s; r_1)P(1s; r_2). \qquad (3.32)$$

Consider a rotation of the radial basis to a new basis. Let **P** be the column vector $[P(1s; r), P(2s; r)]'$ and let $\mathbf{P^*} = \mathbf{OP}$, where **O** is an orthogonal matrix. Then $\mathbf{P} = \mathbf{O'P^*}$ or

$$\left[\begin{array}{c} P(1s; r) \\ P(2s; r) \end{array} \right] = \left[\begin{array}{cc} a & b \\ -b & a \end{array} \right] \left[\begin{array}{c} P^*(1s; r) \\ P^*(2s; r) \end{array} \right].$$

For ^3S it is easy to show that the functional form of the energy expression does not change, that the energy itself is the same whether computed in the original or transformed basis. But it is even simpler to show that the wave function does not change. This is most easily realized by noting that the space part can be written as a determinant which is always invariant under orthogonal transformations.

A similar study for ^1S shows that the form of the energy expression changes. The transformed space part, in fact, is then a linear combination of the space parts of configuration state, $\Phi(1s2s$ $^1S)$ and the function $\{\Phi(1s^2$ $^1S) - \Phi(2s^2$ $^1S)\}/\sqrt{2}$.

Table 3.5. Hartree–Fock calculations for $1s2s$ ^1S for helium.

```
>HF
 Enter ATOM,TERM,Z
 Examples: O,3P,8. or Oxygen,AV,8.
>He,1S,2.
 List the CLOSED shells in the fields indicated (blank line if none)
 ... ... ... ... ... ... ... ... etc.
>
 Enter electrons outside CLOSED shells (blank line if none)
 Example: 2s(1)2p(3)
>1s(1)2s(1)
 There are   2 orbitals as follows:
   1s  2s
 Orbitals to be varied: ALL/NONE/=i (last i)/comma delimited list/H
>all
 Default electron parameters ? (Y/N/H)
>y
 Default values for remaining parameters? (Y/N/H)
>n
 Default values (NO,STRONG) ? (Y/N/H)
>n
 Enter values in FORMAT(I3,1X,L1)
>220,t
 Default values for PRINT, SCFTOL ? (Y/N/H)
>y
 Default values for NSCF, IC ? (Y/N/H)
>y
 Default values for TRACE ? (Y/N/H)
>y
   WEAK ORTHOGONALIZATION DURING THE SCF CYCLE=   F
   ITERATION NUMBER  1
   ----------------
   SCF CONVERGENCE CRITERIA (SCFTOL*SQRT(Z*NWF)) =   2.0D-08
   C( 1s 2s) =    -0.20239   V( 1s 2s) =    -1.67309   EPS = 0.120967
   E( 2s 1s) =    -0.33114   E( 1s 2s) =    -0.33114
             EL        ED           AZ          NORM        DPM
             1s     3.4070067    5.7688840    0.9648931    5.33D-03
      < 1s| 2s>= 1.4D-03
             2s     0.3832448    0.2800888    1.0700990    1.08D-02
             2s     0.3836325    0.2998705    1.0897846    3.04D-03
             1s     3.4126601    5.7650312    0.9634465    8.39D-04
      < 1s| 2s>= 8.0D-04
      . . .
      TOTAL ENERGY (a.u.)
      ----- ------
   Non-Relativistic      -2.16985446    Kinetic      2.16985881
   Relativistic Shift    -0.00011395    Potential   -4.33971326
   Relativistic          -2.16996840    Ratio       -1.999997995
```

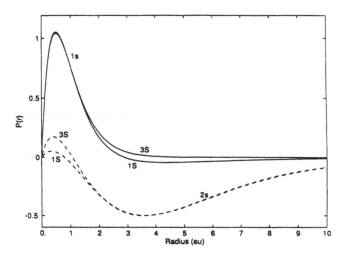

Figure 3.1. Comparison of the $1s2s\ {}^{3}S$ and ${}^{1}S$ Hartree–Fock radial functions.

In the limit of high Z, as will be shown in the next chapter, these two basis states are degenerate and this is the cause of the numerical problems associated with the computation of the Hartree–Fock solution for $1s2s\ {}^{1}S$.

In the $1s2s\ {}^{3}S$ case, the Hartree–Fock radial functions are not unique *unless* another condition is introduced. Koopmans (1933) suggested that the desirable transformation was the one that also minimized the core energy. More precisely, he showed that a transformation of the radial functions transformed the Lagrange multipliers, that the physically desirable solution was the one for which the diagonal energy parameters assumed extreme values, and that this solution was characterized by a zero value for the off-diagonal Lagrange multiplier. Thus the definition of the Hartree–Fock solution requires off-diagonal Lagrange multipliers to be zero whenever the wave function is invariant with respect to rotations of the radial basis.

The Hartree–Fock solution for $1s2s\ {}^{1}S$ can be obtained provided some critical steps are followed. First a calculation should be performed for the rapidly convergent $1s2s\ {}^{3}S$ state and the wave function file wfn.out moved to wfn.inp to be used as initial estimates for the singlet state. Table 3.5 shows part of the calculation. Notice that now one of the default values needs to be changed: the STRONG parameter should be set to true so that after every orbital update, results are considered for orthogonalization. This is called *strong* orthogonality compared to the *weak* orthogonality assumed in earlier examples. Of course, the orbital that is orthogonalized should be less self-consistent than the orbital to which it is orthogonalized. With these two special steps, the calculation converges, although the virial theorem shows that results are not as accurate as found in previous cases.

As soon as input data describing the problem have been obtained, the program rotates the orbital basis for a stationary energy (to first order). The orbitals are updated. If the current orbital is less self-consistent than the other, it is orthogonalized. At the end of an iteration, all orbitals are Schmidt orthogonalized.

The rotation has a significant effect on the radial functions which are shown in figure 3.1. Those for ^3S have the expected hydrogenic form but the rotation greatly reduces the height of the first maximum in the $P(2s; r)$ for ^1S and introduces a node into the $P(1s; r)$ function for this state.

3.6 The general Hartree–Fock equations

Suppose the configuration defining the configuration state γLS has assigned w_a electrons to the $n_a l_a$ subshell and that m subshells are occupied. Then

$$\mathcal{E}(\gamma LS) = \sum_{a=1}^{m} w_a \left[I(a, a) + \left(\frac{w_a - 1}{2} \right) \sum_{k=0}^{2l_a} f_k(l_a) F^k(a, a) \right]$$
$$+ \sum_{a=2}^{m} \left\{ \sum_{b=1}^{a-1} w_a w_b \left[F^0(a, b) + \sum_{k=|l_a - l_b|}^{l_a + l_b} g_k(l_a, l_b) G^k(a, b) \right] \right\}.$$

$$(3.33)$$

This energy expression depends only on the radial integrals. The part of the expression that depends on $P(a; r)$, for example, is the contribution to the energy of the entire $(n_a l_a)^{w_a}$ subshell, the negative of the removal energy of the subshell. Let us denote this quantity by $\bar{\mathcal{E}}((n_a l_a)^{w_a})$. Then the stationary condition for a Hartree–Fock solution applies to this expression but since the variations must be constrained in order to satisfy orthonormality assumptions, Lagrange multipliers need to be introduced. The stationary condition then applies to the functional

$$\mathcal{F}(P(a; r)) = -\bar{\mathcal{E}}((n_a l_a)^{w_a}) + \sum_{b=1}^{m} \delta_{l_a l_b} \lambda_{ab} \langle a | b \rangle. \qquad (3.34)$$

Analysing this energy more carefully, it can be shown that

$$-\bar{\mathcal{E}}((n_a l_a)^{w_a}) = w_a \left(I(a, a) + \left(\frac{w_a - 1}{2} \right) \sum_{k=0}^{2l_a} f_k(l_a) F^k(a, a) \right)$$
$$+ \sum_{\substack{b=1 \\ b \neq a}}^{m} w_a w_b \left[F^0(a, b) + \sum_{k=|l_a - l_b|}^{(l_a + l_b)} g_k(l_a, l_b) G^k(a, b) \right].$$

$$(3.35)$$

Applying the variational conditions to each of the integrals, dividing by $-w_a$, we get the equation

$$\left(\frac{d^2}{dr^2} + \frac{2}{r}[Z - Y(a; r)] - \frac{l_a(l_a + 1)}{r^2} - \varepsilon_{aa}\right) P(a; r)$$

$$= \frac{2}{r} X(a; r) + \sum_{\substack{b=1 \\ b \neq a}}^{m} \delta_{l_a l_b} \varepsilon_{ab} P(b; r)$$

(3.36)

where

$$Y(a; r) = (w_a - 1) \sum_{k=0}^{2l_a} f_k(l_a) Y^k(aa; r) + \sum_{\substack{b=1 \\ b \neq a}}^{m} w_b Y^0(bb; r)$$

(3.37)

$$X(a; r) = \sum_{\substack{b=1 \\ b \neq a}}^{m} w_b \sum_{k=|l_a - l_b|}^{(l_a + l_b)} g_k(l_a, l_b) Y^k(ab; r) P(b; r).$$

Notice that a factor of two arises from the variational properties of the integrals, a factor that becomes four for the $F^k(a, a)$ integrals that represent the 'self-interaction' within a subshell.

3.6.1 Diagonal energy parameters and Koopmans' theorem

The diagonal (ε_{aa}) and off-diagonal (ε_{ab}) 'energy' parameters, are related to the Lagrange multipliers with $\varepsilon_{aa} = 2\lambda_{aa}/w_a$ and $\varepsilon_{ab} = \lambda_{ab}/w_a$. Multiplying equation (3.36) by $P(a; r)$ and integrating, it is easy to show that

$$\varepsilon_{aa} = \frac{2}{w_a} \bar{\mathcal{E}}((n_a l_a)^{w_a}) - (w_a - 1) \sum_{k=0}^{2l_a} f_k(l_a) F^k(a, a).$$

(3.38)

In the special case where $w_a = 1$, ε_{aa} is twice the removal energy, or ionization energy. This is often referred to as Koopmans' theorem but, as seen earlier, if a rotation of the radial basis leaves the wave function unchanged while transforming the matrix of energy parameters (ε_{ab}), the removal energies are extreme values obtained by setting the off-diagonal energy parameters to zero. For multiply occupied shells, ε_{aa} is like an average removal energy, with a correction arising from the self-interaction.

3.6.2 The fixed-core Hartree–Fock approximation

The above derivation has assumed that the solution is stationary with respect to variations in all radial functions. In practice, it may be convenient to assume that certain radial functions are 'fixed' or 'frozen': in other words, these radial functions are assumed to be given. Such approximations are often made about core orbitals and so, this is called a fixed-core HF approximation. Of course, the system of equations defining the solution includes only equations for radial functions that vary.

3.7 Brillouin's theorem

Approximate wave functions can be determined in many different ways. By solving the Hartree–Fock equations, the single-configuration Hartree–Fock approximation has some special properties not possessed by other approximations. One such property is referred to as satisfying Brillouin's theorem (Brillouin 1932, 1934), though we will show that Brillouin's theorem is not always obeyed.

3.7.1 General theory

Let $\Phi^{HF}(\gamma LS)$ be a Hartree–Fock configuration state, where γ denotes the configuration and coupling scheme. With $\Phi^{HF}(\gamma LS)$ are associated the m Hartree–Fock radial functions $P^{HF}(n_1l_1; r)$, $P^{HF}(n_2l_2; r)$, ..., $P^{HF}(n_ml_m; r)$. These radial functions define the 'occupied' orbitals. To this set may be added 'virtual' orbitals that maintain the necessary orthonormality conditions.

First let us consider the case where one of the radial functions $P^{HF}(nl; r)$ is replaced by a radial function $P(n'l; r)$ for a virtual orbital without any change in the coupling of the angular factor. The perturbation of the Hartree–Fock radial function

$$P(nl; r) = P^{HF}(nl; r) + \epsilon P(n'l; r) \tag{3.39}$$

induces a perturbation in the configuration state function

$$\Phi(\gamma LS) = \Phi^{HF}(\gamma LS) + \epsilon F(nl \to n'l) + \mathcal{O}(\epsilon^2) \tag{3.40}$$

which, in turn, leads to a perturbation of the energy functional

$$\mathcal{E}(P^{HF}(nl; r) + \epsilon P(n'l; r))$$
$$= \mathcal{E}(P^{HF}(nl; r)) + 2\epsilon \langle \Phi^{HF}(\gamma LS) | \mathcal{H} | F(nl \to n'l) \rangle + \mathcal{O}(\epsilon^2). \tag{3.41}$$

But the Hartree–Fock energy is stationary to first order in the perturbation of any radial function, provided the perturbation satisfies orthogonality constraints, from which it follows that

$$\langle \Phi^{HF}(\gamma LS) | \mathcal{H} | F(nl \to n'l) \rangle = 0. \tag{3.42}$$

If the function $F(nl \to n'l)$ is a CSF for a configuration γ^*, or proportional to one, then Brillouin's theorem is said to hold between the two configuration states. When nl is singly occupied, $F(nl \to n'l)$ clearly is also a CSF. In the case of equivalent electrons, the answer depends on the coefficients of fractional parentage.

As an example, suppose our reference state is $2p^3\ {}^2P$. Then the $F(2p \to 3p)$ function will be given by a linear combination of the form

$$F(2p \to 3p) = a_1 \Phi \left(2p^2({}^3P)3p\ {}^2P\right) + a_2 \Phi \left(2p^2({}^1D)3p\ {}^2P\right)$$
$$+ a_3 \Phi \left(2p^2({}^1S)3p\ {}^2P\right) \tag{3.43}$$

where the coefficients are proportional to the coefficients of fractional parentage. Thus Brillouin's theorem will not hold for any of the three configuration states in the above equation, only for the linear combination. Generalizing to an arbitrary term of nl^w, it is clear Brillouin's theorem will hold only for terms with a single parent term.

In analysing individual states, it is helpful to remember that in the definition of configuration states, equivalent electrons are first coupled to each other and then to the other electrons. Thus, for the configuration $4s4p^2\ {}^2D$ in Ga I, Brillouin's theorem holds for $\Phi \left(4s\left[4p5p^1D\right]\ {}^2D\right)$, which becomes a linear combination of CSFs when the order of coupling is changed to the normal left to right (see equation (2.12)). At the same time, Brillouin's theorem would not hold for the linear combination representing the spin-polarized configuration state $\Phi \left(4s\left[4p5p^3D\right]\ {}^2D\right)$.

Now let us consider a mono-excitation where the nl electron is moved from one occupied orbital to another. In such cases, the Hartree–Fock orbitals must be perturbed simultaneously. Let us denote the associated perturbation of the CSF as $F(nl \to n'l, n'l \to -nl)$, since in a rotation the sign of the perturbation of the two orbitals is different. At first sight it would appear that Brillouin's theorem could not be satisfied, but the antisymmetry of the CSF plays a significant role. Consider the perturbation $F(1s \to 2s, 2s \to -1s)$ applied to the lithium $1s^2 2s\ {}^2S$ case considered earlier. Then

$$F(1s \to 2s, 2s \to -1s) = \left[a_1 \Phi \left([1s2s]^1S\ 2s\ {}^2S\right) - a_2 \Phi \left([1s^2]^1S\ 1s\ {}^2S\right)\right]. \tag{3.44}$$

A recoupling gives $\Phi \left([1s2s]^1S\ 2s\ {}^2S\right) = \Phi \left(1s2s^2\ {}^2S\right)$ and, in addition $\Phi \left([1s^2]^1S\ 1s\ {}^2S\right)$ is identically zero by the Pauli exclusion principle and so Brillouin's theorem holds for the lithium ground state. In the $1s2s\ {}^3S$ state, neither the $1s \to 2s$ nor the $2s \to 1s$ substitution is allowed: in fact, it can be shown that for this state Brillouin's theorem holds for all mono-excited configurations. The same is not true for $1s2s\ {}^1S$ where the simultaneous perturbation leads to the condition

$$\langle \Phi^{HF}(1s2s\ {}^1S)|\mathcal{H}|[\Phi(1s^2\ {}^1S) - \Phi(2s^2\ {}^1S)]\rangle = 0. \tag{3.45}$$

Thus Brillouin's theorem is not obeyed for either $\Phi(1s^2\ {}^1S)$ or $\Phi(2s^2\ {}^1S)$.

3.7.2 The importance of Brillouin's theorem

The Hartree–Fock solution for $1s2s$ ^3S is a much better approximation than the one for $1s2s$ ^1S. This is directly related to Brillouin's theorem. Let the approximate wave function be a linear expansion

$$\Psi (\gamma LS) = \Phi (\gamma LS) + \sum_{\gamma^*} c_{\gamma^*} \Phi (\gamma^* LS) \qquad (3.46)$$

where $\Phi (\gamma^* LS)$ is a mono-excited CSF. Then, when Brillouin's theorem holds, the non-diagonal elements of the first row/column of the interaction matrix will be zero and the Hartree–Fock energy will be an eigenvalue of the interaction matrix. Thus, the Hartree–Fock approximation *already included* the effect of the singly excited states in the approximate wave function. The same is not true when Brillouin's theorem does not hold for all the singly excited states.

3.8 Term dependence

The $1s$, $2s$, $2p$, ... terminology for electrons in configurations overlooks the fact that the probability distribution of an electron depends also on the coupling, particularly the final LS term. As an example, consider the $1s^22s2p$ configuration in Be which may couple to form either a ^3P or ^1P term. The energy expression differs only in the exchange interaction, $\pm(1/3)G^1(2s, 2p)$, where the $+$ sign refers to ^1P and the $-$ to ^3P. Clearly, the energies of these two terms will differ. What is not quite as obvious is the extent to which the

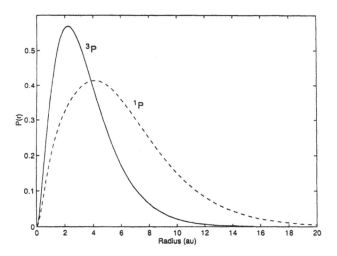

Figure 3.2. A comparison of the $2p$ Hartree–Fock radial functions for the $2s2p$ $^{1,3}P$ states of Be.

$P(2p; r)$ radial functions differ for the two states. The most affected orbital will be the one that is least tightly bound, which in this case is the $2p$ orbital. Figure 3.2 shows the two radial functions. The ^1P orbital is far more diffuse than (not as localized as) the one for ^3P. Such a change in an orbital is called LS-term dependence.

3.9   Iso-electronic sequences and orbital collapse

In an iso-electronic sequence, the only parameter that changes is the nuclear charge†. Calculations for such a sequence provide useful trends, particularly when plotted as a function of $1/Z$, the perturbation parameter from Z-dependent perturbation theory, to be discussed in the next chapter.

An interesting phenomenon that can occur is a very rapid contraction of an orbital which is called *orbital collapse*. This could be an LS-dependent effect, but it can also occur along an isoelectronic sequence of values. This effect is most noticeable in the high-l orbitals. In hydrogen, it can be shown that the mean radius of an orbital is $\langle r \rangle = (1/2)[3n^2 - l(l+1)]$. Thus the higher-$l$ orbitals are more contracted. In neutral many-electron systems, the high-l orbitals have a higher energy and are more diffuse. This is due, in part, to the 'angular momentum barrier', the $l(l+1)/r^2$ term that appears in the definition of the $\mathcal{L}$ operator. In the Hartree model, where the radial equation has the form‡

$$\left(\frac{\mathrm{d}^2}{\mathrm{d}r^2} + \frac{2}{r}[Z - Y(a;r)] - \frac{l_a(l_a+1)}{r^2} - \varepsilon_{aa} \right) P(a;r) = 0 \qquad (3.47)$$

it is possible for $-\frac{2}{r}[Z - Y(a;r)] + \frac{l_a(l_a+1)}{r^2}$ to have two wells, an inner well and an outer shallow well. When the probability distribution of the lowest eigenfunction changes rapidly as a function of increasing nuclear charge from the outer well to the inner well, orbital collapse is said to occur.

3.10   Quantum defects and Rydberg series

Spectra of atoms often exhibit phenomena associated with a Rydberg series, i.e. states where one of the electrons is in an nl orbital, with n assuming a sequence of values. An example is the $1s^2 2snd\ ^3D$ series in Be, $n = 3, 4, 5, \ldots$. For such a series, a useful concept is that of a quantum defect parameter δ. In hydrogen, the ionization energy (IE) is $1/(2n^2)$ au. In complex neutral systems, the effective charge would be the same as for hydrogen at large r. As n increases, the mean radius becomes larger and the probability of the electron being in the hydrogen-like potential increases. Thus one could define an effective quantum

---

† In the HF program it is treated as a REAL variable.
‡ Strictly speaking, in the Hartree model, the definition of $Y(r)$ given by (3.36) must be restricted to $k = 0$.

number $n^* = n - \delta$ such that

$$\text{IE } (nl) = \frac{1}{2(n - \delta)^2}. \tag{3.48}$$

In general, when one is not dealing with neutral systems, the equation must be modified to

$$\text{IE } (nl) = \frac{1}{2} \left[\frac{(Z - N + 1)}{(n - \delta)} \right]^2$$

where N is the number of electrons.

Often the quantum defect parameter is defined with respect to observed data, but it can also be used to evaluate Hartree–Fock energies where $\text{IE} = \varepsilon_{nl,nl}/2$ so that $\varepsilon_{nl,nl} = [(Z - N + 1)/(n - \delta)]^2$. Table 3.6 shows the screening parameter, effective quantum number and quantum defect for the Hartree–Fock $2snd$ ^3D and ^1D terms in Be as a function of n. For the triplet series, the screening is less than 3, so the $Z_{\text{eff}} = Z - \sigma(nl)$ is greater than one. At the same time, for the singlet series, the screening is slightly greater than 3, making the effective nuclear charge less than one. This is the effect of exchange and it is reflected also in the quantum defect parameter which is positive for the triplets and negative for the singlet series. This series is discussed further in section 5.4.

Table 3.6. The screening and quantum defect parameters of the $2snd$ Rydberg series in Be.

	^3D			^1D		
n	$\sigma(nl)$	n^*	$\delta(nl)$	$\sigma(nl)$	n^*	$\delta(nl)$
3	2.960	2.968	0.032	3.012	3.014	−0.014
4	2.972	3.960	0.040	3.008	4.012	−0.012
5	2.979	4.957	0.043	3.006	5.013	−0.013
6	2.983	5.955	0.045	3.005	6.013	−0.013

3.11 Computational aspects

The numerical Hartree–Fock method is based on finite difference approximations. That is, the problem of determining functions is replaced by one of determining values $P_{ja}, j = 1, \ldots, M_a$ such that $P_{ja} \approx P(a; r_j)$, at a discrete set of points, $r_j, j = 1, \ldots, M_a$ called *mesh points* or *grid points*. Differential or integral operators are then replaced by their finite difference analogues. Crucial in this process is the selection of the grid points.

3.11.1 The logarithmic grid

Numerical methods are greatly simplified when the grid points defining the discretized problem are equally spaced with respect to the independent variable. At the same time, in order to achieve a desired accuracy, the points must be sufficiently close together that a polynomial passing through adjacent points (r_j, P_{ja}) is a reasonable approximation to the function $P(a; r)$. In regions where the function is changing rapidly, or where some higher derivative is large, the points must be close together.

The natural choice for the independent variable is the radius r. But in terms of this variable the radial functions change fairly rapidly near the origin, which is also a singular point. As a result, it is necessary to have points close together near the origin and further apart for larger r. Early methods would use equal step sizes h between points with an occasional doubling of step size. The process of doubling h is not difficult but it does increase the complexity of a numerical algorithm. A similar result may be achieved analytically by transforming to a new variable.

A transformation that has been used successfully in the present atomic structure program is

$$\rho = \log(Zr) \Rightarrow r = \frac{e^\rho}{Z} \tag{3.49}$$

along with the transformation of the dependent variable

$$\overline{P}(\rho) = P(r)/\sqrt{r}. \tag{3.50}$$

With this transformation the Hartree–Fock equations become

$$\left[\frac{\mathrm{d}^2}{\mathrm{d}\rho^2} + 2Zr - (l_a + \tfrac{1}{2})^2 - r^2\varepsilon_{aa}\right]\overline{P}(a; \rho)$$

$$= 2r\left[Y(a; r)\overline{P}(a; \rho) + \overline{X}(a; \rho)\right] + \sum_{\substack{b=1 \\ b \neq a}}^{m} \delta_{l_a l_b} r^2 \varepsilon_{ab}\overline{P}(a; \rho) \tag{3.51}$$

with the boundary conditions $\overline{P}(a; \rho) \to 0$ as $\rho \to \pm\infty$. Here $\overline{X}(a; \rho) = X(a; r)/\sqrt{r}$.

This transformation has two advantages. As will be shown in the next subsection, the equations defining $Y^k(ab; r)$ now have a simplified form and can be computed more rapidly. For large atoms this is extremely important since the number of exchange terms increases rapidly with the number of functions. Another advantage arises from the fact that to zero order within perturbation theory, the radial functions are hydrogenic functions with Zr as independent variable. The transformation makes it possible to use the same set of ρ values for all atoms in the periodic table. Only in exceptional circumstances is the grid changed.

A disadvantage of the transformation is the fact that the range $(0, \infty)$ is now transformed to $(-\infty, \infty)$. In order to avoid the point at $-\infty$ it is customary to divide the range of r into two regions, $(0, r_1)$ and (r_1, ∞), and transform only the latter to the ρ variable. The value of r_1 is chosen so that $\rho_1 = \log(Zr_1)$ is a predetermined number, sufficiently small that a series expansion of three terms may be used near the origin. Numerous tests on the hydrogen equation suggest that appropriate values are $\rho_1 = -4$ and $h = 1/16$ for an accuracy of nine digits in a hydrogenic $1s$ orbital energy. Thus the grid points become

$$\rho_j = -4 + (j-1)/16 \quad \text{or} \quad r_j = e^{\rho_j}/Z, \, j = 1, 2, \dots, M. \tag{3.52}$$

In a many-electron system, since the energies of other electrons are smaller, not only because of the principal quantum number but also because of the smaller Z_{eff}, the total energy is expected to be limited by the accuracy of the $1s$ orbital.

In the interval $(0, r_1)$ all functions are approximated by a series expansion about the origin. Some important series are

$$P(nl; r) = a_0(nl)r^{l+1}\left\{1 - \frac{Zr}{l+1} + \alpha r^2 + \mathcal{O}(r^3)\right\} \tag{3.53}$$

$$Y(nl; r) = vr + \mathcal{O}(r^3) \tag{3.54}$$

$$X(nl; r) = a_0(nl)wr^{l+2} + \mathcal{O}(r^{l+3}) \tag{3.55}$$

as $r \to 0$ where

$$\alpha = \frac{Z^2 + (l+1)(\varepsilon/2 + v + w)}{(2l+3)(l+1)} \tag{3.56}$$

$$\varepsilon = \varepsilon_{nl,nl} + \sum_{n'l \neq nl} \varepsilon_{nl,n'l} \frac{a_0(n'l)}{a_0(nl)}. \tag{3.57}$$

The constants v and w depend on the configuration and $a_0(nl)$ is the AZ(nl) parameter included in the program. It is clear from these expansions that the least accurate case always occurs for $l = 0$, where the series converge least rapidly.

3.11.2 Formulae for integration and differentiation

Many classes of formulae can be represented most concisely in terms of the differences of the function values at a set of equally spaced grid points. Of the possible notations, the *central-difference* notation is the most convenient and is the one used throughout.

Let $f_j = f(x_j)$ where x is used to denote either r or ρ. By definition, the central-difference operator is given by

$$\delta f(x) = f(x + h/2) - f(x - h/2). \tag{3.58}$$

Then

$$\delta f_{j+1/2} = f_{j+1} - f_j \tag{3.59}$$

and

$$\delta^2 f_j = \delta f_{j+1/2} - \delta f_{j-1/2}$$
$$= f_{j+1} - 2f_j + f_{j-1}. \tag{3.60}$$

Many atomic properties require the evaluation of an integral, say,

$$I = \int_0^\infty f(x)\, dx \approx \int_0^{x_M} f(x)\, dx \tag{3.61}$$

where x_M is a suitably chosen large value of x, such that $\int_{x_M}^\infty f(x)\, dx$ is negligible. When $f(x)$ is tabulated at equally spaced points $x_j = jh, j = 0, 1, \ldots, M$, where M is even, the integration may be represented by the composite rule

$$\int_0^{x_M} f(x)\, dx = \sum_{j\ \text{odd}} \int_{x_{j-1}}^{x_{j+1}} f(x)\, dx \tag{3.62}$$

where each individual integral is over two adjacent intervals. It can be shown that

$$\int_{x_{j-1}}^{x_{j+1}} f(x)\, dx = 2h\left(f_j + \frac{1}{6}\delta^2 f_j - \frac{1}{180}\delta^4 f_j\right) + \mathcal{O}(h^7). \tag{3.63}$$

This formula uses values of the integrand outside the range of integration and cannot be used for the first and last integral. In these cases the δ^4 term may be omitted leading to the well known Simpson rule

$$\int_{x_{j-1}}^{x_{j+1}} f(x)\, dx = \frac{h}{3}\left[f_{j-1} + 4f_j + f_{j+1}\right] + \mathcal{O}(h^5). \tag{3.64}$$

For certain integrals the accuracy of Simpson's rule is satisfactory.

3.11.3 The functions $Y^k(ab; r)$

By definition

$$Y^k(ab; r) = \int_0^r \left(\frac{s}{r}\right)^k P(a; s) P(b; s)\, ds$$
$$+ \int_r^\infty \left(\frac{r}{s}\right)^{k+1} P(a; s) P(b; s)\, ds. \tag{3.65}$$

Because integrals of this type occur frequently in Hartree–Fock calculations, it is desirable to determine them with maximum efficiency. Direct evaluation, using quadrature, cannot be considered since an integral over the whole range would be required for *each* value of r. These functions can be determined much more rapidly as solutions of a pair of differential equations as suggested by Hartree (1957).

Define

$$Z^k(ab; r) = \int_0^r \left(\frac{s}{r}\right)^k P(a; s)P(b; s)\, ds. \tag{3.66}$$

Then it is easy to show that

$$\frac{d}{dr}Z^k(ab; r) = P(a; r)P(b; r) - \frac{k}{r}Z^k(ab; r) \tag{3.67}$$

and

$$\frac{d}{dr}Y^k(ab; r) = \frac{1}{r}\left[(k+1)Y^k(ab; r) - (2k+1)Z^k(ab; r)\right] \tag{3.68}$$

with the boundary conditions

$$Z^k(ab; 0) = 0 \quad \text{and} \quad Y^k(ab; r) \rightarrow Z^k(ab; r) \quad \text{as} \quad r \rightarrow \infty. \tag{3.69}$$

With the ρ variable, the equations for Z^k and Y^k transform to

$$\frac{d}{d\rho}Z^k = r\overline{P}_a\overline{P}_b - kZ^k \tag{3.70}$$

$$\frac{d}{d\rho}Y^k = (k+1)Y^k - (2k+1)Z^k. \tag{3.71}$$

As was shown by Worsley (1958), both these equations have simple integrating factors, namely $e^{k\rho}$ and $e^{-(k+1)\rho}$, respectively, so that

$$d\left(e^{k\rho}\, Z^k\right) = r\overline{P}_a\overline{P}_b\, e^{k\rho}\, d\rho \tag{3.72}$$

$$d\left(e^{-(k+1)\rho}\, Y^k\right) = -(2k+1)Z^k\, e^{-(k+1)\rho}\, d\rho. \tag{3.73}$$

On integrating these equations we get

$$Z^k_{j+m} = e^{-mhk}\, Z^k_j + \int_{\rho_j}^{\rho_{j+m}} r\overline{P}_a\overline{P}_b\, e^{k(\rho-\rho_{j+m})}\, d\rho, \qquad j = 1, 2, \ldots \tag{3.74}$$

$$Y^k_j = e^{-mh(k+1)}\, Y^k_{j+m} + (2k+1)\int_{\rho_j}^{\rho_{j+m}} Z^k\, e^{-(k+1)(\rho-\rho_j)}\, d\rho, \qquad j = \ldots, 2, 1. \tag{3.75}$$

Note that the calculation of Z^k proceeds by outward integration using known starting values whereas Y^k is obtained by inward integration. Here we have used the fact that $\rho_{j+m} - \rho_j = mh$. In the HF program $m = 2$ is used; the first and last pair of intervals are integrated using Simpson's rule but for other pairs of intervals the more accurate $\mathcal{O}(h^7)$ formula is used. For $k > 0$ the method is stable in that errors at previous points are damped by an exponential factor, but for $k = 0$, errors accumulate for the outward integration of Z^k though the errors in Y^k are again damped.

The Y^k functions can also be thought of as solutions of a boundary value problem. Differentiating the first-order equation for Y^k and using the equation for Z^k to eliminate the latter, we obtain

$$\frac{d^2}{dr^2} Y^k(ab; r) = \frac{k(k+1)}{r^2} Y^k(ab; r) - \frac{(2k+1)}{r} P(a; r) P(b; r) \qquad (3.76)$$

with the boundary conditions

$$Y^k(ab; 0) = 0 \quad \text{and} \quad \frac{d}{dr} Y^k(ab; r) = -\frac{k}{r} Y^k(ab; r) \quad \text{as} \quad r \to \infty. \qquad (3.77)$$

This equation is not stable for outward integration but has been used successfully in basis type methods (Brage and Froese Fischer 1994).

3.11.4 Solution of the radial equation

The radial equation for a single electron outside a core is a linear integro-differential equation of the form

$$y''(r) + \left[f(r) - \varepsilon \right] y(r) = \int_0^\infty K(r, s) \, y(s) \, ds \qquad (3.78)$$

where $y(0) = 0$ and $y(r) \to 0$ as $r \to \infty$. The integral equation part includes the exchange effect as well as off-diagonal energy parameters, if any. A direct discretization would lead to an eigenvalue problem

$$(\mathbf{T} - \varepsilon \mathbf{I} - \mathbf{K})\mathbf{y} = 0 \qquad (3.79)$$

where $\mathbf{T}$ is a band matrix (the band width depending on the order of discretization of y'') and $\mathbf{K}$ is a full matrix related to the kernel $K(r, s)$, but also including factors arising from a quadrature rule. This eigenvalue problem could be solved for a selected eigenvalue using Rayleigh-quotient iteration, but would then require the storage of the full matrix and $\mathcal{O}(M^3)$ floating point operations per iteration. An iterative procedure requiring only $\mathcal{O}(M)$ operations per iteration is used by the HF program.

Let $\{y^{(k)}(r), k = 0, 1, \ldots\}$ be a sequence of approximate solutions. Define

$$g^{(k)}(r) = \int_0^\infty K(r, s) \, y^{(k)}(r) \, ds \qquad (3.80)$$

and

$$\left\{ \frac{d^2}{dr^2} + f(r) - \varepsilon \right\} y^{(k+1)}(r) = g^{(k)}(r). \qquad (3.81)$$

For the latter, the familiar Numerov method based on the first two terms of the discretization

$$\delta^2 y_j = h^2 \left(1 + \frac{1}{12}\delta^2 - \frac{1}{240}\delta^4 + \frac{31}{60480}\delta^6 - \dots \right) y_j'' \tag{3.82}$$

leads to a system of equations

$$(\mathbf{T} - \varepsilon\mathbf{S})\mathbf{y}^{(k+1)} = \mathbf{c}^{(k)} \tag{3.83}$$

where $\mathbf{T} = (t_{ij})$ and $\mathbf{S} = (s_{ij})$ are $(M-1) \times (M-1)$ tridiagonal matrices with

$$t_{ij} = \begin{cases} -2 + (10/12)h^2 f_i, & i = j \\ 1 + (1/12)h^2 f_j, & |i-j| = 1 \qquad f_j \equiv f(r_j) \\ 0, & \text{otherwise} \end{cases}$$

and

$$s_{ij} = \begin{cases} (10/12)h^2, & i = j \\ (1/12)h^2, & |i-j| = 1 \\ 0, & \text{otherwise.} \end{cases}$$

The column vector $\mathbf{c} = (c_1, c_2, \dots, c_{M-1})^t$ has components

$$c_i = \frac{h^2}{12}(g_{i+1} + 10g_i + g_{i-1}), \qquad g_i \equiv g^{(k)}(r_i). \tag{3.84}$$

For simplicity the above assumes boundary conditions of zero at both ends.

Unlike (3.78), (3.81) is no longer an eigenvalue problem and unique solutions to the discretized approximations of (3.83) will exist for values of ε if $(\mathbf{T} - \varepsilon\mathbf{S})$ is non-singular. If the coefficient matrix is singular, unique solutions may still exist if either $\mathbf{c}^{(k)} = 0$ or if $\mathbf{c}$ is orthogonal to the desired eigenvector. In either case, for the solutions of (3.83) to converge to those of (3.78) it is necessary to define $\varepsilon = \varepsilon^{(k)}$ so that the sequence $\varepsilon^{(k)}, k = 0, 1, \dots$ converges to an eigenvalue. A natural choice is the Rayleigh quotient for $y^{(k)}(r)$ (see section 3.3). At the same time, we would like the iterations to converge to normalized eigenfunctions where normalization is defined in terms of numerical integration.

Define $(\mathbf{u}, \mathbf{v})$ to be the numerical approximation to $\int_0^\infty u(r)v(r)\,dr$ using only vectors $\mathbf{u}$ and $\mathbf{v}$ representing values of the functions on the grid. Also define $||\mathbf{u}||^2 = (\mathbf{u}, \mathbf{u})$. The method M1 implemented in the HF program combines these ideas as follows:

For $k = 0, 1, 2, \dots$

    (i) $\varepsilon^{(k)} = \mathbf{y}^{(k)t}\left[\mathbf{T}\mathbf{y}^{(k)} - \mathbf{c}^{(k)}\right] / \mathbf{y}^{(k)t}\mathbf{S}\mathbf{y}^{(k)}$

    (ii) $\left[\mathbf{T} - \varepsilon^{(k)}\mathbf{S}\right]\mathbf{z}^{(k+1)} = \mathbf{c}^{(k)}$

    (iii) $\mathbf{y}^{(k+1)} = \mathbf{z}^{(k+1)}/||\mathbf{z}^{(k+1)}||.$

This method works remarkably well in many cases but when the exchange effect is extremely small, the coefficient matrix is too nearly singular and a second method M2 is used. The latter, as $\mathbf{c} \to 0$ goes over to inverse iteration, a method for finding the eigenvector for the eigenvalue closest to $\varepsilon^{(k)}$.

Method M2 proceeds very much as M1, but also introduces an inverse iteration step

$$\left[\mathbf{T} - \varepsilon^{(k)}\mathbf{S}\right]\mathbf{w}^{(k+1)} = \mathbf{S}\mathbf{y}^{(k)}. \tag{3.85}$$

(The latter is the variational equation for $w(r) = \partial y(r)/\partial \varepsilon$ under the assumption that $f(r)$ and $g(r)$ do not depend on the energy). Then the next estimate is formed as

$$\mathbf{y}^{(k+1)} = \mathbf{z}^{(k+1)} + \beta\mathbf{w}^{(k+1)}; \quad \beta \text{ such that } ||\mathbf{y}^{(k+1)}|| = 1. \tag{3.86}$$

The HF and MCHF programs solve the radial equation on the logarithmic grid, but not directly by linear algebra. Instead outward and inward integration are used, somewhat for historical reasons. Outward integration is stable in the oscillatory region where matrix methods would require pivoting. In the tail region the matrix is diagonally dominant and pivoting is not needed. In this region a system of equations is solved that matches the solution of outward integration, at the same time looking for a point where the boundary condition at large r can be applied. Thus the algorithm is 'adaptive' and the user need not be concerned with the extent of the radial orbital. In both regions a particular solution and solution of the homogeneous equation are obtained. These are then combined so that the three-term recurrence at the join is satisfied.

In order to ensure that the iterative process converges to the desired eigenfunction, node counting is used and the notion of an 'acceptable' solution is applied. Particularly in the early iterations, the solutions obtained by both M1 and M2 may be 'unusual'. In such cases, the Rayleigh quotient value of ε must be replaced by some other value. A somewhat *ad hoc* search for an acceptable solution is initiated. In outward/inward integration, the three-term recurrence is satisfied at all points except the join. In some instances, it has proven useful to relate the residual of the equation at the join to the change in energy required to satisfy this condition.

When orbitals are multiply occupied, the function $f(r)$ also depends on $y(r)$ and the problem is non-linear. In such cases, the SCF iterations may oscillate and accelerating techniques may need to be used as described earlier.

3.12 Exercises

(i) Given the energy expression,

$$\mathcal{E}(1s3d\ ^3D) = I(1s, 1s) + I(3d, 3d) + F^0(1s, 3d) - (1/5)G^2(1s, 3d), \tag{3.87}$$

derive the Hartree–Fock equations for the $1s3d\ ^3D$ state.

(ii) Now suppose that $P(4d)$ is a 'fixed-core' Hartree–Fock radial function for the $1s4d\ ^3D$, where $P(1s)$ is the function of example (i). Show that $\langle P(3d)|P(4d)\rangle = 0$.

(iii) The energy expression for the $1s^22s^2\ ^1S$ ground state of Be is

$$\mathcal{E}(1s^22s^2) = 2I(1s, 1s) + 2I(2s, 2s) + F^0(1s, 1s)$$

$$+ 2F^0(1s, 2s) - G^0(1s, 2s). \qquad (3.88)$$

Show that the expression is invariant under a rotation of the radial basis; i.e. $\mathcal{E}(P(1s),\ P(2s)) = \mathcal{E}(P^*(1s),\ P^*(2s))$, where orbitals are transformed as given by (3.27).

(iv) Using the GENCL program, determine all the allowed terms for the $2p^2$ and $2p^4$ configurations. The latter may be viewed as a configuration with two holes, say $2p^{-2}$. Show that the deviations from the average energy are the same for particles and holes.

(v) Consider the [Xe] $6s$ iso-electronic sequence, for $Z = 55$, 56, 57 and 58, where [Xe] is used to denote the xenon core consisting of $1s^22s^22p^63s^23p^63d^{10}4s^24p^64d^{10}5s^25p^6$. Tabulate the screening parameter and mean radius of the $6s$ orbital as a function of Z.

(vi) Repeat the calculations of the previous exercise for $4f$ rather than $6s$. Does orbital collapse occur? Estimate the effective nuclear charge where collapse occurs. You may wish to perform some calculations at fractional nuclear charges.

(vii) The size of an atom may be defined as the mean radius of the outer electron. Perform calculations for He $1s^2\ ^1S$ ($Z = 2$) and for Xe ($Z = 54$). How do the two sizes compare?

(viii) Prove that Brillouin's theorem holds for all the mono-excited states of closed shell systems.

(ix) Let $P(1s)$, $P(2s)$, $P(2p)$ be Hartree–Fock radial functions for the $1s^22s^22p^2\ ^3P$ ground state of carbon. Let $P(3p)$ be a fixed-core Hartree–Fock function for the excited state $1s^22s^22p3p\ ^3P$ of carbon. Prove that $\langle 1s^22s^22p^2\ ^3P|\mathcal{H}|1s^22s^22p3p\ ^3P\rangle = 0$.

(x) Write a program that numerically solves the radial eigenvalue problem (1.20) for the hydrogen potential $U(r) = -1/r$. Proceed in the following steps:

(a) Consider $r \in [0, 100]$. For simplicity, use an equally spaced grid $r_0, r_1, \ldots, r_M$ with $r_0 = 0$, $r_M = 100$ take $M = 10\,000$.

(b) Start with a trial energy eigenvalue $E = -1$.

(c) Determine r_i such that $E = l(l + 1)/2r_i^2 + U(r_i)$.

(d) Integrate the equation from r_0 to r_{i+1} using the Numerov method based on $\delta^2 y_j = h^2\left(1 + \frac{1}{12}\delta^2\right)y_j''$. For starting values, use $P_0 = 0$ and the exact value of P_1 as given in table 1.1.

(e) Integrate the equation from r_M to r_{i-1}. Again, use the asymptotic behaviour of the exact solution to obtain the starting values P_M and P_{M-1}.

(f) Scale the outward and inward solutions so that they have the same value at r_i.

(g) Calculate the difference in derivative at r_i between the outward and inward solutions. Note the sign on this difference.

(h) Increase the trial energy E with a small amount, say 0.1, and repeat steps (c)–(g) until the sign of the differences in derivative changes. At this stage we know that the exact eigenvalue E has been passed.

(i) Use a secant search to zoom in on the exact eigenvalue for which the outward and inward solutions should have continuous derivatives at r_i. Using the program, verify numerically that the energy for the lowest s, p and d states are given by the expression $E = -1/(2n^2)$ where n is the principal quantum number. For a numerically determined eigenvalue, integrate outward from r_0 to r_M. How does P_i look in the region of the tail?

(xi) Modify the program to allow for a different number of grid points. Determine the eigenvalue for the 1s state using a grid with (a) $M = 100$, (b) $M = 1000$ and (c) $M = 10\,000$.

(xii) Alternatively, show that the use of the Numerov method together with zero boundary conditions at r_0 and r_M leads to a matrix eigenvalue problem. Find the lowest ten eigenvalues. Vary the range from $r \in [0, 100]$ to $r \in [0, 1000]$. How do the ten lowest eigenvalues change?

Chapter 4

Multiconfiguration Hartree–Fock Wave Functions

4.1 Correlation in many-electron atoms

The Hartree–Fock method predicts many atomic properties remarkably well, but when analysed carefully, systematic discrepancies can be observed. Consider the ionization potentials tabulated in table 4.1 compared with the observed. In these calculations, the energy of the ion was computed using the same radial functions as for the atom. Thus no 'relaxation' effects were included. For some systems, like Li and Na, the computed ionization potentials agree very well with the observed, but for others the difference can be large. One would expect a similar pattern of deviations for the Na, Mg, Al, ... sequence where the $3s$ and $3p$ subshells are being filled as for the Li, Be, B, ... sequence where the $2s$ and $2p$ subshells are being filled, but in fact the patterns differ substantially.

Of course, observed data include other effects as well such as relativistic effects, finite mass and volume of the nucleus, but these are small for light atoms. For such systems the largest source of discrepancy arises from the fact that the Hartree–Fock solution is an approximation to the exact solution of Schrödinger's equation. Neglected entirely is the notion of 'correlation in the motion of the electrons'; each electron is assumed to move independently in a field determined by the other electrons. For this reason, the error in the energy was defined by Löwdin (1955), to be the *correlation* energy, i.e.

$$E^{\text{corr}} = E^{\text{exact}} - E^{HF}. \tag{4.1}$$

In this definition, E^{exact} is not the observed energy—it is the exact solution of Schrödinger's equation which itself is based on a number of assumptions.

4.2 Z-dependent perturbation theory

An indication of the important correlation corrections can be obtained from a perturbation theory study of both the Hartree–Fock approximation and the exact

Table 4.1. Observed and Hartree–Fock ionization potentials for the ground states of neutral atoms, in eV.

Atom	Obs.	HF	Diff.
n = 2 shell			
Li	5.39	5.34	0.05
Be	9.32	8.42	0.90
B	8.30	8.43	−0.13
C	11.26	11.79	−0.53
N	14.53	15.44	−0.91
O	13.62	14.45	−0.85
F	17.42	18.62	−1.20
Ne	21.56	23.14	−1.58
n = 3 shell			
Na	5.14	4.96	0.18
Mg	7.65	6.89	0.76
Al	5.98	5.71	0.27
Si	8.15	8.08	0.07
P	10.49	10.66	−0.17
S	10.36	10.10	0.26
Cl	12.97	12.98	−0.01
Ar	15.76	16.08	−0.32

wave function. Hartree (1957) had been interested in the Z-dependence of radial functions, mainly as a means of obtaining good initial estimates. In the following section, we follow closely the approach taken by Layzer *et al* (1964) in the study of the Z-dependent structure of the total energy.

4.2.1 General theory

Let us introduce a new variable $\rho = Zr$, which in effect changes the unit of length. Then the Hamiltonian becomes

$$\mathcal{H} = Z^2 \left(\mathcal{H}_0 + Z^{-1} V \right) \tag{4.2}$$

where

$$\mathcal{H}_0 = \sum_{i=1}^{N} \left(-\frac{1}{2} \nabla_i^2 - \frac{1}{\rho} \right), \tag{4.3}$$

$$V = \sum_{i>j}^{N} \frac{1}{\rho_{ij}} \tag{4.4}$$

and Schrödinger's equation becomes

$$\left(\mathcal{H}_0 + Z^{-1}V\right)\psi = \left(Z^{-2}E\right)\psi. \tag{4.5}$$

If we now assume

$$\psi = \psi_0 + Z^{-1}\psi_1 + Z^{-2}\psi_2 + \dots \tag{4.6}$$

in the ρ unit of length, and

$$E = Z^2\left(E_0 + Z^{-1}E_1 + Z^{-2}E_2 + Z^{-3}E_3 + \dots\right) \tag{4.7}$$

we may insert these expansions in (4.5) to obtain equations for ψ_k and E_k:

$$(\mathcal{H}_0 - E_0)\psi_0 = 0, \tag{4.8}$$

$$(\mathcal{H}_0 - E_0)\psi_1 = (E_1 - V)\psi_0, \tag{4.9}$$

$$(\mathcal{H}_0 - E_0)\psi_2 = (E_1 - V)\psi_1 + E_2\psi_0. \tag{4.10}$$

The solutions of the first equation are products of hydrogenic orbitals.

Let $|(nl)\nu LS\rangle$ be a configuration state function constructed from products of hydrogenic orbitals. Here (nl) represents a set of N quantum numbers $(n_1l_1, n_2l_2, \dots, n_Nl_N)$ and ν any additional quantum numbers such as the coupling scheme or seniority needed to distinguish the different configuration states. Then

$$\mathcal{H}_0|(nl)\nu LS\rangle = E_0|(nl)\nu LS\rangle \tag{4.11}$$

$$E_0 = -\frac{1}{2}\sum_{i=1}^{N}\frac{1}{n_i^2}. \tag{4.12}$$

Since E_0 is independent of the angular quantum numbers, it is now clear that different configurations may lead to the same E_0; that is, E_0 is degenerate. According to first-order perturbation theory for degenerate states (Messiah 1965, chapter 16), ψ_0 then is a linear combination of the degenerate configuration state functions $|(nl')\nu'LS\rangle$; the coefficients are components of an eigenvector of the interaction matrix, $\langle(nl')\nu'LS|V|(nl)\nu LS\rangle$, and E_1 is the corresponding eigenvalue. Then

$$\psi_0 = \sum_{(l')\nu'} c_{(l')\nu'}|(nl')\nu'LS\rangle. \tag{4.13}$$

But only configurations with the same parity π interact and so the linear combination is over all CSFs with the same set of principal quantum numbers and the same parity. This is the set of configurations referred to as the *complex* by Layzer *et al* (1964) and denoted by the quantum numbers $(n)\pi LS$.

The first-order correction ψ_1 is a solution of (4.9) orthogonal to ψ_0. It can be expanded as a linear combination of normalized intermediate configuration state functions $|\gamma_v LS\rangle$ belonging to $\mathcal{H}_0$, but outside the complex. Then

$$\psi_1 = \sum_v \frac{|\gamma_v LS\rangle\langle\gamma_v LS|V|\psi_0\rangle}{E_0 - E_{\gamma_v LS}} \tag{4.14}$$

where $\sum$ denotes both summation over discrete states and integration over continuum states. In the above, $E_{\gamma_v LS} = \langle\gamma_v LS|\mathcal{H}_0|\gamma_v LS\rangle$, and states with $E_{\gamma_v LS} = E_0$, i.e. states belonging to the complex, are omitted. Then

$$E_2 = \langle\psi_0|V|\psi_1\rangle = \sum_v \frac{\langle\gamma_v LS|V|\psi_0\rangle^2}{E_0 - E_{\gamma_v LS}}. \tag{4.15}$$

The first-order correction also determines the third-order energy correction through the relation

$$E_3 = \langle\psi_1|V - E_1|\psi_1\rangle. \tag{4.16}$$

It is clear from these remarks that the structure of ψ_1 is extremely important. An analysis by Layzer (1959) of experimental term energies in isoelectronic sequences showed that for degrees of ionization greater than three, the third-order energy contribution $Z^{-1}E_3$ is usually smaller than the relativistic effects.

4.2.2 Structure of ψ_1

The zero-order wave function ψ_0 describes the many-electron system in a general way, but for many atomic properties the first-order correction may also be important. Substituting equation (4.13) into (4.14) and interchanging the orders of summation, we find

$$\psi_1 = \sum_{(l')v'} c_{(l')v'} \sum_v \frac{|\gamma_v LS\rangle\langle\gamma_v LS|V|(nl')v'LS\rangle}{E_0 - E_{\gamma_v LS}}. \tag{4.17}$$

In other words, the mixing coefficient, $c_{(l')v'}$, is a weight factor in the sum over intermediate configuration state functions interacting with $|(nl')v'LS\rangle$. Consequently, in the general case we may view the first-order correction as a linear combination of first-order corrections. Thus it is sufficient to study in detail only the non-degenerate case.

Let us now assume $\psi_0 = |\gamma LS\rangle$, where $\gamma LS = (nl)vLS$.

The configuration states interacting with γLS are of two types: those that differ by a single electron (single substitution S) and those that differ by two electrons (double substitution D). The former can be further subdivided into

(i) Those that differ from γLS by one principal quantum number but retain the same spin and orbital angular coupling. These configuration states are part of *radial correlation*.

(ii) Those that differ by one principal quantum number and also differ in their coupling. Often the only change is the coupling of the spins, in which case the configuration states are part of *spin polarization.*

(iii) Those that differ in the angular momentum of exactly one electron and are accompanied by a change in orbital angular coupling of the configuration state and possibly also the spin coupling. The latter represent *orbital polarization.*

The sums over CSFs that differ in two electrons can also be classified. Let $\{a, b, c, \ldots\}$ be occupied orbitals in $(nl)vLS$ and $\{v, v', \ldots\}$ be unoccupied or *virtual* orbitals. Then the double replacement $ab \rightarrow vv'$ may generate a CSF in the expansion for ψ_1 that can be classifed according to different correlation effects:

(i) If ab are orbitals for outer electrons the replacement represents outer or *valence correlation.*

(ii) If a is core orbital but b is an outer orbital, the effect represents the polarization of the core and is referred to as *core–valence correlation.*

(iii) If both orbitals are from the core, the replacement represents *core–core correlation.*

4.3 Pair-correlation expansions

Z-dependent perturbation theory is not appropriate for many of the atomic systems of interest; however, it can be a very useful guide for indicating how an initial approximation can be improved. The focus is not on a formal perturbation parameter, but rather on the structure of ψ_0 and its correction, ψ_1. In some instances, a natural choice would be $\psi_0 = \Phi^{HF}(\gamma LS)$, then ψ_1 should be a sum over CSFs interacting with ψ_0. Since the Coulomb operator is a two-electron operator, the CSFs in ψ_1 cannot differ from the Hartree–Fock configuration by more than two electrons.

The sums in (4.14) defining ψ_1 involve infinite sums along with integration over continuum states. In variational methods for bound states, it is not necessary to include continuum functions but even so, in theory, the number of bound orbitals may be infinite. In practice, the set of orbitals defining the CSFs in the ASF will be finite. This set is called the *active set* (AS) of orbitals. Though, given a reference CSF, not all CSFs that are generated by the SD substitutions in fact will interact. Table 4.2 shows a GENCL run for an SD expansion where AS $= \{1s, 2s, 2p, 3s, 3p, 3d\}$ with all but $1s$ in the *virtual set* to which replacements may be made from the occupied set. Four CSFs are generated that do not interact with the Hartree–Fock *reference configuration*, the configuration from which the replacements are made. These are CSFs of the form $2p3p^2(^1D,^3P)\,^2P$ and $2p3d^2(^1D,^3P)\,^2P$ where the coupling of the equivalent electrons results in a zero interaction.

Table 4.2. Generating an SD expansion for $1s^2 2p$ using GENCL.

```
>Gencl
>
              Header    ?
>Lithium
      Closed Shells    ?
>
      Reference Set    ?
>1s(2)2p(1)
                   2    ?
>
          Active Set    ?
>
        Replacements   ?
>sd
          Virtual Set   ?
>2s,2p,3s,3p,3d
   From which shell    ?
>1
     To which shell    ?
>2
          Final Terms   ?
>2P
                   2    ?
>
```

In order to limit the CSFs from SD expansions to interacting ones, it is necessary to take coupling information into account, something not done in the present code. Continuing with the lithium example, to include correlation in the $1s^2$ core, the replacements are from the coupled pair of orbitals, $1s^2 \ ^1S$, to $nln'l \ ^1S$. In other words, the pair of virtual orbitals must couple to the same LS as the occupied pair. The function defined by all such possible replacements is called a *pair-correlation* function (PCF). Note that it is associated with a specific LS-symmetry. Of course, there also are other occupied pairs, namely $1s2p \ ^3P$ and $1s2p \ ^1P$. The CSFs that result from SD replacements from the different pairs, coupled to the remaining subconfiguration, define the first order correction to the zero-order Hartree–Fock approximation. Note that, unlike perturbation theory, where the one-electron orbitals are defined in terms of a zero-order Hamiltonian with a corresponding perturbation potential, in the variational approach we define a zero-order wave function and consider corrections to this wave function.

Z-dependent perturbation theory alerts us to the fact that the zero-order wave function should not always be a single CSF. In fact, other members of the complex should not be ignored, though a simple test may show that for some neutral atoms, certain members of the complex are not important. At the same time, experience has shown that in some instances, CSFs not in the complex are extremely important. Examples will be discussed in chapter 6. In any event, when ψ_0 is an expansion over a set of CSFs, the latter form a *multireference* set and the first-order correction is contained in the set obtained from SD replacements from the multireference set. Note that when coupling information is ignored, some CSFs will be generated that do not interact with ψ_0.

4.4 Complete and restricted active spaces

In general, given an active set of orbitals (AS) and a rule for generating CSFs, a set of CSFs is generated. This set forms a basis for a configuration space for the wave function expansion. When the rule generates all possible CSFs, then the resulting space is referred to as the *complete active space* (CAS). Often the CAS is defined relative to a set of closed subshells common to all configurations. The orbitals for these subshells are said to be *inactive*. For example, the complex for the $3s^2 3p^2 \, ^3P$ ground state of Si is a neon-like core, $1s^2 2s^2 2p^6$, coupled to the $\{N = 4, \pi = \text{even}, \, ^3P\}$ CAS of the active set $\{3s, 3p, 3d\}$. If the AS is extended to $\{3s, 3p, 3d, 4s, 4p, 4d, 4f\}$ or the $n = 4$ AS for short, then all single, double, triple, and quadruple (SDTQ) replacements from the outer four electrons will be generated. Of course, the number of CSFs will increase rapidly, both with the size of the active set and with the number of electrons N outside the closed subshells.

Frequently, the important set of CSFs can be obtained by the union of sets obtained by applying different rules. For example, the SD $n = 6$ set for the Si ground state may be combined with the SDTQ $n = 4$ set, resulting in a restricted active space (RAS). Examples of these ideas will be presented in the next few chapters. Once the CSFs have been determined the multiconfiguration Hartree–Fock (MCHF) method may be applied.

4.5 The MCHF approximation

In the multiconfiguration Hartree–Fock (MCHF) method, the wave function is approximated by a linear combination of orthonormal configuration state functions so that

$$\Psi(\gamma \, LS) = \sum_{i=1}^{M} c_i \Phi(\gamma_i \, LS), \quad \text{where} \quad \sum_{i=1}^{M} c_i^2 = 1. \tag{4.18}$$

Then the energy expression becomes

$$\mathcal{E}(\gamma \, LS) = \sum_{i=1}^{M} \sum_{j=1}^{M} c_i c_j \langle \Phi(\gamma_i \, LS) | \mathcal{H} | \Phi(\gamma_j \, LS) \rangle$$

$$= \sum_{i=1}^{M} \sum_{j=1}^{M} c_i c_j H_{ij}$$

$$= \sum_{i=1}^{M} c_i^2 H_{ii} + 2 \sum_{i>j}^{M} c_i c_j H_{ij} \tag{4.19}$$

where

$$H_{ij} = \langle \Phi \left(\gamma_i \, LS \right) | \mathcal{H} | \Phi \left(\gamma_j \, LS \right) \rangle. \tag{4.20}$$

Because $H_{ij} = H_{ji}$, the sum over i, j may be limited to the diagonals and the lower part of the matrix $\mathbf{H} = (H_{ij})$, called the *interaction* matrix. Let $\mathbf{c} = (c_1, c_2, \ldots, c_M)^t$ be a column vector of the expansion coefficients, also called *mixing coefficients*. Then the energy of the system is

$$E = \mathbf{c}^t \mathbf{H} \mathbf{c}. \tag{4.21}$$

Let $\mathbf{P}$ be the column vector of radial functions $[P(a; r), P(b; r), \ldots]^t$. Since the interaction matrix elements depend on the radial functions, it is clear the energy functional will depend on both $\mathbf{P}$ and $\mathbf{c}$.

In deriving the MCHF equations the energy needs to be reduced further and expressed in terms of the radial functions and $\mathbf{c}$. From the theory of angular momenta for the evaluation of matrix elements of the Hamiltonian, discussed in section 2.2, it follows that

$$H_{ij} = \sum_{ab} w_{ab}^{ij} I(a, b) + \sum_{abcd;k} v_{abcd;k}^{ij} R^k(ab, cd) \tag{4.22}$$

where the sum on ab or $abcd$ is over the orbitals occupied in either configuration state.

Substituting into the energy expression of (4.19), and interchanging the order of summation, we get

$$\mathcal{E}(\gamma LS) = \sum_{ab} w_{ab} I(a, b) + \sum_{abcd;k} v_{abcd;k} R^k(ab, cd) \tag{4.23}$$

where

$$w_{ab} = \sum_{i=1}^{M} \sum_{j=1}^{M} c_i c_j w_{ab}^{ij} \quad \text{and} \quad v_{abcd;k} = \sum_{i=1}^{M} \sum_{j=1}^{M} c_i c_j v_{abcd;k}^{ij}. \tag{4.24}$$

In this form, the energy is expressed as a list of integrals and their contribution to the energy which depends on the mixing coefficients. To minimize the sum over integrals it is useful to take advantage of the symmetry properties of both the $I(a, b)$ and the $R^k(ab, cd)$ integrals. This can be achieved by defining a canonical order. The MCHF program assumes $a \leqslant b$ for the $I(a, b)$ integral, where the inequality applies to the order of a relative to b in $\mathbf{P}$. Similarly, for $R^k(ab, cd)$ it is assumed that $a \leqslant b$, $a \leqslant c$ and $b \leqslant d$.

As in the derivation of the Hartree–Fock equations, the stationary principal must be applied to a functional that includes Lagrange multipliers for all the constraints. Thus

$$\mathcal{F}(\mathbf{P}, \mathbf{c}) = \mathcal{E}(\gamma LS) + \sum_{a \leqslant b} \delta_{l_a l_b} \lambda_{ab} \langle a | b \rangle - E \sum_{i=1}^{M} c_i^2. \tag{4.25}$$

In deriving the stationary conditions with respect to variations in c_i, the most convenient form for $\mathcal{E}(\gamma LS)$ is (4.19) which leads to the secular equation

$$\mathbf{Hc} = E\mathbf{c}. \tag{4.26}$$

Thus the Lagrange multiplier E is the total energy of the system. The requirement of a stationary condition with respect to variations in the radial functions $P(a; r)$ leads to a system of equations, one for each radial function to be varied.

Let $P(a; r)$ be a radial function to be varied for which (4.25) should be stationary.

(i) The variation of $w_{aa} I(a, a)$ becomes

$$-w_{aa} \int_0^\infty \delta P(a; r) \mathcal{L} P(a; r) \, \mathrm{d}r.$$

(ii) The variation of $\sum_{b;k} v_{abab;k} R^k(ab, ab)$, may be expressed as

$$2w_{aa} \int_0^\infty \delta P(a; r) \frac{1}{r} Y(a; r) P(a; r) \, \mathrm{d}r$$

where a factor $2w_{aa}$ has been taken outside the integral and all other constants incorporated into the definition of $Y(a; r)$.

(iii) Similarly, the variation of all other integrals can be expressed in the form $2w_{aa} \int_0^\infty \delta P(a; r) \frac{1}{r} X(a; r) \, \mathrm{d}r$. Note that some contributions may arise from off-diagonal integrals $I(a, b)$, that the Slater integrals may have the radial integral occurring one, two or three times and each position needs to be checked.

The sum of these variations along with those of the orthonormality constraints may be expressed in the form

$$w_{aa} \int_0^\infty \delta P(a; r) Q(r) \, \mathrm{d}r = 0. \tag{4.27}$$

The requirement that this variation be zero for *all* small variations $\delta P(a; r)$ leads to the condition $Q(r) = 0$ or

$$\left(\frac{\mathrm{d}^2}{\mathrm{d}r^2} + \frac{2}{r} [Z - Y(nl; r)] - \frac{l(l+1)}{r^2} - \varepsilon_{nl,nl} \right) P(nl; r)$$

$$= \frac{2}{r} X(nl; r) + \sum_{n' \neq n} \varepsilon_{nl,n'l} P(n'l; r)$$

$$\tag{4.28}$$

where we have assumed the quantum numbers associated with orbital a are nl. The diagonal and off-diagonal energy parameters are related to the Lagrange multipliers

$$\varepsilon_{nl,nl} = \frac{2\lambda_{nl,nl}}{w_{nl,nl}}; \quad \varepsilon_{nl,n'l} = \frac{\lambda_{nl,n'l}}{w_{nl,nl}}. \tag{4.29}$$

Note that with this definition the matrix of diagonal and off-diagonal energy parameters is not symmetric, however:

$$w_{nl,nl}\varepsilon_{nl,n'l} = w_{n'l,n'l}\varepsilon_{n'l,nl}. \tag{4.30}$$

This system of radial equations has exactly the same form as the Hartree–Fock equations except:

- the occupation numbers $w_{nl,nl}$ are not integers but rather *expected* occupation numbers;
- the function $X(nl; r)$ arises not only from the exchange of electrons within a configuration state, but also from interactions between configuration states.

4.6 A non-orthogonal extension

The above discussion has assumed a single orthonormal basis of radial functions. The MCHF atomic structure package incorporates an extension which allows a limited degree of non-orthogonality which sometimes can be used to advantage. Consider the approximation to the ground state of boron, namely $1s^2 2s^2 2p\ ^2P$. This atomic state has an outer $2p$ orbital, but Z-dependent perturbation theory suggests a strong interaction with $2p^3\ ^2P$. The $2p$ orbital representing the latter may be quite different. Thus an MCHF calculation may be over the set of CSFs $\{1s^2 2s^2 2p_1, 1s^2 2p_2^3\}\ ^2P$. Table 4.3 shows the use of GENCL for such a case. Generally, the orbital label consists of two characters for the nl quantum numbers, and optionally a third character that represents a set indicator. Orbitals with the same set indicator are orthonormal but those with different set indicators are non-orthogonal. Orbitals without a set indicator (such as those we have dealt with so far) are members of all sets.

Evaluation of matrix elements when orbitals are non-orthogonal is complex, in general. In order to keep the energy expression relatively simple and avoid the generalized eigenvalue problem, the following restrictions are imposed:

(i)   The orbitals within each CSF are mutually orthogonal.
(ii) There are *at most two subshells* in $\Phi(\gamma_i LS)$ containing spectator electrons whose orbitals are not orthogonal to orbitals $\Phi(\gamma_j LS)$ for all $i, j \leqslant M$.
(iii) If all the spectator electrons with non-orthogonal orbitals have the same l value, then there are *at most two such electrons* in each of $\Phi(\gamma_i LS)$ and $\Phi(\gamma_j LS)$.
(iv) $\langle \Phi(\gamma_i LS) | \Phi(\gamma_j LS) \rangle = \delta_{ij}$.

Table 4.3. An example using GENCL with non-orthogonal orbitals.

```
>Gencl
             Header   ?
>Boron
      Closed Shells   ?
>1s
      Reference Set   ?
>2s(2)2p1(1)
                2   ?
>2p2(3)
                3   ?
>
         Active Set   ?
>
      Replacements    ?
>
        Final Terms   ?
>2P
                2   ?
>
Listing of cfg.inp
 Boron
  1s
  2s( 2) 2p1( 1)
     1S0      2P1       2P0
  2p2( 3)
     2P1
```

The term 'spectator' refers to those electrons not involved directly in the interaction. The purpose of restriction (iii) is to avoid the need for parentage sums in treating the non-orthogonal spectator electrons.

With the above extension, the radial integrals may be multiplied by overlap integrals of the form $\langle a_1|a_2\rangle^p$, referred to as O^1 integrals, and O^2 overlap integrals of the form $\langle a_1|a_2\rangle^{p_1}\langle b_1|b_2\rangle^{p_2}$, where a_1, a_2, b_1, and b_2 are occupied orbitals. With overlap integrals present in the energy expression, the derivation of the radial equation becomes somewhat more intricate, but the underlying theory is the same.

A solution of the MCHF problem requires the simultaneous solution of the secular equation and the variational radial equations. When the latter are assumed to be given, then only the secular problem needs to be solved and the problem is called a *configuration interaction* (CI) problem. If any radial function is optimized, the calculation is called a *multiconfiguration Hartree–*

Fock (MCHF) calculation. The iterative procedure for its solution is the MC-SCF method, described briefly as:

(i)   Initialize the radial functions
(ii) Solve the secular problem for the desired eigensolution
(iii) Until converged

(a) Improve the radial functions
(b) Solve the secular problem for the desired eigensolution.

These stages are best described by means of an example.

4.7   MCHF calculation for $3s^2 3p\ ^2P$ in Al

The variational procedure can only be applied after the energy expression has been obtained. In the MCHF atomic structure package, GENCL determines the configurations to be included in the wave function expansion and NONH derives the energy expression as a data structure for (4.25). In deriving the expression, NONH assumes all electron labels have n as their first character label, the spectroscopic l value ($s, p, d, f, g, h, i, k, l, m$) as their second character, and optionally also a 'set indicator', some symbol designating the set of orthogonal orbitals. Since no set indicator is used, a single orthonormal basis is implied.

Table 4.4 displays the use of GENCL for producing an expansion for the complex associated with the ground state of Al, namely $3s^23p\ ^2P$. All orbitals are orthogonal. The NONH program is then used to obtain an energy expression by performing all spin-angular integrations. These data are needed to compute the interaction matrix (4.21) and also to determine the contribution of each radial integral to the energy (4.23). The latter is done most efficiently if all the integrals contributing to the energy are first transformed to a canonical form, using the symmetry relations, then sorted and a list of unique integrals formed. NONH reports these integrals, and then, for each integral, the coefficients of the integral and the position in the matrix where the integral occurs. Overlap integrals are also reported. This data structure makes it possible to evaluate an integral once and use it wherever it may occur in the matrix. More details may be found in Froese Fischer (1986). The present NONH calculation computed 28 matrix elements (because of symmetry, only the elements on or below the diagonal were evaluated): of these 26 were non-zero, resulting in an interaction matrix that is 92.857% dense. The total number of unique integrals was 24, but the number of F^k, G^k, R^k or L-integrals[†] that appeared in the non-zero matrix elements was $26 + 23 + 31 + 15 = 95$. Thus, on average, each integral appeared about four times.

For brevity, only a section of the Fk portion of the int.lst is displayed in table 4.5. After a header line, following the spectroscopic notation for the

---

† The derivation of the MCHF equations is based on the $\mathcal{L}$-operator and it is convenient to introduce L-integrals where $I(a, b) = -\frac{1}{2}L(a, b)$.

integral, there is an integer that designates the *last* position in the coefficient list of a contribution by the integral to the energy. Thus the coefficients of the current integral are grouped together. The *first* coefficient is either the beginning of the list or the coefficent following the last one of the preceding integral. The coefficient list consists of a coefficient for the associated integral followed by row and column indices designating the position in the interaction matrix. Thus the first integral, F 0(3s, 3s) occurs exactly once with coefficient 1.0 and contributes to the (1, 1) position of the interaction matrix. The second integral, F 0(3s, 3p) contributes to three locations as specified by rows 2, 3 and 4 of the coefficient list. Unlike the F^k and G^k integrals, the R^k and L integrals, maybe also occur with an overlap integral. If the integral is to be multiplied by an overlap integral, an index points to that integral (a zero index indicates the absence of an overlap factor). A positive pointer designates a single overlap matrix (possibly raised to a power), whereas a negative value designates a pair of overlap integrals.

Table 4.4. Example of the use of the program GENCL to generate an active set expansion for the $3s^2 3p$ ^2P state of an Al-like system and NONH for determining the energy expression for the wave function expansion.

```
>GENCL
                Header   ?
>Al
      Closed Shells   ?
> 1s   2s   2p
      Reference Set   ?
>3s(2)3p(1)
                2   ?
>
          Active Set   ?
>3s,3p,3d
Type of set generation ?
>0
      Replacements   ?
>2P
                2   ?
>
      Final Terms   ?
>b                           (for going back to correct an error)
      Replacements   ?
>
      Final Terms   ?
>2P
                2   ?
```

Table 4.4. (continued)

```
>NONH
 FULL PRINT-OUT ? (Y/N)
>n
 STATE  (WITH  7 CONFIGURATIONS):
 ------------------------------
 THERE ARE  3 ORBITALS AS FOLLOWS:
     3s  3p  3d
 THERE ARE 3 CLOSED SUBSHELLS COMMON TO ALL CONFIGURATIONS:
     1s  2s  2p

 CONFIGURATION  1 (OCCUPIED ORBITALS= 2 ): 3s( 2)   3p( 1)
                       COUPLING SCHEME:    1S0      2P1
                                                            2P0
 CONFIGURATION  2 (OCCUPIED ORBITALS= 3 ): 3s( 1)   3p( 1)   3d( 1)
                       COUPLING SCHEME:    2S1      2P1       2D1
                                                            1P0       2P0
 CONFIGURATION  3 (OCCUPIED ORBITALS= 3 ): 3s( 1)   3p( 1)   3d( 1)
                       COUPLING SCHEME:    2S1      2P1       2D1
                                                            3P0       2P0
 CONFIGURATION  4 (OCCUPIED ORBITALS= 1 ): 3p( 3)
                       COUPLING SCHEME:    2P1
 CONFIGURATION  5 (OCCUPIED ORBITALS= 2 ): 3p( 1)   3d( 2)
                       COUPLING SCHEME:    2P1      1S0
                                                            2P0
 CONFIGURATION  6 (OCCUPIED ORBITALS= 2 ): 3p( 1)   3d( 2)
                       COUPLING SCHEME:    2P1      1D2
                                                            2P0
 CONFIGURATION  7 (OCCUPIED ORBITALS= 2 ): 3p( 1)   3d( 2)
                       COUPLING SCHEME:    2P1      3P2
                                                            2P0
 ALL INTERACTIONS ? (Y/N)
>y
   28 matrix elements        26 non-zero matrix elements
   92.8571 % dense           NF= 26 NG= 23 NR= 31 NL= 15
   Total number of terms= 95 The total number of integrals =  24
```

Generally the results from the calculations of tables 4.4 and 4.5 are not of interest in themselves. They do however, provide information needed by the MCHF program. For the latter it also is helpful to first perform a Hartree–Fock calculation. In this way, the more complex MCHF calculation will start with good initial estimates for the core, in fact, it may be desirable for some reason to keep the core fixed.

Table 4.6 shows a calculation where HF provides the initial estimates (wfn.out is moved to wfn.inp) and then only the $n = 3$ active set is optimized.

Table 4.5. The F^k portion of the int.1st file.

```
Al
F  0( 3s,  3s)     1
F  0( 3s,  3p)     4
F  0( 3s,  3d)     6
F  0( 3p,  3p)     7
F  0( 3p,  3d)    12
F  0( 3d,  3d)    15
F  2( 3p,  3d)    20
F  2( 3d,  3d)    23
F  4( 3d,  3d)    26
*
     1.00000000F  1  1
     2.00000000F  1  1
     1.00000000F  2  2
     1.00000000F  3  3

  .   .   .
             *
```

Having obtained initial estimates (an estimate of $3d$ was not found in the input file, so a screened hydrogenic estimate was computed by the program), the calculation sets up the interaction matrix and finds an eigenvector with a large component in the first position, then starts iteration 1. The first phase involves a rotation analysis (none in this case since the core orbitals are fixed), and then computes off-diagonal energy parameters. Each orbital is updated or improved in turn. Strong orthogonality is imposed and each new orbital is orthogonalized immediately to the core. The second phase consists of the CI problem where the interaction matrix is computed and a single eigenvector obtained.

Two convergence criteria are applied. The total energy should converge to a relative change determined by the CFGTOL parameter, but the self-consistent field process applied to the orbitals should also converge to a desirable accuracy. In MCHF calculations, it is possible to have orbitals that contribute very little to the total energy and the orbitals may not stabilize to a high degree of self-consistency. Thus the change in the orbital is weighted by $\sqrt{w_{aa}}$. However, this constraint is relaxed throughout the MC-SCF process, since ultimately it is the stationary property of the energy that is important. In the present calculation, the total energy convergence criterion of 7.5×10^{-10} has been met and was the most severe criterion. The orbitals are well within the self-consistency criterion of 5.7×10^{-5}.

The final results are summarized at the end, as well as in the SUMMRY file. Note that in this calculation, the virial theorem is not exactly -2; this is associated with the fact that not all orbitals were varied.

Table 4.6. Table showing the stages of an MCHF calculation for $3s^2 3p \ ^2P$ of Al.

```
>HF                              !Start with a Hartree-Fock Calculation
 Enter ATOM,TERM,Z
 Examples: O,3P,8. or Oxygen,AV,8.
>Al,2P,13.
 List the CLOSED shells in the fields indicated (blank line if none)
 ... ... ... ... ... ... ... ... etc.
> 1s  2s  2p  3s
 Enter electrons outside CLOSED shells (blank line if none)
 Example: 2s(1)2p(3)
>3p(1)
      ... etc.
    TOTAL ENERGY (a.u.)
    ----- ------
    Non-Relativistic    -241.87670717   Kinetic      241.87673123
    Relativistic Shift    -0.41796905   Potential   -483.75343840
    Relativistic        -242.29467622   Ratio         -1.999999901
 Additional parameters ? (Y/N/H)
>n
>mv wfn.out wfn.inp             !Move wfn.out to wfn.inp
>MCHF                           !Perform the MCHF calculation
 ATOM, TERM, Z in FORMAT(A,A,F) :
>Al,2P,13.
 There are   6 orbitals as follows:
   1s  2s  2p  3s  3p  3d
 Enter orbitals to be varied:
      (ALL,NONE,SOME,NIT=,comma delimited list)
>=3
 Default electron parameters ? (Y/N)
>y
 Default values (NO,REL,STRONG) ? (Y/N)
>y
 Al    2P        13.   220 6  3  7  F  T
       WAVE FUNCTIONS NOT FOUND FOR   3d
 Default values for other parameters ? (Y/N)
>y
         WEAK ORTHOGONALIZATION DURING THE SCF CYCLE=   F
         ACCELERATING PARAMETER FOR MCHF ITERATION  = 0.00
         SCF CONVERGENCE TOLERANCE (FUNCTIONS)      = 1.00D-07
         NUMBER OF POINTS IN THE MAXIMUM RANGE      = 220
         RELATIVISTIC DIAGONAL  ENERGY CORRECTIONS  =   F

         TOTAL ENERGY =     -241.915884727
    WEIGHTS
 1  0.9641  2  0.1751  3  0.1541  4 -0.1106  5 -0.0514  6 -0.0213
 7 -0.0272
```

Table 4.6. (continued)

```
ITERATION NUMBER   1
----------------
CONVERGENCE CRITERIA:ENERGY  (CFGTOL)            =   1.0D-10
                  :FUNCTION(SCFTOL*SQRT(Z*NWF))=   8.8D-07
   E( 3s 1s) =    -0.00065   E( 1s 3s) =     0.00001
   E( 3s 2s) =    -0.00277   E( 2s 3s) =     0.00001
   E( 3p 2p) =    -0.00448   E( 2p 3p) =     0.00001
          EL          ED           AZ           NORM         DPM
< 1s| 3s>=-1.9D-05
< 2s| 3s>=-3.6D-04
          3s       0.8467147     5.3970729    1.0043018    3.92D-03
< 2p| 3p>= 6.2D-03
          3p       0.5456788    12.9479743    0.9678655    4.23D-02
          3d       1.0540099     1.3816417    0.9778572**  9.65D-02

   TOTAL ENERGY =      -241.920208609
WEIGHTS
  1  0.9613  2  0.1875  3  0.1536  4 -0.1137  5 -0.0545  6 -0.0223
  7 -0.0287
LEAST SELF-CONSISTENT FUNCTION IS 3p: WEIGHTED MAXIMUM DPM=4.36D-02


   ITERATION NUMBER   2
   ----------------
   CONVERGENCE CRITERIA:ENERGY  (CFGTOL)            =   1.4D-10
                     :FUNCTION(SCFTOL*SQRT(Z*NWF))=   1.8D-06
   E( 3s 1s) =    -0.00101   E( 1s 3s) =     0.00001
   E( 3s 2s) =    -0.00481   E( 2s 3s) =     0.00001
   E( 3p 2p) =    -0.00607   E( 2p 3p) =     0.00001
          EL          ED           AZ           NORM         DPM
< 1s| 3s>=-1.3D-05
< 2s| 3s>=-2.2D-04
          3s       0.8351237     5.3671214    1.0023457    2.07D-03
< 2p| 3p>= 1.8D-03
          3p       0.5537035    13.3150424    0.9887112    9.27D-03
          3d       1.0303276     1.5201562    0.9971922**  7.78D-03

   TOTAL ENERGY =      -241.920404991
WEIGHTS
  1  0.9606  2  0.1899  3  0.1541  4 -0.1146  5 -0.0545  6 -0.0226
  7 -0.0290
LEAST SELF-CONSISTENT FUNCTION IS 3p: WEIGHTED MAXIMUM DPM=9.59D-03
                . . .
```

Table 4.6. (continued)

```
ITERATION NUMBER  7
----------------
CONVERGENCE CRITERIA:ENERGY  (CFGTOL)             =   7.5D-10
                   :FUNCTION(SCFTOL*SQRT(Z*NWF))=  5.7D-05
   E( 3s 1s) =    -0.00116   E( 1s 3s) =    0.00001
   E( 3s 2s) =    -0.00547   E( 2s 3s) =    0.00001
   E( 3p 2p) =    -0.00611   E( 2p 3p) =    0.00001
             EL          ED              AZ          NORM        DPM
< 1s| 3s>=-3.7D-09
< 2s| 3s>=-5.7D-08
        3s    0.8321410         5.3565340    1.0000006    4.68D-07
< 2p| 3p>= 6.0D-07
        3p    0.5543190        13.4282909    0.9999972    2.51D-06
        3d    1.0193740         1.5477546    1.0000001**  4.44D-07

   TOTAL ENERGY =      -241.920412414
WEIGHTS
  1  0.9604  2  0.1903  3  0.1543  4 -0.1149  5 -0.0545  6 -0.0227
  7 -0.0291

ENERGY (a.u.)
------
   Total              -241.920412414
   Potential          -483.837234533
   Kinetic             241.916822120
   Ratio                 2.000014841
```

4.8   Properties of MCHF wave functions

The MCHF wave function has many of the same properties as the HF wave function though the interpretation must now be modified.

4.8.1   Diagonal energy parameter

In the Hartree–Fock approximation the diagonal energy parameter was directly related to the binding energy of an electron in the case of a singly occupied orbital. When a subshell was multiply occupied, the energy parameter could be related to the binding of the subshell, though some corrections for self-interactions needed to be included.

In MCHF, the diagonal energy of an orbital singly occupied in one configuration can also be shown to be related to the binding energy. But frequently an orbital appears in many configurations and the diagonal energy parameter no longer is a meaningful parameter. If it is singly occupied in all configurations, as $3d$ in $\{2s^2 3d, 2p^2(^1S)3d\}$, then the binding energy is relative to a correlated core with mixing coefficients given by the MCHF approximation.

Often orbitals are referred to as *spectroscopic*, in that the form of the orbital is similar to a Hartree–Fock orbital and its diagonal energy is closely related to a binding energy. Otherwise, the orbital is said to be a *correlation orbital*. Frequently the diagonal energy parameter of a correlation orbital is large (note in this case that $\varepsilon_{3d,3d}$ is almost twice as large as that for $\varepsilon_{3p,3p}$). This has the effect of contracting the orbital so that it can correct the wave function in the region of space occupied by the spectroscopic orbitals of the main configurations.

Examples of such situations will be found in the next two chapters.

4.8.2 Generalized Brillouin's theorem

The MCHF wave function is stationary with respect to perturbations of the radial functions of the form $P(nl; r) \rightarrow P(nl; r) + \epsilon P(n'l; r)$. If the associated perturbation of Ψ, to first order, is a function proportional to $F(nl \rightarrow n'l)$ then, by the stationary principle,

$$\langle \Psi^{\mathrm{MCHF}} | \mathcal{H} | F(nl \rightarrow n'l) \rangle = 0.$$

If $F(nl \rightarrow n'l)$ happens to be a CSF, then that CSF has been included in the MCHF approximation, at least to first order. This is the generalization of Brillouin's theorem. When the generalized Brillouin theorem holds *to all orders*, the perturbation has been included to all orders and will not change the wave function. In the following chapter we will show how, in some simple cases, this condition can be exploited in selecting the form of the wave function expansion and improve the convergence rate.

4.8.3 Stationary condition with respect to rotations

Whenever an orthonormal radial basis is present, the energy must be stationary with respect to a rotation of the basis. We have already seen in the case of a single configuration state that a wave function may be invariant under such a rotation, in which case Koopmans' theorem was used to select the basis that would minimize the energy of the ion. This solution was characterized by having a zero off-diagonal energy parameter. In an active space wave function expansion, when orbitals of the same symmetry are present, a rotation of the radial basis corresponds to a transformation within the complete active space and though the mixing coefficients will vary, the wave function and energy are unchanged. Thus the MCHF wave function is not unique. A degree of freedom may be removed by selecting that solution for which a certain expansion coefficient is zero. Then, by a generalized Brillouin theorem (GBT), it can be shown that the effect of a rotational perturbation is zero to all orders in the expansion of the energy. This technique of eliminating a CSF in order for the MCHF problem to have a unique solution and relying on Brillouin's theorem is referred to as the *GBT approach*. The MCHF program, however, when it determines that the energy is stationary with respect to a rotation of a pair of orbitals, will set the associated Lagrange multiplier to zero, as in the HF case.

These and other properties will be investigated in greater detail in subsequent chapters.

4.9   Computational aspects

In most situations, the dominant component of the wave function has spectroscopic orbitals similar to Hartree–Fock orbitals. Consequently, the numerical techniques for solving the radial equation for these orbitals require the radial functions to be 'acceptable', in that they have the correct number of nodes. The MCHF program assumes that orbitals for the first CSF are expected to be spectroscopic, and that all others are correlation orbitals. The latter do not need to be 'acceptable' and are allowed to assume any form. They are obtained using method 1, without any checking of 'acceptability'. Also, it has been found that, generally, the best convergence is obtained if all radial functions are updated once, in turn, using strong orthogonality, after which the expansion coefficients are updated. The latter requires that the interaction matrix be computed and a new mixing coefficients determined.

The method used for updating expansion coefficients is one that attempts to find the eigenvector for which the first component is the largest contributor. It assumes the radial functions for this CSF are reasonably accurate and may be described as follows.

Assume an approximate, normalized vector $\mathbf{c}$ is known. Let $\mathbf{H}$ and $\mathbf{c}$ be partitioned so that

$$\mathbf{H} = \begin{bmatrix} H_{11} & \mathbf{H}_1^t \\ \mathbf{H}_1 & \mathbf{H}' \end{bmatrix}, \quad (1/c_1)\mathbf{c} = \begin{bmatrix} 1 \\ \mathbf{c}' \end{bmatrix}.$$

Until E has converged:

(i)   Compute $E = \mathbf{c}^t\mathbf{H}\mathbf{c}$;

(ii)   Solve $\mathbf{H}_1 + \mathbf{H}'\mathbf{c}' = E\mathbf{c}'$, for $\mathbf{c}'$;

(iii)   Set $\mathbf{c} = \alpha \begin{bmatrix} 1 \\ \mathbf{c}' \end{bmatrix}$ where $\alpha = 1/\sqrt{1 + \mathbf{c}''\mathbf{c}}$.

This method works reasonably well for ground state systems and for many excited states.

It may at first sight seem desirable to find the nth eigenfunction of the interaction matrix if the atomic state is the nth in the observed spectrum. But this can be highly misleading. Consider a calculation for $2s^2 2p^2 3d\ ^4P$ of nitrogen (with $1s$ inactive) which happens to be the fourth ^4P. Clearly, an MCHF calculation will not yield the eigenvalue for this state as the fourth until an adequate basis is present for the lower states, namely $2s^22p^23s$, $2s2p^4$, $2s^22p^24s$. By optimizing on the dominant component without regard to the position of the eigenvalue, calculations can be performed for multiply excited states. Examples will be presented in later chapters. However, the method is not without its difficulties when several configurations are strongly interacting, but procedures that follow a specific eigenvalue tend to have even more problems.

The cfg.inp file produced by GENCL contains no mixing coefficients. When such a file is read, the program creates an initial guess in which the **c'** is assumed to be zero. In the case of highly degenerate states (often associated with excited states, it is helpful to provide an assist. If one solution has been obtained with expansion coefficients, say (c_1, c_2), for the lower state, then we know the higher state must be orthogonal with expansion coefficients close to $(c_2, -c_1)$. Though convergence may be obtained when the largest component is not in the first position, it is generally safer to do so.

The method used here is not one designed for the large, sparse, eigenvalue problem. For such cases, better performance is obtained using methods such as Davidson's algorithm (Davidson 1975, 1989) though with this algorithm it becomes necessary to track a specific eigenvalue.

A complete description of the numerical MC-SCF procedures may be found in Froese Fischer (1986).

4.10   Exercises

(i)   Derive an expression for the contribution to the energy from the $3d$ orbital in the wave function expansion over the configuration states $\{1s^22s^23d, 1s^22p^2(^1S)3d\}$ and relate it to the binding energy.

(ii) Run the NONH program for non-orthogonal orbital expansion of table 4.3 and, from the int.lst derive the energy expression for the approximation.

(iii) Perform an MCHF calculations for a wave function expansion over the CSFs $\{1s2p_1, 2s2p_2, 2p_h3s, 2p_h3d\}$ for He, where $1s, 2s, 2p_h$ are fixed hydrogenic functions (default screening parameter of zero), $1s, 2s, 3s$ are orthonormal but $2p_h, 2p_1, 2p_2$ are non-orthogonal. Show that

(a)   The diagonal energy parameters of $3s, 2p_2, 3d$ are all the same, except for numerical inaccuracies.

(b)   The difference in the diagonal energy parameter of $2p_1$ and $2p_2$ is twice the difference of the energies of the two targets, namely $4*(1-1/4)=3$.

(iv) Determine the complex for $3s^23p^2$ ^3P.

(v)   Determine the complex for $2s^22p3p$ ^3P. Hint: Use NONH and perform a CAS expansion. Delete all CSFs not in the complex.

(vi) Repeat the GENCL run of table 4.2. Then perform a NONH run to determine the CSFs that interact with the reference configuration.

Chapter 5

Two-Electron Systems

To illustrate many of the points made in the earlier chapters, we will start by discussing two-electron systems. This includes not only helium and helium-like systems, but also systems that in a first approximation can be modelled as a core and two valence electrons (such as beryllium and magnesium). These are the simplest systems in which we can study correlation, but they do exhibit many interesting challenges and examples of complex problems. Let us start by applying some general methods to the helium-like systems.

5.1 Non-uniqueness of the wave function

The CAS expansion for a two-electron system can be written

$$\Psi(\gamma LS) = \sum_{nl,n'l'} c_{nn'}^{(ll')} \Phi(nln'l'LS) \tag{5.1}$$

where the sum on nl and $n'l'$ is over orbitals in a given active set. This expansion is invariant with respect to rotations of orbitals with the same symmetry, and, as discussed in section 4.8.3, some condition needs to be imposed in order to obtain a uniquely defined solution. Many conditions are possible. From the point of maximal numerical stability in the MC-SCF proceedure the orbital rotations should result in an expansion where as many as possible of the coefficients $c_{nn'}^{(ll')}$ are identically zero. Such an expansion is said to be in *reduced* form.

5.2 The reduced form

To investigate the reduced form we order the CSFs in expansion (5.1) according to their spin-angular symmetry

$$\Psi(\gamma LS) = \sum_{ll'} \left(\sum_{nn'} c_{nn'}^{(ll')} \Phi(nln'l'LS) \right) \tag{5.2}$$

where the first sum is over different possible pairs of ll' quantum numbers. In the example of the $1s2p\ ^3P$ term we get

$$\Psi(^3P) = \sum_{nn'} c_{nn'}^{(sp)} \Phi(nsn'p\ ^3P) + \sum_{nn'} c_{nn'}^{(pd)} \Phi(npn'd\ ^3P)$$
$$+ \sum_{nn'} c_{nn'}^{(df)} \Phi(ndn'f\ ^3P) + \dots \tag{5.3}$$

and for $1s^2\ ^1S$ we get

$$\Psi(^1S) = \sum_{nn'} c_{nn'}^{(ss)} \Phi(nsn's\ ^1S) + \sum_{nn'} c_{nn'}^{(pp)} \Phi(npn'p\ ^1S)$$
$$+ \sum_{nn'} c_{nn'}^{(dd)} \Phi(ndn'd\ ^1S) + \dots \ . \tag{5.4}$$

Each of these sums on nn' corresponds to a pair-correlation function (PCF). Let us examine the structure of these PCFs a little more closely. We start by looking at the case where $l \neq l'$ for which each of the above sums on nn' can be written, in terms of unantisymmetrized functions, as (see section 2.1.4)

$$\sum_{n \geqslant m, n' \geqslant m'} c_{nn'}^{(ll')} \Phi(nln'l'\ LS)$$
$$= \sum_{n \geqslant m, n' \geqslant m'} c_{nn'}^{(ll')} \mathcal{A}\left[\frac{1}{r_1 r_2} P(nl; r_1) P(n'l'; r_2)(q_1' q_2' | ll' LS\rangle^u \right]. \tag{5.5}$$

Here $m = l + 1$ and $m' = l' + 1$ are the lowest quantum numbers for a given l or l'. Interchanging the order of antisymmetrization and summation, this sum can be written as

$$\mathcal{A}\left[\frac{1}{r_1 r_2} \left(\sum_{n \geqslant m, n' \geqslant m'} c_{nn'}^{(ll')} P(nl; r_1) P(n'l'; r_2) \right) (q_1' q_2' | ll' LS\rangle^u \right]. \tag{5.6}$$

The sum on n, n' is now over only the radial factors and may be expressed simply in matrix vector form

$$\sum_{n \geqslant m, n' \geqslant m'} P(nl; r_1) c_{nn'}^{(ll')} P(n'l'; r_2) = \mathbf{P}_l^t \mathbf{C} \mathbf{P}_{l'} \tag{5.7}$$

where the $\mathbf{P}$ are column vectors of radial functions and $\mathbf{C}$ is the coefficient matrix. We can always find two orthogonal matrices, $\mathbf{O}_1$ and $\mathbf{O}_2$, that diagonalize the coefficient matrix

$$\mathbf{D} = \mathbf{O}_1 \mathbf{C} \mathbf{O}_2^t \tag{5.8}$$

which leads to

$$\mathbf{P}_l' = \mathbf{O}_1 \mathbf{P}_l \quad \mathbf{P}_{l'}' = \mathbf{O}_2 \mathbf{P}_{l'}. \tag{5.9}$$

If we assume that $l' > l$, this leads to the relation

$$\sum_{n \geqslant m, n' \geqslant m'} c_{nn'}^{(ll')} P(nl; r_1) P(n'l'; r_2) = \sum_{n \geqslant m} d_n^{(ll')} P'(nl; r_1) P'(n'l'; r_2) \qquad (5.10)$$

where, in the latter sum, $n' = n + m' - m$. Inserting (5.10) into (5.5) we obtain the *reduced form* of the pair-correlation function

$$\sum_{n \geqslant m, n' \geqslant m'} c_{nn'}^{(ll')} \Phi(nln'l' \ LS) = \sum_{n \geqslant m} d_n^{(ll')} \Phi'(nln'l' \ LS) \qquad (5.11)$$

where $\Phi'(nln'l' \ LS)$ is a CSF defined in terms of the transformed radial functions. As an example, we can again look at $1s2p \ ^3P$. The reduced form of the pair correlation expansion, where now $n' = n + 1$, is given by

$$\Psi(1s2p \ ^3P) = \sum_{n \geqslant 1} d_n^{(sp)} \Phi(nsn'p \ ^3P) + \sum_{n \geqslant 2} d_n^{(pd)} \Phi(npn'd \ ^3P) + \ldots . \qquad (5.12)$$

It is important to note that the p orbitals of the $npn'd$ CSFs are different and not orthonormal to the p's of the $nsn'p$ CSFs, due to the independent transformations of the terms of the two sums.

For the PCF with $l = l'$ we start with (5.5). The action of the antisymmetrization operator on the factors

$$P(nl; r_1) P(n'l; r_2) (q_1' q_2' | llLS)^u$$

gives

$$\mathcal{A} \left[P(nl; r_1) P(n'l; r_2) (q_1' q_2' | llLS)^u \right]$$
$$= \frac{1}{\sqrt{2}} \{ P(nl; r_1) P(n'l; r_2) (q_1' q_2' | llLS)^u - P(n'l; r_1) P(nl; r_2) (q_2' q_1' | llLS)^u \}. \qquad (5.13)$$

But now $(q_2' q_1' | llLS)^u = (-1)^{L+S+1} (q_1' q_2' | llLS)^u$ and so

$$\mathcal{A} \left[P(nl; r_1) P(n'l; r_2) (q_1' q_2' | llLS)^u \right]$$
$$= \frac{1}{\sqrt{2}} \left\{ P(nl; r_1) P(n'l; r_2) + (-1)^{(L+S)} P(n'l; r_1) P(nl; r_2) \right\} (q_1' q_2' | llLS)^u. \qquad (5.14)$$

Consequently

$$\sum_{n \geqslant m, n' \geqslant m'} c_{nn'}^{(ll')} \Phi(nln'l' \ LS) = \frac{1}{r_1 r_2} \mathbf{P}_l^t(r_1) \mathbf{C} \mathbf{P}_l(r_2) (q_1' q_2' | llLS)^u \qquad (5.15)$$

where $\mathbf{C}$ is symmetric if $L + S$ is even and antisymmetric if $L + S$ is odd. The PCF is now in a form that allows an orbital transformation. Let us start with the symmetric case. For a symmetric matrix there exists an orthogonal matrix **O** and a diagonal matrix **D** such that

$$\mathbf{D} = \mathbf{OCO}'. \tag{5.16}$$

Defining the transformed orbital basis $\mathbf{P}'_l = \mathbf{OP}_l$, called the *natural orbital basis*, and inserting into (5.15) we finally obtain the expansion

$$\sum_{n\geqslant m,n'\geqslant m} c^{(l)}_{nn'}\Phi(nln'l\ LS) = \sum_{n\geqslant m} d^{(l)}_n \Phi'(nlnl\ LS). \tag{5.17}$$

In the antisymmetric case a real orthogonal matrix can transform **C** into a 2×2 block diagonal form, resulting in the expansion

$$\sum_{n\geqslant m,n'\geqslant m'} c^{(l)}_{nn'}\Phi(nln'l\ LS) = \sum_{n\geqslant m,\Delta n=2} d^{(l)}_n \Phi'(nl(n+1)l\ LS) \tag{5.18}$$

where $\Delta n = 2$ implies that n increases by 2 from one term to the next in the sum.

5.2.1 Practical considerations

In an actual calculation the unique solution corresponding to the reduced form may be targeted by including, in the expansion that enters the MCHF program, only those CSFs that should have non-zero coefficients. This will activate the rotation analysis described in the previous chapters as an important tool in the self-consistent field procedure. As an example we consider $1s2p\ ^3P$. When using the active set of orbitals with n quantum numbers less than or equal to four, the following CSFs should be included in the cfg.inp file; $\{1s2p_1, 2s3p_1, 3s4p_1, 2p_23d_2, 3p_24d_2, 3d_34f_3\}\ ^3P$. Here orbitals with the same set indicator are orthonormal while those with different set indicators are non-orthogonal.

5.2.2 The $1s^2\ {}^1S$ state

There are basically two different ways of increasing the set of orbitals in a systematic way, for a reduced pair expansion. In equation (5.2) the outer summation is on l, l', leading to what is called an *l*-expansion, but by interchanging the order of summation so the outer summation is on n, n' we obtain an *n*-expansion. For the latter expansion, we include all reduced pairs generated by all orbitals with principal quantum number less than a given maximum n. In table 5.1 we give results for the $1s^2\ {}^1S$ of helium using the *n*-expansion technique, while in table 5.2 we give results from an *l*-expansion.

Since this is our first example, let us look in more detail at the MCHF calculations performed to arrive at these results. In each step we assume that the atomic state function has the form of an expansion in natural orbitals and

Table 5.1. Energies of an MCHF expansion by n for the He ground state.

n	E (au)
1	−2.8616800
2	−2.8976736
3	−2.9018406
4	−2.9029094
5	−2.9033003
6	−2.9034761
⋮	
Exact[a]	−2.9037243535...

[a]Accad *et al* (1971).

Table 5.2. Energies of an MCHF expansion in partial waves for the He ground state.

$n\,l$	s	p	d	f	g
1	−2.8616800				
2	−2.8779968	−2.8985828			
3	−2.8788709	−2.9001777	−2.9022770		
4	−2.8789901	−2.9004263	−2.9026193	−2.9031173	
5	−2.8790161	−2.9004836	−2.9027063	−2.9032262	−2.9033842
6	−2.8790237	−2.9005008	−2.9027337	−2.9032627	−2.9034278
ΔE_l		−0.0214771	−0.0022329	−0.0005290	−0.0001651

then determine the radial functions for this form. By Brillouin's theorem, the coefficients of the off-diagonal CSFs can be shown to be zero. For example, a simple MCHF calculation over the configuration states, $\{1s^2,\ 2s^2,\ 2p^2\}\ {}^1S$, yields an energy of −2.8976736 au. and, since the energy is stationary with respect to the rotation of the $1s$, $2s$ orbitals, the interaction with $\Phi\left(1s2s\ {}^1S\right)$ is zero.

Table 5.3 shows an MCHF calculation for such an expansion. From the initial estimates (hydrogenic in this case) the program determines the interaction matrix and computes an eigenvector. A simple procedure is used, one that tends to converge to the eigenvector with a large component in the first position. Thus, in the present case, the $1s^2$ configuration should appear first, though the order of the other configuration states is not important. Having obtained estimates of expansion coefficients, the program determines the stationary conditions and performs a small rotation (the EPS parameter is restricted to a maximum of 0.025 in magnitude). The Lagrange multipliers are determined and orbitals updated as in a Hartree–Fock program. The SCF and diagonalization procedure are repeated until convergence or until the maximum number of iterations has been reached.

In MCHF calculations, it usually is advisable to start with a Hartree–Fock calculation for the dominant term in the expansion ($1s^2$, in this case) and then add new orbitals to the calculation. With m orbitals of the same symmetry, the variational procedure needs to satisfy $m(m - 1)/2$ stationary conditions with respect to rotations in that symmetry. When orbitals do not contribute significantly to the energy, the stationary conditions are not well defined, and the SCF procedure may not converge. For this reason, an n-expansion is most practical, namely a series of calculations in which the maximum principal quantum number is increased by unity each time. Then the initial estimates for orbitals $P(ml; r)$ for which $m < n$ are those from the previous MCHF calculation and only the new orbitals with $m = n$ are estimated as screened hydrogenic functions. Furthermore, during the first MCHF calculation, only the new orbitals are varied, all others being kept fixed. For very large calculations, with many Lagrange multipliers, it may be necessary to use accelerating parameters.

However, l-expansions are interesting theoretically. Such an expansion is essentially a partial wave expansion. Starting with $l = 0$, the maximum principal quantum number n is increased until some level of convergence has been reached. Then the maximum l is increased and new orbitals introduced. Table 5.2 shows results for helium of such a procedure where the column designates the largest l and the row the largest n for that l, the largest n for lower l being $n = 10$: note that the calculation has proceeded by columns. The interesting row is the last row. Let us refer to these energies as $E_l, l = 0, \ldots$. Define $\Delta E_l = E_l - E_{l-1}$, which represents the contribution to the energy from the lth partial wave. From a perturbation theory analysis, using hydrogenic functions, Schwartz (1963) has shown that $\Delta E_l = \mathcal{O}((l + \frac{1}{2})^{-4})$. For the ground state of helium, assuming more fully converged partial waves, values of l up to 100 are required for seven decimal places of accuracy. This example shows the slow rate of convergence of the configuration model, but fortunately the $1s^2$ subshell is the most severely affected in this manner because these electrons are in closer proximity to each other than any other two electrons in a many-electron system.

5.2.3 The $1s2p$ $^{3,1}P$ states

Our second example is $1s2p$ ^3P and ^1P in helium. The computational approach where partial waves are added systematically is quite efficient for these cases and we show the results in tables 5.4 and 5.5. The main complication, compared to the ground state, is the introduction of non-orthogonal orbitals. It is also interesting to observe how the ^3P converges faster than the ^1P.

Table 5.3. MCHF calculation for the He $1s^2$ ^1S ground state with a wave function expansion over the $\{1s^2, 2s^2, 2p^2\}$ ^1S set of configuration states.

```
ATOM, TERM, Z in FORMAT(A,A,F) :
>He,1S,2.
 There are   3 orbitals as follows:
   1s  2s  2p
 Enter orbitals to be varied:
   (ALL,NONE,SOME,NIT=,comma delimited list)
>all
 Default electron parameters ? (Y/N)
>y
 Default values (NO,REL,STRONG) ? (Y/N)
>y
 Default values for other parameters ? (Y/N)
>y
 TOTAL ENERGY =  -2.752426275: WEIGHTS 1 0.9995 2 -0.0201 3  0.0261
   ITERATION NUMBER  1
   ----------------
 CONVERGENCE CRITERIA:ENERGY  (CFGTOL)          =  1.0D-10
                     :FUNCTION(SCFTOL*SQRT(Z*NWF))=  2.4D-07
 C( 1s 2s) =      0.73175  V( 1s 2s) =      2.88126  EPS =-0.025000
 E( 2s 1s) =     19.40677  E( 1s 2s) =      0.00783
             EL        ED            AZ           NORM        DPM
     < 1s| 2s>=-1.7D-01
             1s      1.6577883     4.4911114    3.8789792    1.66D-01
     < 1s| 2s>= 3.9D-01
             2s      9.7986379     7.2202350    1.5223809**  7.71D-01
             2p      6.7088545     5.0973929    1.3892430**  5.61D-01
     < 1s| 2s>=-1.1D-10
 TOTAL ENERGY =        -2.888659245
 WEIGHTS 1  0.9956  2 -0.0647  3  0.0679
 LEAST SELF-CONSISTENT FUNCTION IS 1s: WEIGHTED MAXIMUM DPM=2.34D-01
 ...
   ITERATION NUMBER  9
   ----------------
 CONVERGENCE CRITERIA:ENERGY  (CFGTOL)          =  1.5D-09
                     :FUNCTION(SCFTOL*SQRT(Z*NWF))=  6.3D-05
 C( 1s 2s) =     -0.00001  V( 1s 2s) =     10.11922  EPS = 0.000001
 E( 2s 1s) =     10.43887  E( 1s 2s) =      0.04026
             EL        ED            AZ           NORM        DPM
     < 1s| 2s>= 5.4D-07
             1s      1.9083810     4.7633932    1.0000001    6.59D-07
     < 1s| 2s>=-1.8D-06
             2s      6.0364787     7.3328356    0.9999948**  1.18D-05
             2p      6.9293358     7.5860153    0.9999817**  5.16D-05
 TOTAL ENERGY =  -2.897673560 WEIGHTS 1 0.9962 2 -0.0619 3  0.0620
```

Table 5.4. Total energies (in au) for an *l*-expansion, reduced pair-correlation approach for $1s2p \ ^1P$ of helium.

l, l':	sp	pd	df	fg	gh
	−2.1224642	−2.1236610	−2.1237831	−2.1238165	−2.1238289
	−2.1225873	−2.1236854	−2.1237933	−2.1238206	−2.1238307
	−2.1225938	−2.1236886	−2.1237955	−2.1238218	−2.1238314
	−2.1225960	−2.1236895	−2.1237962	−2.1238223	−2.1238317
	−2.1225965	−2.1236897	−2.1237964	−2.1238224	
ΔE_l		−0.0010932	−0.0001067	−0.0000265	−0.0000093
Exact[a]					−2.1238431

[a] Accad *et al* (1971).

Table 5.5. Total energies (in au) for an *l*-expansion, reduce pair-correlation approach for $1s2p \ ^3P$ of helium.

l, l':	sp	pd	df	fg	gh
	−2.1314371	−2.1330572	−2.1331471	−2.1331595	−2.1331623
	−2.1323375	−2.1331078	−2.1331537	−2.1331609	−2.1331626
	−2.1323673	−2.1331133	−2.1331550	−2.1331613	−2.1331628
	−2.1323706	−2.1331144	−2.1331554	−2.1331614	
	−2.1323712	−2.1331146	−2.1331555		
ΔE_l		−0.0007434	−0.0000409	−0.0000059	−0.0000014
Exact[a]					−2.1331642

[a] Accad *et al* (1971).

5.2.4   The $1s2s \ ^1S$ state

One of the most difficult calculations from the point of view of stability of SCF iterations is the HF calculation for the $1s2s \ ^1S$ state of helium. Once obtained, the solution is disappointing in that the energy is too low! In this case, a rotation of the orbital basis introduces the linear combination of configuration states, $\Phi(1s^2 \ ^1S) - \Phi(2s^2 \ ^1S)$, which is also the diagonal form of this state as discussed in the previous section. The latter form is exceedingly unstable and the SCF process has never been applied successfully to this form. From the point of view of Z-dependent perturbation theory, it can be shown that $\Phi(1s2s \ ^1S)$ and $\left[\Phi(1s^2 \ ^1S) - \Phi(2s^2 \ ^1S)\right]/\sqrt{2}$ are degenerate, and that degenerate perturbation theory needs to be applied.

The difficulty arises because of the orthogonality constraint.   A HF calculation for $1s2s' \ ^1S$, where $\langle 1s|2s' \rangle \neq 0$, has a non-degenerate perturbation expansion. However, since the variational procedure relies on an energy expression, it then becomes necessary to derive an energy expression for a non-orthogonal case and include the overlap integral. This may be difficult in general. An alternative is to express the non-orthogonal CSF in terms of an orthogonal basis. Let $|\bar{2s}\rangle = |2s'\rangle - \langle 2s'|1s\rangle |1s\rangle$ be an unnormalized orbital, i.e. orthogonal to $1s$. Let the $2s$ be the normalized version; $|2s\rangle = b|\bar{2s}\rangle$. Then $|1s2s'\ ^1S\rangle = c_1|1s^2\ ^1S\rangle + c_2|1s2s\ ^1S\rangle$, which defines a stable multiconfiguration Hartree–Fock approximation. Also, by the Hylleraas–Undheim–MacDonald theorem (Hylleraas and Undheim 1930, MacDonald 1933), the energy is an upperbound to the second ^1S state. A more accurate MCHF calculation would start with this combination, and retain the diagonal reduced form for the remaining correlation effect. The expansion differs from the expansion for the ground state in that the $2s^2$ configuration state has been replaced by $1s2s$ which is, in fact, the dominant term in the expansion.

It turns out that the l-expansion is quite unstable for this case. In table 5.6 we therefore show results from an n-expansion, reduced calculation.

Table 5.6. Resulting total energies (in au) for different steps in a reduced form calculation for $1s2s\ ^1S$ of helium.

CSFs included	Total energy
$1s2s,\ 1s^2$	-2.1434743
$+2p^2$	-2.1438019
$+3s^2, 4p^2, 3d^2$	-2.1456870
$+4s^2, 4p^2, 4d^2, 4f^2$	-2.1458729
$+5s^2, 5p^2, 5d^2, 5f^2, 5g^2$	-2.1459327
$+6s^2, 6p^2, 6d^2, 6f^2, 6g^2, 6h^2$	-2.1459531
$\vdots$	
Exact[a]	-2.1459740

[a]Accad *et al* (1971).

5.2.5 The $1s2s\ ^3S$ state

The $1s2s\ ^3S$ atomic state function differs from the ^1S in that the radial factors for the partial wave expansions are now antisymmetric. In this case, orthogonal matrices exist such that $\mathbf{D}^l = \mathbf{O}\mathbf{C}^l\mathbf{O}^t$ where now $\mathbf{D}^l$ has a 2×2 block diagonal form, leading to an expansion over the configuration states $\{1s2s, 3s4s, 5s6s, \ldots, 2p3p, 4p5p, \ldots, \text{etc}\}\ ^3S$. Note that, as in the case of the ground state, each orbital appears in only one CSF, and each CSF requires two orbitals.

Unlike the ground state of helium (see table 5.2) where Schwartz (1963)

Table 5.7. Resulting total energies in different steps of an l-expansion, reduced-pair correlation approach for $1s2s\ ^3S$ of helium.

l, l':	ss	pp	dd	ff	gg
	−2.1742508	−2.1751669	−2.1752187	−2.1752265	−2.1752283
	−2.1742644	−2.1751717	−2.1752202	−2.1752270	
	−2.1742648	−2.1751719	−2.1752204		
ΔE_l		−0.0009071	−0.0000485	−0.0000066	−0.0000013
Exact[a]					−2.1752294

[a] Accad *et al* (1971).

showed a $\mathcal{O}((l + \frac{1}{2})^{-4})$ dependence of ΔE_l, for the $1s2s\ ^3S$ the dependence is $\mathcal{O}((l + \frac{1}{2})^{-6})$. The faster rate of convergence is seen in table 5.7. The simple calculation is already seen to be accurate to 1×10^{-6} hartrees.

The simplicity of the MCHF method makes it well suited to these fairly accurate calculations for lowly excited states of two-electron systems, but its real strength is more apparent for other cases.

5.2.6 Doubly excited states in helium

In chapter 10 we will discuss quasi-bound states, above the first ionization limit, in different systems. In most cases, these will interact with the continuum, which causes autoionization. In many cases the position of these autoionizing states can be determined with quite high accuracy, by assuming that they are bound. We thereby disregard the continuum and only compute the discrete part of the wave function.

A good example of this is the $2s2p\ ^3P$ term in helium. To a certain level the reduced expansion is stable for this state, and we can with high accuracy arrive at the expansion in table 5.8. This corresponds to an excitation energy of 58.313 eV above the $1s^2\ ^1S$ ground term of helium, in excellent agreement with the experimental values of 58.312 ± 0.006 eV (Berry *et al* 1972), 58.30 ± 0.03 eV (Hicks and Cromer 1975) and 58.29 ± 0.03 eV (Gelebart *et al* 1976). In chapter 10 we will return to the interesting problem of computing the width of such resonance states.

The expansion in table 5.8, not including the last CSF, is something of a limit for this approach, since including $4s_14p_1$ CSF gives instabilities. The $4s_1$ approaches a hydrogenic function and the $4p_1$ simulates a continuum-like solution. To solve this we relax the orthogonality condition, and include a $4s_14p_3$ CSF.

Table 5.8. An MCHF expansion for the $2s2p$ ^3P term of He I. Also given is the total energy (E_{TOT}) and excitation energy (E_{exc}) as a function of the length of the expansion.

CSF	c_i	E_{TOT} (au)	E_{exc} (eV)
$2s_1 2p_1$	0.9917619	−0.7536079	58.4996
$3s_1 3p_1$	−0.0548442	−0.7551786	58.4569
$2p_2 3d_2$	−0.1143651	−0.7601703	58.3211
$3p_2 4d_2$	−0.0066678	−0.7602695	58.3184
$4p_2 5d_2$	−0.0025272	−0.7602907	58.3178
$3d_3 4f_3$	0.0098318	−0.7604116	58.3145
$4d_3 5f_3$	0.0017987	−0.7604224	58.3142
$4f_4 5g_4$	−0.0027411	−0.7604397	58.3138
$5g_5 6h_5$	0.0010749	−0.7604437	58.3137
$6h_6 7i_6$	−0.0005089	−0.7604449	58.3136
$4s_1 4p_3$	0.0127035	−0.7604589	58.3132
Experiment[a]			58.312 ± 0.006

[a]Berry *et al* (1972).

5.3 Rydberg series

Since there is no difference in the MCHF equations for different terms in a Rydberg series, such as the $2snd$ ^3D series in beryllium, we need to find a stable approach to 'get the right solution'. The most obvious is node counting, since we assume that, at least in simple cases, the number of nodes of the orbital with principal quantum number n is $n - l - 1$.

In principle the Hylleraas–Undheim–MacDonald theorem (discussed in section 1.5.4) indicates that we need to include all lower terms (with principle quantum number less than n) to satisfy the upper bound principle for the energy eigenvalue of the $2snd$ ^3D solution. By using the GBT it is immediately clear that this is not necessary. A lower Rydberg term, such as $2s3d$ ^3D, is just a mono-excited CSF from our CSF associated with the largest weight, that is the $2snd$ ^3D. The interaction between them will therefore be zero.

However, in some cases instabilities might occur and force us to consider the inclusion of lower Rydberg terms. This is more due to the fact that the eigenvalues are exceedingly close for higher members of a Rydberg series, and the initial guess might lead us to the wrong solution. The node counting here becomes an art, due to the small inner oscillations of high-n orbitals.

Let us look a little more closely at our example, the $2snd$ ^3D in neutral beryllium. We use an approach with the following reduced expansion of CSFs:

$$\{2snd_1, 3sn'd_2, 2p_23p_2, 2p_34f_3, 3d_44d_4, 3d_55g_5\}^3D.$$

The resulting binding energies, relative to the $1s^22s$ ^2S limit, are presented in

Table 5.9. Frozen-core MCHF results for binding energies (in cm^{-1}) for members of the $2snd$ ^3D Rydberg series of beryllium. The limit energy ($E = -14.2773948$ au) is taken as from a Hartree–Fock calculation for $2s$ ^2S of Be$^+$.

	Binding energies			
Term	Hydrogen[a]	HF	MCHF	Experiment
$3d$	12192.1	12453.3	13142.4	13138.35[b]
$4d$	6858.0	6996.9	7251.5	7250.41[b]
$5d$	4389.1	4466.2	4589.0	4588.31[b]
$6d$	3048.0	3094.3	3163.1	3162.57[b]
$7d$	2239.4	2269.1	2311.4	2311.17[c]
$8d$	1714.5	1734.7	1762.8	1762.74[c]
$9d$	1354.7	1369.0	1388.5	1389.5[b]
$10d$	1097.3	1107.8	1121.9	1122.1[b]
$11d$	906.8	914.8	925.4	924.0[b]

[a]Hydrogenic, $1/2n^2$, value.
[b]From Johansson (1962).
[c]From Holmström and Johansson (1969).

table 5.9, while we give the mixing coefficients in table 5.10. Here we use a frozen-core approximation, where the $1s$ radial function is kept fixed from the Hartree–Fock for the limit.

The approach is stable for all $n \leqslant 11$, except for $n = 7$. Here we encounter an instability for the $2p_2 3p_2$ CSF, whose orbitals we did not manage to optimize as for the other members of the series. To overcome this we use the radial $2p_2$- and $3p_2$-functions from the calculation for the $n = 6$ term and kept them fixed. At the end, after all other CSFs had been added, we included a $4p_2 5p_2$ CSF to correct for the small difference in this contribution to the reduced form between $n = 6$ and 7 terms.

5.4 Rydberg series with perturber

Many examples of Rydberg series are not as simple as the one described above. Different perturbers, that is terms which do not belong to the series but interact strongly with it, will influence the structure and cause irregularities. We will discuss this in more detail in the next chapter, but here let us give an example from two-electron systems, namely the ^1D series in beryllium. This will illustrate an important stepwise procedure for convergence which is useful in many situations.

The lowest member of this series turns out to have the largest weight associated with the $2p^2$ ^1D CSF. It is quite straightforward to perform this

Table 5.10. CSF composition for different members of the $2snd$ ^3D Rydberg series of beryllium.

	c_i coefficients					
Term	$2snd_1$	$3sn'd_1$	$2p_23p_2$	$2p_34f_3$	$3d_44d_4$	$3d_55g_5$
$3d$	0.99202	0.01444	−0.09791	−0.07775	0.00321	0.00592
$4d$	0.99726	0.01115	−0.05764	−0.04478	0.00150	0.00345
$5d$	0.99872	0.00829	−0.03957	−0.03042	−0.00094	0.00236
$6d$	0.99929	0.00639	−0.02946	−0.02252	−0.00067	0.00176
$7d$[a]	0.99956	0.00508	−0.02349	−0.01752	−0.00051	0.00137
$8d$	0.99971	0.00418	−0.01872	−0.01424	−0.00041	0.00112
$9d$	0.99980	0.00351	−0.01560	−0.01185	−0.00034	0.00093
$10d$	0.99986	0.00300	−0.01326	−0.01006	−0.00029	−0.00079
$11d$	0.99989	0.00260	−0.01145	−0.00868	−0.00025	−0.00068

[a]Including two pp CSFs. The fixed $2p_23p_2$ has $c_i = -0.02349$ and $4p_25p_2$ has $c_i = -0.00042$.

calculation, if the expansion is increased stepwise. We illustrate this in table 5.11. In the first, Hartree–Fock, step, we only include the $2p^2$ ^1D CSF. The resulting orbitals are moved from the wfn.out to the wfn.inp file, and we add the $2s3d_1$ ^1D CSF. Again we optimize all orbitals, except for the $1s$, and use them as an input for step 3 (where five new CSFs are added).

If we now move to the second term in the ^1D series, this kind of approach is unstable. The reason for the instability is the strong interaction between the $2p^2$ and $2s3d$ CSFs, leading to a distribution between CSFs that is close to fifty–fifty. Since the solutions have to be orthogonal, the relative phases of the two CSFs have to be different. That is, simply, since the sign of the c_i coefficents is opposite in the lowest member of the series (0.7615716 and −0.6424744) it should be equal in the second. The switch of sign is a 'sudden' change, and has to be forced on our solution. The first steps get the phase wrong since the $2p^2$ ^1D in a Hartree–Fock approximation is far above $2s3d$ ^1D, while in a fully correlated model the atomic state function with a dominant $2p^2$ ^1D CSF is the lowest. Switching the order does switch the sign, and the correct phase for the second atomic state with $2s3d$ ^1D character can only be obtained once the localized $2p^2$ ^1D component has been sufficiently correlated.

A possible approach therefore is to first compute a correlated $2p^2$ ^1D, without the $2snd$ ^1D series being included, much in the same way the $2s2p$ ^3P of helium was computed without the interaction with the continuum. Let us label this model as truncated. After this we turn our attention to the $2snd$ ^1D series. For the $2s3d$ ^1D the first step is to include a fixed $2s$ and $3d_1$, obtained from a frozen-core Hartree–Fock calculation. By merging the .c and .w files

Table 5.11. Frozen-core, MCHF calculations for $2p^2$ ^1D in neutral beryllium. For each step the new CSFs are shown, with their final c_i coefficients, together with the total energy in au.

Step	Energy	CSF	c_i
HF	-14.3027281	$2p_2^2$	0.7615716
2	-14.3556190	$2s3d_1$	-0.6424744
3	-14.3597595	$3p_2^2$	-0.0643771
		$2p_34f_3$	-0.0373705
		$3d_4^2$	0.0364892
		$3s4d_1$	0.0163561
		$3d_55g_5$	0.0030702
4	-14.3599482	$4p_2^2$	-0.0056754
		$4f_4^2$	-0.0049263
		$4d_4^2$	0.0044462
		$3p_35f_3$	-0.0023875
		$4s5d_1$	-0.0016899
		$4d_55g_5$	-0.0010421

from this and the truncated calculation, and adding $2s4d_1$ ^1D CSF, we get input files for the next step. In this we first optimize only the $4d_1$, to include the effect of the rest of the Rydberg series. Next we optimize all the orbitals, except for the $1s$, $2s$ and $3d_1$. After this has converged, we include the $3s3d_3$ and $4s4d_3$ to represent correlation with a contracted d orbital. This can be thought of as representing the $3dns$ series.

For the $2s4d$ state, the $4d_1$ orbital is optimized alone in a 'Hartree–Fock-like' calculation with the expansion

$$\{2s4d_1, 2s3d_1\}^1D$$

keeping $1s$, $2s$ and $3d_1$ fixed. After this the calculation proceeds like the one for $2s3d$ ^1D. After merging the Hartree–Fock-like and truncated .c and .w file, and adding a $2s5d_1$ CSF, we have an input for the next stage of the calculations. In contrast to the $2s3d$ the c_i coefficients of $2s3d_1$ and $2s4d_1$ should have the same sign. If the MCHF code gets the pair of values 1.0 and 0.0 for these two coefficients, it converges to the $3d$ state. By giving a non-zero, positive value for the c_i coefficient of $2s3d_1$, as input for the optimization of $5d_1$, the calculations converges to the $2s4d$ term.

For completeness we also use this method for the lowest, $2p^2$ ^1D, where we use the same CSF expansion as for $2s3d$, keeping the $2s$ and $3d_1$ orbitals fixed as from a frozen-core Hartree–Fock calculation.

This is just one example of a Rydberg series, strongly affected by one single perturber. We will return to other examples in the next chapter.

Table 5.12. Binding energies (cm^{-1}) for different members of the $2snd$ ^1D series in beryllium, from different models. CSF composition from final calculations. The energy of the $2s$ ^2S limit is taken from a Hartree–Fock calculation $(E = -14.2773948$ au$)$.

	$2p^2$	$2s3d$	$2s4d$
$E(HF)$	5559.6	12107.4	6815.6
$E(trunc.)$	9073.0		
$E(MCHF)$	18099.6	10656.0	6369.9
$E(exp.)$	18309.6	10763.8	6411.2

CSF	c_i		
$2p_2^2$	0.760695	−0.408857	−0.219782
$3p_2^2$	−0.064234	0.033853	0.019201
$3d_2^2$	0.036488	−0.030185	−0.019834
$4d_2^2$	0.004407	−0.004000	−0.002564
$4f_2^2$	−0.005202	0.004974	0.000898
$5f_2^2$	−0.012113	0.001326	0.003416
$2p_34f_3$	−0.037662	−0.056714	−0.045716
$3p_35f_3$	−0.002396	0.002727	0.001740
$2s3d_1$	0.517140	0.842319	0.099646
$2s4d_1$	0.382136	−0.343228	0.942128
$3s3d_3$	0.030766	−0.014265	−0.006684
$4s4d_3$	0.002422	−0.001393	0.000891
$2s5d_1$	—	—	−0.226368

5.5   The GBT method

As was discussed in the previous sections a unique solution to the two-electron MCHF problem can be obtained by deleting a number of CSFs from the CAS expansion, retaining only the reduced forms of the pair correlation functions. In certain cases this also introduces non-orthogonal orbitals. Expansions based on reduced forms are optimal from the point of numerical stability, but they are difficult to generalize to many-electron systems. Therefore, we will discuss another, simpler method for obtaining a unique solution. In this method, known as the GBT method, *one* CSF is excluded (requiring the orbital rotations to be such that the expansion coefficient for the CSF is zero) from the CAS expansion for each *pair* of orbitals $(nl, n'l)$ with a common l-quantum number. The general rule is to exclude the CSF, $\Phi_x(nl \to n'l)$, that is obtained by a single-orbital, coupling-preserving substitution $P(nl; r) \to P(n'l; r)$, from the CSF, $\Phi_m(nl)$, that among all those with nl occupied has the largest weight.

As an example let us look at $1s^2$ ^1S. When using the active set with maximum $n = 2$, that is $\{1s, 2s, 2p\}$, we have one orbital pair to investigate, $(1s, 2s)$. As we already pointed out, the $\Phi_m(1s)$ in this case is the $1s^2$ ^1S, and

Table 5.13. Results for the $1s^2$ ^1S ground state of helium and $2s2p$ ^3P state of beryllium, from active space calculations, without (CAS) and with (GBT) deletion according to GBT. n denotes the maximum quantum number in the active set and NIT the number of iteration for the two approaches. Other definitions are given in the text. A calculation is considered converged when the total energy has stabilized to the ninth decimal place.

					NIT		
n	nl	$n'l$	Φ_m	Φ_x	CAS	GBT	Energy (au)
			$1s^2$ ^1S of helium				
1					4	4	-2.8616800
2	$1s$	$2s$	$1s^2$	$1s2s$	15	8	-2.8976736
3	$1s$	$3s$	$1s^2$	$1s3s$			
	$2s$	$3s$	$2s^2$	$2s3s$			
	$2p$	$3p$	$2p^2$	$2p3p$	16	11	-2.9018406
4	$1s$	$4s$	$1s^2$	$1s4s$			
	$2s$	$4s$	$2s^2$	$2s4s$			
	$3s$	$4s$	$3s^2$	$3s4s$			
	$2p$	$4p$	$2p^2$	$2p4p$			
	$3p$	$4p$	$3p^2$	$3p4p$			
	$3d$	$4d$	$3d^2$	$3d4d$	33	19	-2.9029094
			$2s2p$ ^3P of beryllium				
2					6	6	-14.5115018
3	$2s$	$3s$	$2s2p$	$2p3s$			
	$2p$	$3p$	$2s2p$	$2s3p$	12	5	-14.5182649
4	$2s$	$4s$	$2s2p$	$2p4s$			
	$3s$	$4s$	$3s3p$	$3p4s$			
	$2p$	$4p$	$2s2p$	$2s4p$			
	$3p$	$4p$	$3s3p$	$3s4p$			
	$3d$	$4d$	$2p3d$	$2p4d$	14	5	-14.5186321

the $\Phi_x(1s \rightarrow 2s)$ is therefore $1s2s$ ^1S. When going to $n = 3$, we will have to consider three additional rotating pairs, $(1s, 3s)$, $(2s, 3s)$, and $(2p, 3p)$. The different Φ_m and Φ_x in this case are given in table 5.13. In this table we also show the number of MCHF iterations required for the complete active space calculation with and without deleting CSFs. Note that in this case the GBT expansion is identical to the reduced form.

As a second example, let us look at the lowest odd term of beryllium, the $1s^2 2s2p$ $^3P^o$. We will treat this as a two-electron system, by keeping the $1s^2$ core frozen. Table 5.13 again gives the strategy for this state.

In both the examples discussed so far, the orbitals occupied in the main CSF have belonged to the same complex. The natural way of performing these calculations in a systematic way was to extend the current active set of orbitals

Table 5.14. Complete active space, GBT calculations for $1s2p\ ^3P$. In the table, N_{CSF} gives the number of CSFs included in the expansion.

Active set	nl	$n'l$	Φ_m	Φ_x	N_{CSF}	Energy (au)
$1s, 2p$					1	-2.1314371
$+2s, 3p, 3d$	$1s$	$2s$	$1s2p$	$2s2p$		
	$2p$	$3p$	$1s2p$	$1s3p$	4	-2.1330205
$+3s, 4p, 4d, 4f$	$1s$	$3s$	$1s2p$	$2p3s$		
	$2s$	$3s$	$2s3p$	$3p3s$		
	$2p$	$4p$	$1s2p$	$1s4p$		
	$3p$	$4p$	$2s3p$	$2s4p$		
	$3d$	$4d$	$3p3d$	$3p4d$	10	-2.1331356

with all orbitals whose principal quantum numbers are the next higher n. In some cases this is not the preferred approach, and will not give a simple way of applying the GBT. This is seen most readily from an example, the $1s2p\ ^3P$ term of helium, where it is most natural to start with the $\{1s, 2p\}$ set, defining the Hartree–Fock case. After that we extend the set with $\{2s, 3p, 3d\}$ and so on. This makes it possible to apply the GBT, and the number of orbitals for each symmetry is balanced. The strategy for this state is described in table 5.14.

5.6 Exercises

(i) We will investigate GBT for the ground state of helium.
  (a) Solve the MCHF problem for the $\{1s^2, 1s2s, 2s^2\}^1S$ expansion.
  (b) Solve the MCHF problem for the $\{1s^2, 2s^2\}^1S$ expansion. Observe that the energy is the same.
  (c) Use the `wfn.out` from the latter calculation as input for an MCHF calculation, with the expansion $\{1s^2, 1s2s, 2s^2\}^1S$. Do not optimize any orbitals (put NIT = 0). What is the resulting c_i coefficient for the second CSF? Explain your results.

(ii) Derive the reduced form of the two partial waves of (5.12). What would it be for the $1s^22s2p\ ^3P$ state of Be?

(iii) Perform an MCHF calculation for Be over the two configuration states $1s^22s2p_1$ and $1s^22p_23d$ and plot the two p orbitals.

(iv) Repeat the calculation of table 5.6 (at least up to $n = 5$) and plot the radial functions. What do you observe about correlation orbitals?

(v) Reproduce the results of tables 5.11 and 5.12 for the lowest states. Note that the relative sign of the $2p_2^2\ ^1D$ and $2s3d_1\ ^1D$ expansion coefficients are different. Plot the radial functions and explain the difference.

Chapter 6

Correlation in Many-Electron Systems

In chapter 4 we defined correlation as the difference between the exact and Hartree–Fock solutions to the Schrödinger equation. To classify different types of correlation we looked at the structure of the first-order correction and the CSFs used to represent this correction. In the first part of this chapter we will discuss the major contributions to the wave function which will represent the zero-order wave function. The remaining part of the chapter is devoted to various improvements. Most important for our discussion is the contribution from correlation within the valence subshells and between valence and core subshells.

6.1 Zero-order wave functions

The complex is the set of CSFs with the same set of principal quantum numbers and parity as the reference configuration. According to Z-dependent perturbation theory (see section 4.2), the complex should be part of the zero-order wave function that describes the dominant correlation effects in the atom. But this theory is valid for high Z and may not be practical for neutral atoms or ions of low degrees of ionization. Even within the complex, there may be CSFs that are not major contributors. To exemplify these issues, in table 6.1 we give some examples of the mixing coefficients from MCHF calculations including the complex as part of a larger calculation. In the case of silicon, the $3s^2 3p^2$ ^1S term (rather than the ground state) was chosen to minimize the length of the expansion associated with the complex. The notation $3d^4$ $_4^1S$ is the spectroscopic notation for designating the seniority which in a cfg.inp file would be designated as 1S4. Note that in all cases considered, some members of the complex are *not* important contributors to the wave function. Thus it is not necessary to think of them as part of the zero-order wave function.

Table 6.1. The mixing coefficients from MCHF calculations including the complex of CSFs for different systems, together with an estimate of the 'rest' of CSFs percentage ($\sum c_i^2$) from a large calculation.

Mg $3s^2$ ^1S		Mg $3s3d$ ^1D		Al $3s^2 3p$ ^2P		Si $3s^2 3p^2$ ^1S	
CSF	c_i	CSF	c_i	CSF	c_i	CSF	c_i
$3s^2$	0.9640	$3s3d$	0.8789	$3s^2 3p$	0.9604	$3s^2 3p^2$	0.9415
$3p^2$	0.2640	$3p^2$	0.4769	$3p^3$	0.1903	$3s^2 3d^2$	0.1779
$3d^2$	−0.0305	$3d^2$	0.0052	$3s3p(^1P)3d$	−0.1543	$3s3p^2 3d$	0.0342
rest	0.19%	rest	0.74%	$3s3p(^3P)3d$	0.1149	$3p^4$	0.2201
				$3p3d^2(^1S)$	−0.0545	$3p^2(^1S)3d^2$	−0.0079
				$3p3d^2(^1D)$	−0.0227	$3p^2(^1D)3d^2$	−0.0167
				$3p3d^2(^3P)$	−0.0291	$3p^2(^3P)3d^2$	−0.0178
				rest	1.6%	$3d^4$ $(_0{}^1S)$	−0.0074
						$3d^4$ $(_4{}^1S)$	−0.0003
						rest	3.15%

6.1.1   Important substitutions

In many systems important correlation effects are described by CSFs both within and outside the complex. A strategy for finding the CSFs, to be included in the zero-order wave function, is to look at specific orbital substitutions that have been found by experience to be important. For systems where the complex is big, this is a valuable method for finding the largest contributions within the complex. Below we will look at some important orbital substitutions.

6.1.1.1 $s \leftrightarrow d$

The importance of this substitution, even outside the complex, is well illustrated by the strong mixing between $2s2p$ ^1P and $2p3d$ ^1P in beryllium. In a two-configuration MCHF calculation with non-orthogonal orbitals the $2s2p$ ^1P term has the expansion

$$0.9716|2s2p_1 \ ^1P\rangle - 0.2367|2p_2 3d \ ^1P\rangle.$$

Another example of the $s \leftrightarrow d$ substitution is the strong mixing between $3s^2 3p$ ^2P and $3s3p(^{1,3}P)3d$ ^2P in aluminum-like systems, illustrated in table 6.1.

6.1.1.2 $s^2 \leftrightarrow p^2$

Basically all substitutions of two equivalent orbitals, with two others, are important. As examples we found strong mixing of $1s^2$ ^1S with $2p^2$ ^1S in helium and of $3s^2 3p$ ^2P with $3p^3$ ^2P in aluminum.

6.1.1.3 $p^2 \leftrightarrow sd$

We have already seen one example of this notoriously important substitution, in the $2snd$ ^1D series of beryllium (table 5.12). For the lowest member of the series, $2snd$ ^1D and $2p^2$ ^1D are mixed almost equally. In the homologous magnesium-like systems, this substitution represents one of the strongest mixings known, which is persistent along the whole isoelectronic sequence. In table 6.2 we give the mixing coefficients of the $3s3d$ ^1D and $3p^2$ ^1D terms in a number of magnesium-like systems.

Table 6.2. The mixing coefficients from a simple MCHF calculation for the lowest ^1D in magnesium-like systems, including only the $n = 3$ complex.

CSF	c_i coefficients				
	Mg I	Al II	S V	Ar VII	Fe XV
$3p^2$	0.4769263	0.7460571	0.8626667	0.8798048	0.8978611
$3s3d$	0.8789281	0.6651281	0.5039948	0.4535076	0.4386277
$3d^2$	0.0051713	0.0316772	0.0423723	0.0416417	0.0380949

6.1.1.4 $d^2 \leftrightarrow pf$

A less pronounced, but still important, substitution is apparent in the strong mixing between $3d^2$ and $3p4f$ in magnesium-like ions. We show in table 6.3 the mixing coefficients of the lowest ^3F term in some magnesium-like ions, obtained from limited MCHF calculations. The CSFs included are

$$\{3d^2, 3p4f, 4d^2, 4f^2\}^3F .$$

Table 6.3. Mixing coefficients for the lowest ^3F term in some magnesium-like ions.

CSF	c_i coefficients				
	Al II	Si III	P IV	S V	Ar VII
$3d^2$	0.1245811	0.2797952	0.7886900	0.9711470	0.9949187
$3p4f$	0.9921618	0.9599922	−0.6141674	0.2357899	0.0960150
$4d^2$	−0.0094191	−0.0113483	−0.0238715	−0.0245391	−0.0184865
$4f^2$	−0.0023910	−0.0009215	0.0140243	0.0259717	0.0240044

6.1.2   Long- and short-range interaction

Looking at the $3p4f\ ^3F - 3d^2\ ^3F$ and $3s3d\ ^1D - 3p^2\ ^1D$ mixings there are some important differences in the behaviour along the isoelectronic sequence. The latter mixing is smoothly varying (at least after Mg I), and persists along the sequence, while the former shows an almost resonance type of behaviour around P IV. This is an example of different ranges of an interaction†, which has given rise to the labels *long-* and *short-range* interactions (Froese Fischer 1980). Just as in our examples, it is generally true that long-range behaviour is associated with an interaction within a complex, while short-range behaviour arises from a more accidental degeneracy effect. The long- and short-range behaviour can be understood by looking at the expression for mixing coefficients of interacting CSFs along an isoelectronic sequence.

6.1.3   Close degeneracy and plunging configurations

In a two-configuration model, where $\Phi(\gamma_1 LS)$ interacts with a dominant configuration state function $\Phi(\gamma_0 LS)$, the mixing coefficient for the former is

$$c_1 \approx \frac{\langle \Phi(\gamma_1 LS)|\mathcal{H}|\Phi(\gamma_0 LS)\rangle}{E(\gamma_0 LS) - E(\gamma_1 LS)}, \tag{6.1}$$

where $E(\gamma_0 LS)$ and $E(\gamma_1 LS)$ are the energies associated with the individual CSFs. The numerator in the above expression is large if the orbitals of the two CSFs occupy roughly the same region in space. This is the case when the CSFs belong to the same complex. The example of $3p4f\ ^3F - 3d^2\ ^3F$ illustrates another possibility for strong mixing, namely a small denominator due to a close energy degeneracy of the two CSFs. Let us look at some rules and examples that will help us to predict when close degeneracies might occur.

Neglecting the non-central part of the electron–electron Coulomb repulsion, the energy $E(\gamma LS)$ of a CSF can be estimated in the central-field approximation (see section 1.4.1). Here the electrons are moving in an effective central potential $U(r)$, and the energy of the configuration is given as the sum of the orbital energies. Assume that the atom has N electrons and a nuclear charge Z. Close to the nucleus the electrons see the full nuclear charge, and the effective potential goes as $-Z/r$. Far away the nucleus is screened by the electrons, in which case the potential goes as $-(Z - N + 1)/r$. For highly charged ions, where N is small compared to Z, the central potential is almost a pure Coulomb potential, and the orbital binding energies‡ depend primarily on the n quantum numbers. For neutrals and ions of small charge the effective potential differs appreciably from a Coulomb potential, and the orbital binding energies depend also on the l quantum numbers. The dependence on l is due to the different forms of

---

† Configurations are sometimes said to interact, meaning that there is a mixing between the corresponding CSFs.
‡ In the central-field model the orbital binding energy is the negative of the orbital energy.

the orbitals and their penetration into the core, arising from the centrifugal term $l(l+1)/r^2$ in the radial one-electron equation (1.20). For s orbitals the centrifugal term is zero, and the corresponding electron has a finite probability of being in the vicinity of the nucleus where it is affected by the full nuclear charge. This leads to a large binding energy. For a p orbital the centrifugal term to some extent prevents the electron from penetrating the core as effectively, leading to smaller binding energies as compared to the s orbitals. For orbitals with higher l-quantum numbers the centrifugal term becomes increasingly important yielding successively smaller binding energies. A qualitative rule is that in the beginning of an iso-electronic sequence the binding energy depends mainly on the value of $n + l$. If this value is equal for two orbitals the one with lowest n quantum number has the largest binding energy. Thus, at the beginning of an iso-electronic sequence the orbitals ordered according to their bindings energies are

$$1s, 2s, 2p, 3s, 3p, 4s, 3d, 4p, 5s, 4d, 5p, 6s, 4f, 5d, 6p, 7s, 5f, 6d, 7p, 8s, \ldots .$$

At the end of the sequence the binding energies are determined by the n quantum numbers, and we get the ordering

$$1s, 2s, 2p, 3s, 3p, 3d, 4s, 4p, 4d, 4f, 5s, 5p, 5d, 5f, 5g, 6s, 6p, 6d, 6f, 6g, \ldots .$$

Somewhere in the middle of a sequence there are many cases of nearly equal binding energies, and thereby also substitutions of the form $nd \leftrightarrow (n+1)s$, $nf \leftrightarrow (n+1)p$, that lead to close energy degeneracies of the corresponding CSFs. Severe cases are usually in the singly ionized spectra, where a number of configurations are degenerate. An example is the $\{3d^7, 3d^6 4s, 3d^5 4s^2\}$ set of configurations† in the Mn I isoelectronic sequence.

A consequence of the change in relative binding energies of the orbitals is that configurations with more than one excited electron, relative to the ground configuration, have a large excitation energy in the beginning of the sequence, even if all excitations are within a complex. For high-Z members of the sequence only the n quantum number is important, and the configurations are grouped according to complexes. Somewhere along the sequence the picture has to change from one extreme to the other, and the energy level structure has to be rearranged. Therefore the doubly excited configurations, belonging to the same complex as the ground term, are *plunging* relative to the mono-excited *normal* configurations. This will become clearer when looking at an example.

6.1.4   Example 1. Magnesium-like ions

We have already encountered examples of plunging configurations. The $3p^2\ ^1D$ term of magnesium-like ions is above the ionization limit for Mg I, well above the whole Rydberg series of normal, $3snd\ ^1D$ terms. Already in Al II it has

† Sometimes also designated as $(3d + 4s)^7$.

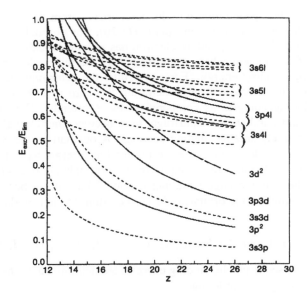

Figure 6.1. Relative, average excitation energies as a function of nuclear charge, Z, for some magnesium-like ions.

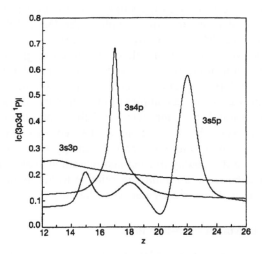

Figure 6.2. The absolute value of the weight of the $3p3d$ CSF in the expansion for $3s3p$, $3s4p$ and $3s5p$ ^1P as a function of Z in the Mg-like isoelectronic sequence.

plunged down to become the lowest ^1D term. As a matter of fact, the magnesium sequence exhibits some other good examples of plunging configurations, as illustrated in figure 6.1. In this figure we show the average Hartree–Fock excitation energies of a number of configurations as a function of nuclear charge. The energies are normalized by the ionization energy in each ion. The dashed lines give the normal, $3snl$ configurations, while the solid lines depict members of the displaced, $3pnl$ configurations. The second displaced configuration, represented by the long-dashed curve, is $3d^2$. It is clear from this figure how the displaced configurations are plunging through the normal ones, creating perturbations and strong interactions. As an example, we look at $3p3d$ ^1P.

To show the influence on the 'normal' $3snp$ ^1P series, of this plunging configuration, we depict in figure 6.2 the absolute values of the mixing coefficients for $\Phi(3p3d$ $^1P)$ in the expansion of the three normal terms, $3s3p$, $3s4p$ and $3s5p$ ^1P. The representation for the first two is

$$\{3s_1np_1, 3p_23d_2\}^1P \quad \text{for} \quad n = 3, 4.$$

To get convergence for all members of the isoelectronic sequence we need to use a different approach for the $3s5p$ ^1P term. Here we start by using the $1s, 2s, 2p, 3s, 3p_1$ orbitals from the calculation for $3s3p$ ^1P. The core and the $3p_1$ orbitals will be kept fixed for the rest of the calculation. A $4p_1$ is optimized on the $\{3s4p_1, 3s3p_1\}$ ^1P expansion. Finally, the $3s, 5p_1, 3p_2$ and $3d_2$ are optimized on the expansion

$$\{3s5p_1, 3s3p_1, 3s4p_1, 3p_23d_2\}^1P .$$

Note that the $3p_2$ orbital of the $3p3d$ CSF is neither orthogonal to nor the same as the np_1 of the $3snp$'s.

The $3p3d$ plunges through the $3s4l$ complex around Cl VI ($Z = 17$), causing a strong interaction with the $3s4p$ ^1P. Before that it has passed through the $3s5l$ complex in P IV ($Z = 15$), indicated by the first local maximum in the $3s5p$ curve. The structure of the $3s5p$ curve is then complicated by the strong interaction between $3s4p$ and $3p3d$ in a sort of secondary effect. Finally, the $3p4d$ passes through $3s5p$ in Ti XI ($Z = 22$), giving rise to the last maximum. This illustrates the flexibility of the MCHF method, since the $3d_2$ correlation orbital here will have more of a $4d$ character. The same is true for high Z in the $3s4p$ ^1P, where the long-range interaction with $3p4d$ dominates. This, together with the smooth behaviour of the $3s3p$ curve, illustrates the long-range $3snp$–$3pnd$ interaction, in contrast to the short-range one between the $3snp$ and $3pn'd$ ($n' < n$).

It is also interesting to observe how the second displaced configuration $3d^2$ falls faster and passes through both the normal and the first $3pnl$ series. The crossing with the third configuration in the $3p4l$ series $3p4f$, gives the strong mixing in P IV, already observed in table 6.3.

Another interesting effect is the $3d$–$4s$ degeneracy. For low ionization, such as Mg I, the $3d$ electron has a smaller binding energy than the $4s$, due to the latter's more extensive penetration into the core. For larger Z the opposite is true, and somewhere around the second spectrum the two are about equal. In figure 6.1 we see examples of this, where $3s3d$ is above $3s4s$ in Mg I, but already below in Si III (a closer look shows that the ^3D is below, while the ^1D is above in this ion).

It is clear that even a very crude calculation, such as the one on which figure 6.1 is based, can be very useful for predicting the existence of perturbations and for determining what CSFs need to be included in order to take close degeneracy into account. It is also clear from this figure, that the calculations become more simple, and stable, for high Z, where the level structure is more separated, while for low ionization stages the picture is much more complicated.

6.1.5 Example 2. Gallium-like ions

As a second example of the effect of plunging configurations, and its interplay with important substitutions, we will look at the $4s^2 5s$ ^2S state of gallium-like ions, which for low Z is the first excited term. By using the rules outlined above we know that we should include the important $4s^2 \rightarrow 4p^2$ and $4s^2 \rightarrow 4d^2$ substitutions. A reasonable starting point is therefore the following expansion

$$\{4s_1^2 5s_1, 4p^2 4s_2, 4d^2 4s_3\}{}^2S.$$

This simple calculation will illustrate the power of the non-orthogonal MCHF approach. We have labelled the two correlation s orbitals, $4s_2$ and $4s_3$, with $n = 4$. This only affects the initial estimate, since the MCHF code does not count nodes for correlation orbitals. If the $s^2 \rightarrow p^2, d^2$ substitutions represent the most important correlation, these two orbitals should resemble a $5s$. This is actually the case for Ga I, as can be seen in part (a) of figure 6.3. In this ion the $4s_3$ and the outer part of $4s_2$ are perfect matches to the $5s_1$. The picture changes however, and if we look at part (b), which shows the s orbitals for Ge II ($Z = 32$), and even more extreme part (c), for As III ($Z = 33$), we see especially how the $4s_3$, but also the $4s_2$, becomes more and more $4s$-like. This is linked to the plunging $4s4p^2$ configuration, which is very close to the $4s^2 5s$ in As III. Already for Se IV the order of the two atomic states has changed, and the $4s4p^2$ has become the lowest member of the ^2S Rydberg series.

6.2 First-order wave functions

The zero-order wave function should include the dominant CSFs. For some purposes the zero-order wave function is adequate, but in many cases higher-order corrections are important. Defining an active set of orbitals, the zero-order wave function can be improved by including CSFs generated by orbital

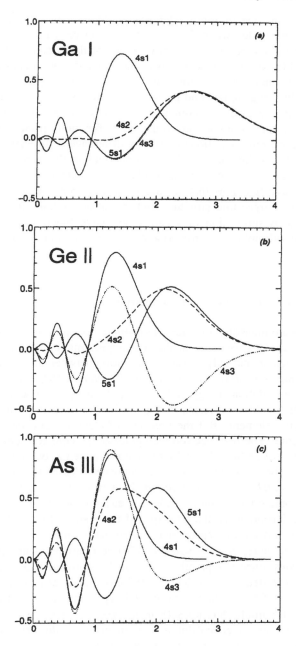

Figure 6.3. The radial s-functions from MCHF calculations using the expansion $\{4s_1^2 5s_1,$ $4s_2 4p^2,\ 4s_3 4d^2\}\ ^2S$, for (a) Ga I, (b) Ge II and (c) As III.

replacements from one or more reference configurations (frequently from all CSFs in the zero-order wave function) to the active set. By enlarging the active set in a systematic way, the convergence of the computed values can be monitored. Which orbital replacements to include and how to enlarge the active set depends on a number of factors, such as the complexity of the system, the atomic property being studied and the desired accuracy. Another very important consideration is the available computer resources. In the following sections we will discuss corrections to zero-order wave functions from the perspective of obtaining good representations of energy differences. By energy differences we mean not only differences in *LS* term energies in the same atom, but also ionization energies. In the remaining chapters of the book we will discuss approximate wave functions that are aimed at describing other properties such as isotope shifts, hyperfine structure and transition probabilities. In this we will be guided by the notion of a zero-order wave function and a first-order correction, but will not necessarily adhere to it strictly as in perturbation theory.

6.2.1   Valence and core–valence correlation

Frequently we are interested in determining energy separations between different levels. In these cases we may, in the first approximation, define closed inner subshells as inactive, and only consider correlation between the outer, or valence, electrons, correlation referred to as *valence correlation* (VC). The rationale for this is that the correlation energy in the core, although large in an absolute sense, largely cancels for different terms in a given ion, or even for terms in different ions of the same element. But the presence of electrons outside a core has an effect on the core. The first electron may polarize the core considerably, but this polarization will often be reduced by the second since the two electrons will try to avoid each other and prefer to be on opposite sides of the core. The latter is referred to as *dielectronic polarization*. However, examples have been found where dielectronic polarization may also increase the polarization (Norcross and Seaton 1976), depending on the total spin of the system. In the MCHF scheme, the effect of polarization is represented as correlation between the outer valence electrons and the core, or *core–valence* (CV) correlation. Generally, for a few valence electrons, energy separations are much improved if also core–valence correlation is included. Core–valence correlation, as described in section 4.2.2, is represented by CSFs obtained by orbital replacements from the zero-order wave function (or multireference set) of the type $ab \rightarrow vv'$ where a and b are respectively core and valence orbitals. The inclusion of this correlation reduces the energy and, with respect to a fixed core, increases the binding of the outer electrons to the core. In the case of a single electron, it increases the binding of the electron which is reflected in a *contraction* of the orbital. The contraction of the valence orbitals often has a large effect on different properties, and we

Table 6.4. Scheme for active set, GBT calculations for $4s^2 4p$ ^2P and $4s4p^2$ ^2D of Ge II.

		$4s^2 4p$ ^2P		$4s4p^2$ ^2D	
AS	$nl, n'l$	$\Phi_m(nl)$	$\Phi_x(nl \to n'l)$	$\Phi_m(nl)$	$\Phi_x(nl \to n'l)$
$n = 5$	$4s, 5s$	$4s^2 4p$	$4s5s(^1S)4p$	$4p^2 4s$	$4p^2 5s$
	$4p, 5p$	$4s^2 4p$	$4s^2 5p$	$4p^2 4s$	$4p5p(^1D)4s$
	$4d, 5d$	$4s4d(^3D)4p$	$4s5d(^3D)4p$	$4d4s^2$	$5d4s^2$
	$4f, 5f$	$4s4d(^1D)4f$	$4s4d(^1D)5f$	$4p4f(^3D)4s$	$4p5f(^3D)4s$
$n = 6$	$4s, 6s$	$4s^2 4p$	$4s6s(^1S)4p$		
	$4p, 6p$	$4s^2 4p$	$4s^2 6p$		
	$4d, 6d$	$4s4d(^3D)4p$	$4s6d(^3D)4p$		
	$4f, 6f$	$4s4d(^1D)4f$	$4s4d(^1D)6f$		
	$5s, 6s$	$4s5s(^1S)5p$	$4s6s(^1S)5p$		
	$5p, 6p$	$4p^2(^3P)5p$	$4p^2(^3P)6p$		
	$5d, 6d$	$4s5d(^3D)5p$	$4s6d(^3D)5p$		
	$5f, 6f$	$5f^2(^1S)4p$	$5f6f(^1S)4p$		
	$5g, 6g$	$4s4f(^1F)5g$	$4s4f(^1F)6g$		

will refer to this as a core–valence effect†.

6.2.2 Valence correlation in Ge II

As an example of valence correlation we consider the ground $4s^2 4p$ and first excited configurations $4s4p^2$ in Ge II. Since there are only three valence electrons, we use a CAS expansion to describe the correlation‡. As was discussed in the previous chapter, the CAS wave function is not unique. To obtain a unique solution the GBT approach can be used to delete certain CSFs. To identify which deletions should be done in the different steps, we have to use a correct order of the orbitals. This is clear from looking at the $4s^2 4p$ ^2P and table 6.4. In the first step we include the $n = 4$ orbitals. When extending the orbital set to include $n = 5$, we use a generalized form of the analyses discussed in chapter 5. For the $(4s, 5s)$ pair we find that, of course, $\Phi_m(4s)$ is $4s^2 4p$ ^2P and the corresponding $\Phi_x(4s \to 5s)$ is represented by $4s5s(^1S)4p$ ^2P. To generate these states we need to give the orbitals in the order

$$\{4s, 5s, 4d, 5d, 4f, 5f, 5g, 4p, 5p\}$$

† Formally this is not always a correct definition. If we use a frozen-core approach, opening up the core also represents 'relaxation' effects, that is the change in the core subshells between different ASFs.

‡ The size of the CAS expansions grows quickly with respect to the increasing active set of orbitals, and in order to perform some of the larger calculations referred to in this section, we need to increase different dimensions in the enclosed MCHF package.

when running the GENCL code. In general, when the first subshell has an occupancy larger than one, we order the orbitals in the active set by starting with all the orbitals of the same l-symmetry as the first subshell (in this case the s orbitals) and ending with orbitals of the same l-symmetry as the last subshell (in this case p orbitals). Orbitals with l-quantum numbers not present in the main CSF are put in between, since the correlation in the subshell with more than one electron is expected to be larger than between this and the other subshell. We therefore expect the most important correlation to arise from substitutions out of this subshell. As an example, the most important contribution from $4d$, representing the important $4s \rightarrow 4d$ substitution, is from the CSF $4s4d(^3D)4p$ ^2P with $c_i = -0.144$, while the $4s4d(^1D)4p$ ^2P has $c_i = 0.011$. If we instead had used the $4s, 4p, 4d, 4f$ ordering, we would have got 0.119 and -0.082 for the two c_i coefficients. A choice with one dominant CSF gives a more stable approach when applying GBT. We can clearly claim that the $4s5d(^3D)4p$ ^2P is the correct CSF to delete.

Table 6.5. Binding and excitation energies (in cm^{-1}) for different terms in Ge II and Ge III, from different active set, GBT calculations.

			AS			
E of:	Rel. to:	HF	$n = 4$	$n = 5$	$n = 6$	Exp.[a]
		Binding energies				
$4s^2 4p$ ^2P	$4s^2$ ^1S	122311	123969	125854	126043	127343
$4s4p^2$ ^4P	$4s4p$ ^3P	135415	136788	137268	137299	139140
$4s4p^2$ ^2D	$4s4p$ ^3P	106774	120916	123417	123636	126732
		Excitation energies				
$4s^2 4p$ ^2P	$4s4p^2$ ^4P	36332	44205	45685	45860	52710
$4s^2 4p$ ^2P	$4s4p^2$ ^2D	63939	60078	59536	59521	65117
$4s^2$ ^1S	$4s4p$ ^3P	49435	57025	57099	57108	63328

[a] Sugar and Musgrove (1993).

When moving to the excited term $4s4p^2$ ^2D we also need to consider the ordering of the 'occupied' orbitals. As it stands, we cannot directly identify how to apply GBT for the $4p \rightarrow 5p$ substitution, and whether we should delete the $4s4p(^3P)5p$ or $4s4p(^1P)5p$ CSF. The obvious solution is to reorder the orbitals, and write the main CSF as $4p^2(^1D)4s$ ^2D, so that the problem is recast in a similar manner to the one above. Now it is clear that $4p5p(^1D)4s$ should be deleted, while $4p5p(^3D)4s$ should be retained. The complete active space, GBT calculations for these two terms are illustrated in table 6.4. For the step where $n = 6$ orbitals are included, the calculation for $4s4p^2$ ^2D is stable without applying GBT.

It is quite fruitful to look at the binding energies for different terms, in order to analyse our results. This links our calculations to a system with one fewer electron, which in most cases is simpler. As we will see from an example we will be able to understand deviations from experimental values for certain properties by using such a procedure.

If we return to Ge II, we can conclude that the ground term, $4s^2 4p \ {}^2P$, is based on the $4s^2 \ {}^1S$ of Ge III, while the first excited, $4s4p^2 \ {}^4P$, has the $4s4p \ {}^3P$ as a parent term. For the $4s4p^2 \ {}^2D$ the parent is a linear combination of the $4s4p \ {}^3P$ and ${}^1P$, and both can serve as reference points. If we therefore extend our active set, GBT calculations to these three terms in Ge III, up to $n = 6$, we arrive at the results in table 6.5.

It is interesting to notice that binding energies in all three cases are quite accurate (error of just a few $1000 \ \mathrm{cm}^{-1}$), while the excitation energies deviate by about $7000 \ \mathrm{cm}^{-1}$, and are rather constant for the three examples in table 6.5. The reason for the deviation can more or less almost entirely be referred to as an inaccuracy in the determination of the 'distance between the two limits', $4s^2 \ {}^1S$ and $4s4p \ {}^3P$. In the next section we will turn to this two-electron system and investigate the effect of core–valence correlation.

6.2.3   Core–valence correlation in Ge III and Ge IV

It is in principle quite simple to include core–valence correlation in an MCHF calculation, since this concept has a straightforward interpretation in the choice of CSFs. In the case of the terms of Ge III the core–valence correlation of the $3d$ subshell is included via CSFs of the form†

$$1s^2 \ldots 3p^6 3d^9 n_1 l_1 n_2 l_2 n_3 l_3 \quad \text{where} \quad n_1, n_2, n_3 > 3 \,.$$

The choice of core orbitals is not obvious. If selected as the bare core, the $3d^{10} \ {}^1S$ ground state of Ge V in our case, Brillouin's theorem will hold for all core excitations of the bare core, but not when additional electrons are present. However, conceptually it is an attractive choice since all terms,

$$1s^2 \ldots 3p^6 3d^{10} n l n' l' \ LS, \quad \text{where} \quad n, n' \geqslant 4,$$

will have the same core and differences in energy can be interpreted relative to this core. In a formal sense, by fixing core orbitals in this manner rather than choosing Hartree–Fock core orbitals for each state in question, we are introducing some *core rearrangement* relative to a strict definition of correlation.

Returning to Ge III and looking at the different terms we perform CAS-GBT calculations for valence correlation, similar to the ones discussed above. The main difference is the different core used. In the earlier calculations, which were concentrated on the Ge II system, we used a core optimized on the ground

† This expansion may include more than precisely the first-order core–valence correction since this would require that we define the zero-order wave function.

Table 6.6. Example of a GENCL run for core-polarization in Ge III and a listing of the cfg.inp file produced.

```
                  Header   ?
>Ge III
        Closed Shells   ?
>1s 2s 2p 3s 3p
        Reference Set   ?
>3d(10)4s(2)
                   2   ?
>
          Active Set   ?
>3d,4s,4p
                Type of set generation ?
>1
        Replacements   ?
>
          Final Terms   ?
>1S
                 2   ?
>

cfg.inp file produced:

Ge III
 1s  2s  2p  3s  3p
 3d(10)  4s( 2)
     1S0      1S0       1S0
 3d(10)  4p( 2)
     1S0      1S0       1S0
 3d( 9)  4s( 1)  4p( 2)
     2D1      2S1      1D2       1D0       1S0
```

state of Ge II, $4s^2 4p$ ^2P, while to be able to clearly investigate the effect of core–valence correlation, here we need to use a core optimized on the $3d^{10}$ ^1S of Ge V. We also do CAS-GBT calculations for five different terms of Ge III, including only valence correlation and using orbitals up to $n = 6$. The results are given in table 6.7.

As a second step we included also the core–valence correlation type of CSF. Again we extended the active set, step by step, but allow at most one excitation from the $3d$ subshell. GENCL permits the generation of the required CSFs from *one* core subshell, $3d^{10}$ in the present case. This is shown in table 6.6. In the generation of CSFs from the active set of orbitals, when the Type of set generation is greater than zero, this number (1, 2 or 3), represents the number

Table 6.7. Results from active set, core–valence correlation calculations for different terms in Ge III and Ge IV, binding energies, transition energies, and term splittings.

Term	HF	VC[a]	+CV[b]	+RS[c]	Exp.[d]
		Binding energies			
Ge III $4s^2$ ^1S	254298	263362	270669	274633	274693
Ge III $4s4p$ ^3P	205662	207054	210709	211356	211365
Ge III $4s4p$ ^1P	166940	175941	182013	182184	182820
Ge III $4p^2$ ^3P	200299	202040	207454	208467	208369
Ge III $4p^2$ ^1D	189582	207916	210903	212604	212890
Ge IV $4s$ ^2S	354003	354003	363159	368959	368720
Ge IV $4p$ ^2P	277341	277341	284189	286041	285548
		Transition energies and term splittings			
Ge III ^1S–$^3P^o$	48636	56308	59960	63277	63328
Ge III $^3P^o$–^3P	82025	81676	82668	85807	86168
Ge III $^3P^o$–$^1P^o$	38722	31113	28696	29172	28545
Ge III ^3P–^1D	−10717	5876	3449	4137	4521
Ge IV ^2S–$^2P^o$	76662	76662	78970	82918	83172

[a] Including only valence correlation.
[b] Including core–valence correlation.
[c] Including relativistic shifts (see chapter 7).
[d] Experiment, see http://aeldata.phy.nist.gov/nist_beta.html

of holes that may be created in the filled subshell of the first orbital in the list of virtual orbitals. Thus, in the present case, a maximum of one hole is allowed. For the two-electron system Ge III the number of CSFs increases very rapidly with the orbital set and the examples included in this chapter cannot be handled with the standard values of parameters in the supplied MCHF package.

The calculations are performed in three steps, with successively larger active sets. The first two, $n = 4$ and 5, are stable, and GBT does not need to be applied. In the third step only the orbitals with $n = 6$ were optimized, since the contribution is small and a full optimization unstable.

We can see from table 6.7 that core–valence correlation has a major influence on the binding energy of $4s^2$ ^1S, increasing it by more than 7000 cm^{-1}. The effect is considerably smaller in $4s4p$ ^3P and therefore corrects the transition energy between these two terms by 3652 cm^{-1}. We can also see from this table that the term splittings (between different terms of the same configuration) is corrected by the inclusion of core–valence correlation. This is a general result, and observed in many semiempirical studies, where different Slater integrals are automatically scaled by a factor of 0.7–0.9 (Cowan 1981). This is basically equivalent to scaling the term splittings. Here we see that the cause of this 'error' in term splitting can be attributed to omitted core–valence correlation in

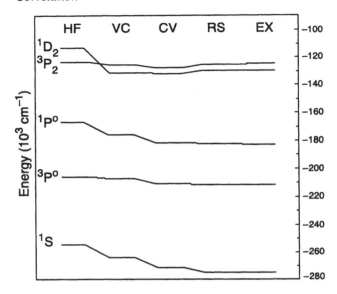

Figure 6.4. The energies of different terms of Ge III, relative to the $4s\ ^2S$ limit of Ge IV. Results from different MCHF approximations (HF = Hartree–Fock, VC = valence correlation, CV = core valence, RS = relativistic shifts) and experiment (EX).

the semi-empirical calculations.

It is clear from table 6.7 that there exists a remaining discrepancy between the core–valence correlation results and experiment (about 4600 cm^{-1} for the $4s^2\ ^1S$). This can be attributed to relativistic effects, a topic that will be dealt with in the next chapter. For completeness we have added the results when the *relativistic shifts* are included.

If we backtrack and remember our original problem, $4s^24p\ ^2P$ and $4s4p^2\ ^4P$ of Ge II, we deduce that the error in the transition energy between these two can be derived more or less entirely from an inaccuracy in the distance between their two limits ($4s^2\ ^1S$ and $4s4p\ ^3P$). We have now shown that this discrepancy can be attributed largely to core–valence correlation and relativistic shifts. The same should be true for the Ge II term. It would, of course, be possible to use the same method to include core–valence correlation in the calculations for Ge II, but this would leave us with five open subshells and a very fast-growing CSF expansion.

Finally, in figure 6.4 we show the energies of different terms of Ge III relative to the $4s\ ^2S$ limit, from different approaches. In this figure it is clear how the largest effect, both of core–valence correlation and relativistic shifts, is present in the ground term. We can also see how core–valence correlation affects the term splitting. The strong correlation, in this case valence correlation, in the

$4p^2\ {}^1D$ is also apparent from the cross-over in energy between the Hartree–Fock and valence correlation results.

6.2.4 Variation of core–valence correlation

We found that core–valence correlation was very important for the different germanium ions, so let us now investigate how it varies with different properties of the core. We will look at a number of different one- and two-electron systems, to try to get a qualitative idea of how core–valence correlation depends on the system.

It is always a question as to whether we can assume that a subshell is part of the core, and does not have to be treated as a valence subshell. It is safe to assume that closed shells belong to the core, so the first example of cores is $1s^2$. For lithium- and beryllium-like ions it is the only core subshell.

When a subshell is closed, but there are open subshells in the same shell, the complex needs to be considered with care. As an example, it is not a good approximation to assume that $2s^2$ belongs to the core in boron-like ions, in spite of the fact that the ground term is $1s^2 2s^2 2p$, since the zero-order wave function would also contain $1s^2 2p^3$. Clearly the 'core' in the latter is $1s^2$. At the same time the first excited term in these systems is $1s^2 2s 2p^2\ {}^4P$. For all isoelectronic sequences that originate at the first row of the periodic table (together with

Table 6.8. The contributions to the binding energies of $ns\ {}^2S$ and $np\ {}^2P$, and the transition energy, ΔE, between the two, in different ions. HF = Hartree–Fock, CV = including core–valence correlation, RS = including relativistic shifts, EX = experiment, α_d = dipole polarizability in au.

Ion	Core	Term	HF	CV	RS	EX
Be II	$1s^2$	$E_b({}^2S)$	146174	147071	147095	146883
	$\alpha_d = 0.05182$	$E_b({}^2P)$	113989	114979	114981	114954
		ΔE	32185	32092	32114	31929
Mg II	$2p^6$	$E_b({}^2S)$	118641	121358	121550	121268
	$\alpha_d = 0.4698$	$E_b({}^2P)$	84222	85748	85786	85507
		ΔE	34419	35610	35764	35761
Ca II	$3p^6$	$E_b({}^2S)$	91090	96118	96563	95752
	$\alpha_d = 3.254$	$E_b({}^2P)$	67877	70791	70933	70412
		ΔE	23213	25327	25630	25340
Zn II	$3d^{10}$	$E_b({}^2S)$	132872	141711	144128	144891
	$\alpha_d = 2.296$	$E_b({}^2P)$	89567	94922	95455	95828
		ΔE	43305	46788	48673	49063

the lithium and beryllium sequences this includes the boron, carbon, nitrogen, oxygen and fluorine sequences) this is the only true core subshell. We will look at Be I and II to test the importance of the $1s^2$ core.

The next example of a core is $1s^2 2s^2 2p^6$. The simplest examples of systems with this core are sodium- and magnesium-like ions. Our examples will be Mg I and II.

Again, it is not a good approximation to consider the $3s^2$ as a part of the core if $3p$ is open. In contrast, the closed $3p^6$ can be included in the core, which we will see by looking at Ca I and Ca II. After this study, we consider the zinc- and copper-like systems in order to investigate the influence of the $3d^{10}$ as the outermost core subshell.

In all our examples we will use a method parallel to the one used for the germanium ions above. Three successively larger active sets will be used, with $n_{max} = n, n + 1$ and $n + 2$, where n is the principal quantum number of the valence orbitals. In the case of instabilities, we might only optimize the $n + 2$ set in the last step.

We start by looking at the one-electron systems and show in table 6.8 the binding energies of ns ^2S and np ^2P, together with the transition energies between these two, in different approximations.

According to this table, core–valence correlation contributes about 0.6–0.9% to the binding energies of both $2s$ and $2p$ in Be II, about 2% for the two levels in Mg II, about 4–5% for $4s$ and $4p$ in Ca II and finally about 6% to the two binding energies in Zn II. Even more important is the change in the

Table 6.9. The same as in table 6.8, but for two-electron systems and the ns^2 ^1S and $nsnp$ ^3P terms. VC = including valence correlation; CV = including also core–valence.

Ion	Core/α_d	Term	HF	VC	CV	RS	EX
Be I	$1s^2$	$E_b(^1S)$	64879	74974	75072	75078	75192
	$\alpha_d = 0.05182$	$E_b(^3P)$	51366	52930	52923	52916	53212
		ΔE	13513	22044	22149	22161	21980
Mg I	$2p^6$	$E_b(^1S)$	53181	60654	61531	61595	61671
	$\alpha_d = 0.4698$	$E_b(^3P)$	38376	39800	39606	39572	39780
		ΔE	14806	20853	21925	22023	21891
Ca I	$3p^6$	$E_b(^1S)$	41120	47505	49280	49340	49306
	$\alpha_d = 3.254$	$E_b(^3P)$	32644	34088	33899	33832	34043
		ΔE	8475	13417	15381	15568	15263
Zn I	$3d^{10}$	$E_b(^1S)$	60492	67971	73825	75066	75767
	$\alpha_d = 2.296$	$E_b(^3P)$	39846	41149	42728	42614	43070
		ΔE	20647	26822	31096	32452	32696

contribution from core–valence correlation to the $^2S - {}^2P$ transition energy. In Be II it is negligible, while in Mg II, Ca II and Zn II it is about 1200, 2100 and 3400 cm^{-1}, respectively. This corresponds to 3, 8 and 7% of the total transition energy in the three heavier ions. This shows clearly how the core subshells get more polarizable with increasing n quantum number. An indication of this behaviour can also be seen from the theoretical *electric dipole polarizabilities*, α_d (sometimes called electric dipole susceptibilities) for the cores in these elements, which are given in table 6.8. This is defined as the constant of proportionality between the induced dipole moment in the core, and the external field. In our case the external field is produced by the valence electrons.

As a second example, let us consider the two-electron systems Be I, Mg I, Ca I and Zn I, and their first two terms $ns^2\ {}^1S$ and $nsnp\ {}^3P$. The results for these are given in table 6.9.

We can again observe the successive increase in importance of core–valence correlation when moving from the $1s^2$ core in Be I, to the $3d^{10}$ in Zn I. Here it is even more obvious that the polarization of the $1s^2$ core is of minor importance to the properties investigated, while in Zn I the polarization of the $3d^{10}$ core constitutes as much as 40% of the total correlation contribution to the $4s^2\ {}^1S -$ $4s4p\ {}^3P$ excitation energy.

It is also interesting to investigate the importance of core–valence correlation as a function of Z, in an isoelectronic sequence. For trends in the

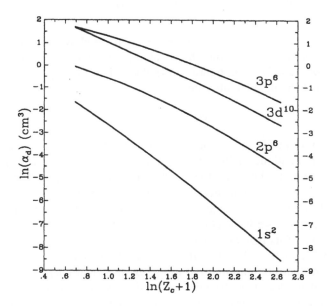

Figure 6.5. The $\ln\alpha_d$ as a function of the logarithm of the core charge for some different cores.

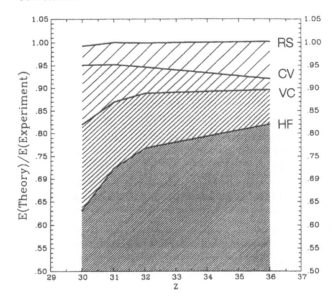

Figure 6.6. Different contributions to the excitation energy of $4s4p$ ^3P in zinc-like ions, as a function of the nuclear charge, normalized to the experimental value. HF = Hartree–Fock value, VC = including valence correlation, CV = including core–valence correlation and RS = including relativistic shifts.

dipole polarizability, we refer to figure 6.5. In it we show $\ln(\alpha_d)$ as a function of $\ln(Z_c + 1)$, where Z_c is the charge of the core and values of α_d are from Johnson *et al* (1983). This figure shows that it is a fairly good approximation to assume that the polarizability decreases as

$$\alpha_d(1s^2) \propto Z_c^{-p}.$$

For the helium-like $(1s^2)$ core the value of p is largest, about 3.5, while for the neon-like $(2p^6)$ and nickel-like $(3d^{10})$ the p varies from 2.0 to 2.5. The slowest decline is shown by the argon-like $(3p^6)$ core, with p ranging from about 1.25 in the beginning of the sequence, to slightly over 2 at the end.

We select the zinc-like sequence, which has a nickel-like core, to investigate a little bit further. We have already computed the $4s^2$ $^1S - 4s4p$ ^3P transition energy for two of its members, Zn I and Ge III. We extend these calculations to include Ga II and Kr VIII. The different contributions to this transition energy are shown in figure 6.6, normalized to the experimental energy difference.

Again it is clear that the relative importance of core–valence correlation decreases along the sequence. At the neutral end it contributes almost 15% to the total excitation energy, while for Kr VII the corresponding value is only slightly more than 2%. At the same time the valence correlation approaches a constant fraction of the total transition energy, and the relativistic contribution increases. We will consider Z-dependence in more detail shortly.

6.2.5 Filled and almost filled shells

The examples so far have dealt with only a few electrons outside filled shells. Not discussed are cases where there are many electrons outside the core. Particularly difficult are cases where the number of 'core' electrons changes as in the $3d^5 4s$ ^7S – $3d^4 4s(^6D)4p$ ^7P transition in Cr I. An early paper (Froese Fischer 1971) showed that matrix elements for interactions between the CSFs of $(3d + 4s)^{n+2}$ depended significantly on how the orbitals were chosen. Botch *et al* (1981) undertook a broad, general study of the major differential valence correlation effects of the low-lying $3d^n 4s^2$, $3d^{n+1}4s$, and $3d^{n+2}$ configurations in the first row of transition metals. They concluded that the major correlation effects arise from the $4s^2 - 4p^2$ angular correlation that we mentioned earlier. The second, was from *radial correlation*. They proposed that $3d^{n+1}$ be modelled as $3d^n 3d'$, ie n equivalent electron and one non-equivalent, non-orthogonal orbital $3d'$, and referred to this as the *projected Hartree–Fock* method. In our MCHF scheme, $3d' = a_1 3d + a_2 4d$ where $4d$ is now the orthogonal project of $3d'$ on $3d$. Then, in the MCHF formalism,

$$|3d^n 3d'\rangle = c_1 |3d^{n+1}\rangle + c_2 |3d^n 4d\rangle.$$

This was a simple method for improving term energy separation. It is clear that the notion of 'equivalent electrons' in broad terms explains the spectra quite well, but intuitively, when the tenth $3d$ electron is added to a $3d^9$ subshell it seems plausible that this electron should not have a radial distribution the same as all the other $3d$'s. In fact, a model that allows each electron to have its own radial function would describe radial correlation much better, but is difficult to treat computationally except in simple cases (Froese 1966).

However, the early investigations omitted an important correlation concept. In Z-dependent perturbation theory, it is possible to describe pair-correlation functions within two-electron systems. Correlation within a many-electron system was then a sum of pair-correlation functions with some corrections. When the number of equivalent electrons, say nl^{q+1}, changes to nl^q, then the number of (nl, nl) pair-correlation functions changes from $(q + 1)q/2$ to $q(q - 1)/2$. This change in correlation *within* the group of equivalent electrons, which usually is stronger than correlation between different electrons, can only be captured through SD replacements for the group of equivalent electrons.

Let us consider a smaller system which illustrates some of these points, namely the $1s^2 2s^2 2p^6$ ^1S – $2p^5 3s$ ^1P transition in Ne I. In later chapters we will see that it is important to have the same core orbitals, that we can only safely have two orbitals different in each state when considering matrix elements between states. Thus, to have a common set of $1s, 2s, 2p$ orbitals we choose the latter as Hartree–Fock orbitals for $1s^2 2s^2 2p^5$ ^2P. As a reference set, we

Table 6.10. The $2p^6$ ^1S – $2p^53s$ ^1P transition energy in Ne I from fixed-core MCHF calculations. In (i), common $1s, 2s, 2p$ orbitals were obtained from a HF calculation for $1s^22s^22p^5$; $1s, 2s$ were treated as inactive and SD replacements included from $\{2p^6, 2p^53p\}$ ^1S and $2p^53s$ ^1P. In (ii), only $1s$ was inactive, different $1s, 2s, 2p$ orbitals were obtained from HF calculations for the two states under consideration, and SD replacements made for $2s^22p^6$ ^1S and $2s^22p^53s$ ^1P.

Type	$E(^1S)$ (au)	$E(^1P)$ (au)	Diff (cm^{-1})
HF	−128.5470959	−127.9862116	123096
Calculation (i)			
$n = 3$	−128.6791504	−128.0978760	127571
$n = 4$	−128.7313895	−128.1225089	133630
$n = 5$	−128.7446059	−128.1322608	134391
Calculation (ii)			
$n = 5$	−128.8391734	−128.2272919	134289
$n = 6$	−128.8468764	−128.2346093	134373
Expt			134269

then choose $\{2p^6, 2p^53p\}$ ^1S for the ground state and $2p^53s$ ^1P for the excited state. SD replacements are now made from these reference states. Table 6.10 shows first the Hartree–Fock result with different orbitals for the core. Notice that the energy separation is too small by 11 000 cm^{-1}, a fairly large amount. Then pair-correlation effects are included in the form of SD replacements which incorporate both the correlation within the subshell of equivalent electrons and between the non-equivalent electrons. The $n = 3$ active set calculation is already 4 500 cm^{-1}, better than the independently optimized HF value. Expanding the orbital set leads to a rapidly converging transition energy close to the observed. This calculation has lead to a satisfactory transition energy, but the quality of the wave function is uncertain. The $2s$ and $2p$ orbitals have similar mean radii. Though no excitations are possible within the ^1S complex, there is a $2s2p^63p$ CSF in the complex of ^1P. The second set, obtained from more powerful codes, uses independently optimized core orbitals, keeping them fixed in the MCHF calculation. With $1s^2$ inactive, SD replacements replacements are made and we have obtained a similar transition energy. This confirms that for transition energies, $2s$ can be treated as part of the inactive core, that our choice of fixed, common core orbitals provided results of the same quality as the independently optimized choice.

6.3   Z-dependence of atomic properties

One way of arriving at the Z-dependence of different atomic properties is to use hydrogenic formulae. These become more accurate for higher members of an isoelectronic sequence, and are useful when trying to form a qualitative picture of any ion. Many examples of the empirical formulae were given in the review article by Edlén (1964) and in the book by Cowan (1981).

The starting point is that the binding energy of a hydrogen-like term can be written as

$$E(nl) = \frac{(Z - \sigma)^2}{2n^2} \tag{6.2}$$

where σ is a screening constant arising from the presence of the core. We also need to use the fact that a hydrogenic function varies according to

$$P^Z(r) = Z^{1/2} P^H(Zr) \tag{6.3}$$

where P^H is the function for hydrogen, and that the one-electron moments scale as

$$\langle r^n \rangle^Z = \frac{\langle r^n \rangle^H}{Z^n}. \tag{6.4}$$

That is, as Z increases, the mean radius of the orbital decreases.

It is easy to show, from these relations, that for transition energies, E_t, we get

$$E_t(\Delta n = 0) \propto (Z - \sigma) \quad \text{and} \quad E_t(\Delta n \neq 0) \propto (Z - \sigma)^2. \tag{6.5}$$

Term splitting and correlation is governed by Slater integrals. From the formulae above they will vary as

$$R^k \propto (Z - \sigma). \tag{6.6}$$

To first order, the energy shift due to the interaction with a certain configuration is given by $\delta \approx (R^k)^2/\Delta$, where Δ is the diagonal energy difference between the two CSFs. From the discussion above, we get that the correlation energy

$$\delta(\Delta n = 0) \propto (Z - \sigma) \tag{6.7}$$

for interaction within the complex ($\Delta n = 0$), and

$$\delta(\Delta n \neq 0) \propto \text{constant} \tag{6.8}$$

for interaction with configuration outside the complex.

In most cases core–valence correlation is correlation with CSFs outside the complex. It will therefore decrease faster, relative to the transition energy, than valence correlation that will contain the important and constant contribution from correlation within the complex. We will return to this $(Z - \sigma)$-dependence for other properties, but can already state that the relativistic effects (including the shifts) increase rapidly as a function of $(Z - \sigma)$, the theoretical limit being

$(Z - \sigma)^4$. They will therefore be of increasing importance for higher ionization stages and for larger systems.

Z-dependence is also useful in interpreting the effect of correlation itself on properties other than energy. We have already seen in table 6.9 that the effect of core–valence with respect to valence correlation is to increase the binding energy of an outer electron. That implies a decrease in the screening parameter of (6.2) or, equivalently, an increase in the effective nuclear charge. From this it follows that there is a contraction in the orbital which decreases the mean radius and increases effects that depend on $1/r^3$. We will see the latter in the next chapter on relativistic effects and investigate the former more thoroughly in connection with transition probabilites in chapter 9.

6.4   Exercises

(i)   The example in table 6.1 for $3s3d$ ^1D used the same orbitals in both $3s3d$ and $3d^2$. Show that, even when different orbitals are used, the contribution (as a percentage of the wave function expansion) remains less than 0.025%.

(ii)   The approach outlined for the gallium-like ions Ga I, Ge II and As III, in section 6.1.5, is unstable for higher Z. An alternative approach is to start with Hartree–Fock for $4s_1^2 5s_1$ ^2S, then do a frozen-core Hartree–Fock for $4s_2 4p^2$ ^2S. Add the following expansion

$$\{4p^2 5s_2, 4p5p(^1S)4s_2, 4p5p(^1S)5s_2, 5p^2 4s_2, 5p^2 5s_2\} \, ^2S$$

to the three CSFs mentioned in the text.

• Do MCHF calculations for a few members of the sequence by optimizing all orbitals except for the ones occupied in $4s_2 4p^2$.
• Study the CSF weights as a function of Z, along the isoelectronic sequence.

(iii)   The configuration $2p^6$ has 15 pairs of $(2p, 2p)$ electrons whereas $2p^5$ has only 10 such pairs. Compare the strength of the interaction $\langle 2p^6|\mathcal{H}|2p^4 3s^2\rangle$ with that of $\langle 2p^5|\mathcal{H}|2p^3 3s^2\rangle$ and show that the square of the ratio (assuming radial integrals are the same) is 1.5, the ratio of the number of pair-correlation pairs.

(iv)   Consider the $3d^5$ ^6S – $3d^4 4p(^6D)4p$ ^6P transition energy in Cr II. Unlike the Ne ^1S – Ne ^1P considered in section 6.2.5, in Cr II the mean radii of $3s$ and $3p$ are considerably smaller than that of $3d$ and can be more safely treated as belonging to a core. Try two types of calculation: one where the $1s, \ldots, 3d$ are common, one where individual cores are used. In each case use SD replacements to a virtual set. Compare your transition energy with the observed value of 48534 cm^{-1} = 0.221138 au.

Chapter 7

Relativistic Effects

7.1 Introduction

The theory developed in the preceeding chapters is based on non-relativistic quantum mechanics. Although this is a good approximation for light atoms, relativistic effects must be included even in these atoms or ions in order for the predictions of the theory to be in detailed agreement with experiment. When moving on to heavy ions, or highly ionized systems, the importance of relativistic effects increases quickly. For a general computational method, whose aim is to be valid for a large variety of systems, it is therefore essential to take relativistic effects into account.

There are some ambiguities in the definition of what is a 'relativistic effect', an effect neither included in a general non-relativistic treatment, nor belonging to QED contributions. For example, it might be questioned whether the interaction between an electron's spin and orbital angular momenta is a relativistic effect, or whether the Breit interaction, which describes the correction to the Coulomb potential for a model with an exchange of a photon between two electrons, is a QED effect or a part of the proper relativistic treatment of the Coulomb interaction. We will avoid these controversies by defining as relativistic effects the difference between the Dirac and the Schrödinger equation for a given system.

To include relativistic effects in a rigorous way the Dirac equation needs to be solved for a many-electron system, a difficult and time-consuming task. For most problems of interest, involving outer electrons, it is sufficient to follow another route, by considering only the lowest-order relativistic corrections to the ordinary Schrödinger equation. These corrections can be derived from the relativistic many-electron equation by expanding in powers of α ($\alpha = 1/c$, where c is the speed of light). The resulting Hamiltonian, which is correct to order α^2, is known as the *Breit–Pauli* Hamiltonian and is the basis for the approach we follow.

In order not to burden the treatment in this chapter with too many details, the relativistic many-electron theory underlying the Breit–Pauli approximation is presented in appendix B. This appendix also contains the derivation of some of the terms in the Breit–Pauli Hamiltonian together with the physical interpretation of the operators.

7.2   The Breit–Pauli Hamiltonian

The Breit–Pauli Hamiltonian can be written

$$\mathcal{H}_{BP} = \mathcal{H}_{NR} + \mathcal{H}_{RS} + \mathcal{H}_{FS}, \tag{7.1}$$

where $\mathcal{H}_{NR}$ is the ordinary non-relativistic many-electron Hamiltonian. The *relativistic shift* operator $\mathcal{H}_{RS}$ commutes with L and S and can be written

$$\mathcal{H}_{RS} = \mathcal{H}_{MC} + \mathcal{H}_{D1} + \mathcal{H}_{D2} + \mathcal{H}_{OO} + \mathcal{H}_{SSC}, \tag{7.2}$$

where $\mathcal{H}_{MC}$ is the *mass correction* term

$$\mathcal{H}_{MC} = -\frac{\alpha^2}{8} \sum_{i=1}^{N} (\nabla_i^2)^\dagger \nabla_i^2 \tag{7.3}$$

and $\mathcal{H}_{D1}$ and $\mathcal{H}_{D2}$ are the one- and two-body *Darwin terms*

$$\mathcal{H}_{D1} = -\frac{\alpha^2 Z}{8} \sum_{i=1}^{N} \nabla_i^2 \left(\frac{1}{r_i} \right), \tag{7.4}$$

$$\mathcal{H}_{D2} = \frac{\alpha^2}{4} \sum_{i<j}^{N} \nabla_i^2 \left(\frac{1}{r_{ij}} \right). \tag{7.5}$$

$\mathcal{H}_{SSC}$ is the *spin–spin contact* term

$$\mathcal{H}_{SSC} = -\frac{8\pi\alpha^2}{3} \sum_{i<j}^{N} \left(s_i \cdot s_j \right) \delta \left(r_i \cdot r_j \right) \tag{7.6}$$

and finally $\mathcal{H}_{OO}$ is the *orbit–orbit* term

$$\mathcal{H}_{OO} = -\frac{\alpha^2}{2} \sum_{i<j}^{N} \left[\frac{p_i \cdot p_j}{r_{ij}} + \frac{r_{ij} \left(r_{ij} \cdot p_i \right) p_j}{r_{ij}^3} \right]. \tag{7.7}$$

The *fine-structure* operator $\mathcal{H}_{FS}$ describes interactions between the spin and orbital angular momenta of the electrons, and does not commute with L and S but only with the total angular momentum $J = L + S$. The fine-structure operator consists of three terms

$$\mathcal{H}_{FS} = \mathcal{H}_{SO} + \mathcal{H}_{SOO} + \mathcal{H}_{SS}. \tag{7.8}$$

Here $\mathcal{H}_{SO}$ is the *nuclear spin-orbit* term

$$\mathcal{H}_{SO} = \frac{\alpha^2 Z}{2} \sum_{i=1}^{N} \frac{1}{r_i^3} \boldsymbol{l}_i \cdot \boldsymbol{s}_i \tag{7.9}$$

$\mathcal{H}_{SOO}$ is the *spin–other-orbit* term

$$\mathcal{H}_{SOO} = -\frac{\alpha^2}{2} \sum_{i<j}^{N} \frac{\boldsymbol{r}_{ij} \times \boldsymbol{p}_i}{r_{ij}^3} \left(\boldsymbol{s}_i + 2\boldsymbol{s}_j\right) \tag{7.10}$$

and $\mathcal{H}_{SS}$ is the *spin–spin* term

$$\mathcal{H}_{SS} = \alpha^2 \sum_{i<j}^{N} \frac{1}{r_{ij}^3} \left[\boldsymbol{s}_i \cdot \boldsymbol{s}_j - 3\frac{\left(\boldsymbol{s}_i \cdot \boldsymbol{r}_{ij}\right)\left(\boldsymbol{s}_j \cdot \boldsymbol{r}_{ij}\right)}{r_{ij}^2} \right]. \tag{7.11}$$

7.3 Breit–Pauli wave functions

The Breit–Pauli Hamiltonian commutes with the total angular momentum operator $\boldsymbol{J}$, and thus the corresponding wave functions should be eigenfunctions of $\boldsymbol{J}^2$ and J_z. In the multiconfiguration approximation the Breit–Pauli wave functions are obtained as linear combinations

$$\Psi(\gamma J M_J) = \sum_{i=1}^{M} c_i \Phi(\gamma_i L_i S_i J M_J), \tag{7.12}$$

where $\Phi(\gamma L S J M_J)$ are LSJ coupled CSFs, that is

$$\Phi(\gamma L S J M_J) = \sum_{M_L M_S} \langle L M_L S M_S | L S J M_J \rangle \Phi(\gamma L M_L S M_S). \tag{7.13}$$

Since neither L nor S are good quantum numbers CSFs with different LS need to be included in the expansion 7.12, and we have a mixing of different LS terms. In this case the wave function is given in the so-called *intermediate coupling*.

In the present codes, the radial functions building the CSFs are taken from a previous non-relativistic MCHF run and only the expansion coefficients are optimized. As described in section 1.5.2, this leads to the matrix eigenvalue problem

$$\mathbf{Hc} = E\mathbf{c}, \tag{7.14}$$

where $\mathbf{H}$ is the Hamilton matrix with elements

$$H_{ij} = \langle \gamma_i L_i S_i J M_J | \mathcal{H}_{BP} | \gamma_j L_j S_j J M_J \rangle. \tag{7.15}$$

Thus, the problem of finding the eigenvalues and eigenfunctions of the Breit–Pauli Hamiltonian reduces to the evaluation of matrix elements between LSJ coupled CSFs and a matrix diagonalization for each J value.

It should be noted that the Breit–Pauli Hamiltonian is a first-order perturbation correction to the non-relativistic Hamiltonian and could give wrong results if treated in higher-order perturbation theory (Bethe and Salpeter 1957).

7.4 Fine-structure levels

To get some insight into the nature of the relativistic energy corrections for a many-electron system we consider the simple case where the expansion (7.12) contains only one term. For this expansion we have

$$E = E_{NR} + E_{RS} + E_{FS}, \tag{7.16}$$

where

$$E_{NR} = \langle \gamma LSJM_J | \mathcal{H}_{NR} | \gamma LSJM_J \rangle \tag{7.17}$$

is the ordinary non-relativistic energy and

$$E_{RS} = \langle \gamma LSJM_J | \mathcal{H}_{RS} | \gamma LSJM_J \rangle \tag{7.18}$$

and

$$E_{FS} = \langle \gamma LSJM_J | \mathcal{H}_{FS} | \gamma LSJM_J \rangle \tag{7.19}$$

are the relativistic energy corrections from the relativistic shift and fine-structure contributions, respectively.

The relativistic shift operators all commute with L and S, and thus E_{RS} is independent of J (and M_J) and represents a shift of the non-relativistic LS term energy E_{NR}. The fine-structure energy may be written

$$E_{FS} = E_{SO} + E_{SOO} + E_{SS} \tag{7.20}$$

where E_{SO}, E_{SOO} and E_{SS} are the energies corresponding to the spin–orbit, spin–other-orbit and spin–spin operators, respectively. These energies all depend on the J quantum number and lead to a splitting of the non-relativistic LS term energy E_{NR} into *fine structure levels*. Using the rules of addition of angular momenta, the possible values of J corresponding to given values of L and S are

$$|L - S|, |L - S| + 1, \ldots, L + S - 1, L + S \tag{7.21}$$

and the number of levels in the term is given by the multiplicity $2S + 1$ if $L \leqslant S$ and $2L + 1$ if $L < S$. In figure 7.1 the splitting of the LS term $1s^2 2s^2 2p^2 \ {}^3P$ into the three different fine-structure levels is shown schematically.

Let us investigate the fine structure in more detail. Glass and Hibbert (1978) show that in a many-electron system the matrix elements of the different fine-structure operators have different dependence on the quantum numbers involved.

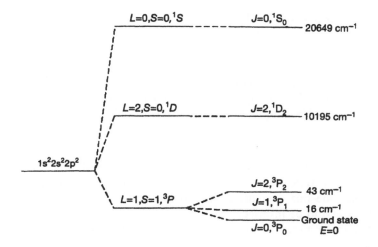

Figure 7.1. The fine structure and term splitting of the $1s^22s^22p^2$ configuration in carbon.

The spin–orbit and spin–other-orbit are both products of rank one spin and spatial tensor operators. Formula (A.55) therefore gives

$$E_{SO} = \langle \gamma LSJM_J | \mathcal{H}_{SO} | \gamma LSJM_J \rangle \propto (-1)^{L+S+J} \left\{ \begin{array}{ccc} L & L & 1 \\ S & S & J \end{array} \right\}$$

(7.22)

$$E_{SOO} = \langle \gamma LSJM_J | \mathcal{H}_{SOO} | \gamma LSJM_J \rangle \propto (-1)^{L+S+J} \left\{ \begin{array}{ccc} L & L & 1 \\ S & S & J \end{array} \right\}.$$

(7.23)

The spin–spin is a scalar product of two rank two tensor operators, and from the Wigner–Eckart theorem we get

$$E_{SS} = \langle \gamma LSJM_J | \mathcal{H}_{SS} | \gamma LSJM_J \rangle \propto (-1)^{L+S+J} \left\{ \begin{array}{ccc} L & L & 2 \\ S & S & J \end{array} \right\}. \quad (7.24)$$

Now, using the explicit expressions for the $6 - j$ symbols

$$(-1)^{L+S+J} \left\{ \begin{array}{ccc} L & L & 1 \\ S & S & J \end{array} \right\} \propto J(J+1) - L(L+1) - S(S+1) \quad (7.25)$$

$$(-1)^{L+S+J} \left\{ \begin{array}{ccc} L & L & 2 \\ S & S & J \end{array} \right\} \propto \frac{3}{4}C(C+1) - L(L+1) - S(S+1),$$

(7.26)

where $C = J(J+1) - L(L+1) - S(S+1)$, it follows that the fine-structure energies can be written

$$E_{SO} = \{J(J+1) - L(L+1) - S(S+1)\}\zeta_{SO}(\gamma LS) \quad (7.27)$$

$$E_{SOO} = \{J(J+1) - L(L+1) - S(S+1)\}\zeta_{SOO}(\gamma LS) \qquad (7.28)$$

and

$$E_{SS} = \left\{ \tfrac{3}{4}C(C+1) - L(L+1) - S(S+1) \right\} \zeta_{SS}(\gamma LS), \qquad (7.29)$$

where $\zeta_{SO}(\gamma LS)$, $\zeta_{SOO}(\gamma LS)$ and $\zeta_{SS}(\gamma LS)$ are factors independent of J.

If we neglect the spin–spin term it is easy to see that the energy difference between two neighbouring fine-structure levels J and $J - 1$ is

$$\Delta E_{FS} = 2\zeta J \qquad (7.30)$$

where $\zeta = \zeta_{SO}(\gamma LS) + \zeta_{SOO}(\gamma LS)$. This is the so-called *Landé interval rule* for the fine structure. If ζ is positive, and the fine-structure energy increases with J, then the fine structure is said to be normal. If ζ is negative the fine structure is said to be inverted.

Often the spin–spin term contributes significantly and cannot be neglected. In this case the Landé interval rule breaks down, and the fine structure shows a more irregular behaviour. This may also be the case when CSFs with different L and S, all coupling to the same total J, are included in the expansion (7.12).

7.5   Computational aspects

The matrix elements of the individual relativistic and non-relativistic operators in the Breit–Pauli Hamiltonian can be written in terms of the radial integrals shown in table 7.1. As an example, for the one- and two-electron Darwin operators we have

$$\langle \gamma LSJM_J | \mathcal{H}_{D1} | \gamma LSJM_J \rangle = \sum_{ab} v_{ab} D(a, b) \qquad (7.31)$$

and

$$\langle \gamma LSJM_J | \mathcal{H}_{D2} | \gamma LSJM_J \rangle = \sum_{abcd;k} w_{abcd;k} X^k(ab, cd), \qquad (7.32)$$

where v_{ab} and $w_{abcd;k}$ are angular coefficients. The sum on ab and $abcd$ is over orbitals occupied in either configuration.

In the MCHF atomic structure package the configuration expansion (7.12) is assumed to be defined in the file cfg.inp. The program BREIT reads the configuration expansion and derives angular coefficients needed for the evaluation of the different Breit–Pauli matrix elements. The angular data, organized in a form similar to the one from the NONH program, is then written to a file int.lst. At this point the cfg.inp could be copied (or moved) to name.c and, if radial functions have been determined, they could be stored as name.w. The program CI can then be used to compute and store the radial integrals, read the angular data on the file int.lst and construct the Hamilton matrix. As a second step the CI program finds a selected number of eigenvalues and the corresponding eigenvectors to the Breit–Pauli matrix and outputs energy eigenvalues and the expansion coefficients as specified by the

Table 7.1. Radial integrals for different Breit–Pauli operators.

(a) Relativistic shift or non-fine-structure operators

$\mathcal{H}_{MC}$:

$$C(a, b) = \frac{\alpha^2}{4} \int_0^\infty \left(\frac{d^2}{dr^2} - \frac{l_a(l_a + 1)}{r^2} \right) P(a; r) \left(\frac{d^2}{dr^2} - \frac{l_b(l_b + 1)}{r^2} \right) P(b; r) \, dr$$

$\mathcal{H}_{D1}$:

$$D(a, b) = \frac{\alpha^2}{4} Z \left(\frac{P(a; r)}{r} \right)_{r=0} \left(\frac{P(b; r)}{r} \right)_{r=0}$$

$\mathcal{H}_{D2}$:

$$X^k(ab, cd) = \frac{\alpha^2}{4} \int_0^\infty \int_0^\infty P(a; r_1) P(b; r_2) \frac{\delta(r_1 - r_2)}{r_1^2} P(c; r_1) P(d; r_2) \, dr_1 \, dr_2$$

$\mathcal{H}_{SSC}$: same as $\mathcal{H}_{D2}$

(b) Fine-structure operators

$\mathcal{H}_{SO}$:

$$Z(a, b) = \frac{\alpha^2}{4} \int_0^\infty P(a; r) \frac{1}{r^3} P(b; r) \, dr$$

$\mathcal{H}_{SS}$:

$$N^k(ab, cd) = \frac{\alpha^2}{4} \int_0^\infty \int_0^\infty P(a; r_1) P(b; r_2) \frac{r_2^k}{r_1^{k+3}} e(r_1 - r_2)$$
$$\times P(c; r_1) P(d; r_2) \, dr_1 \, dr_2$$

where $e(x) = \begin{cases} 1, & x > 0 \\ 0, & x \leqslant 0 \end{cases}$

$\mathcal{H}_{SSO}$: $N^k(ab, cd)$ and $V^k(ab, cd)$ where

$$V^k(ab, cd) = \int_0^\infty \int_0^\infty P(a; r_1) P(b; r_2) \frac{r_<^k}{r_>^{k+3}} r_2 \left(\frac{\partial}{\partial r_1} - \frac{1}{r_1} \right)$$
$$\times P(c; r_1) P(d; r_2) \, dr_1 \, dr_2$$

---

user of the program. Eigenvectors of one or more levels are then stored in name.l and possibly also name.j if fine-structure operators have been selected.

7.6 Fine structure in helium

Of special interest is the fine structure of the $1s2p\ ^3P$ of neutral helium, and we will use it as an illustration of the use of the BREIT and CI programs.

We start by performing non-relativistic MCHF calculations, according to earlier sections. The first step is a Hartree–Fock calculation for $1s2p\ ^3P$. The corresponding wfn.out will be saved in a file named hf.w. Since HF does not produce a cfg.out, either GENCL or an editor can be used to create a file containing the single CSF for this case. The result should be stored as hf.c.

Table 7.2. Breit–Pauli, CI calculation for the helium $1s2p$ ^3P, including only spin-orbit from the J-dependent operators.

```
>cp hf.c cfg.inp
>Breit
 Indicate the type of calculation
 0 => non-relativistic Hamiltonian only;
 1 => one or more relativistic operators only;
 2 => non-relativistic operators and selected relativistic:
>2
 Is full print-out requested? (Y/N)
>y
 All relativistic operators ? (Y/N)
>n
 Spin-orbit ? (Y/N)
>y
 Spin-other-orbit ? (Y/N)
>n
 Spin-spin ? (Y/N)
>n
 THE TYPE OF CALCULATION IS DEFINED BY THE FOLLOWING PARAMETERS -
      BREIT-PAULI OPERATORS            IREL  = 2
      PHASE CONVENTION PARAMETER       ICSTAS = 1
 STATE  (WITH  1 CONFIGURATIONS):
 ------------------------------
 THERE ARE  2 ORBITALS AS FOLLOWS:
      1s  2p
 CONFIGURATION  1 ( OCCUPIED ORBITALS= 2 ):  1s( 1)  2p( 1)
                           COUPLING SCHEME:    2S1    2P1
                                                      3P0
 All Interactions? (Y/N):
>y
                        MULTIPLYING FACTOR    TYPE OF INTEGRAL
 (CONFIG   1/HO /CONFIG    1)
                          1.00000000         I  ( 1s, 1s)
                          1.00000000         I  ( 2p, 2p)
 (CONFIG   1/SO /CONFIG    1)
                          6.00000000         S-O( 2p, 2p)
 (CONFIG   1/Rij/CONFIG    1
                          1.00000000         F 0( 1s, 2p)
                         -0.33333333         G 1( 1s, 2p)
 MATRIX ELEMENTS CONSTRUCTED USING THE SPHERICAL HARMONIC PHASE
 CONVENTION OF CONDON AND SHORTLEY, THEORY OF ATOMIC STRUCTURE
 ------------------------------------------------------------------
>CI
  Name of State
>hf
```

Table 7.2. (continued)

```
  Is this a relativistic calculation ? (Y/N) :
>y
  Is mass-polarization to be included ? (Y/N) :
>n
   ATOM = He      Z =  2.
The size of the matrix is     1
 Enter the approximate number of eigenvalues required
>1
   1 EIGENVALUES FOUND
   -2.13154378  1s.2p_3P
   1.000000
  Maximum and minimum values of 2*J ?
>4,0
  Do you want the matrix printed? (Y or N)
>n
   1 EIGENVALUES FOUND
   -2.13154  1s.2p_3P
   1.000000
   1 EIGENVALUES FOUND
   -2.13154  1s.2p_3P
   1.000000
   1 EIGENVALUES FOUND
   -2.13154  1s.2p_3P
   1.000000
```

To investigate the importance of different effects, we will include the different fine-structure operators one by one in our calculations. In the first step we only include the spin–orbit operator, and run the BREIT and CI codes as shown in table 7.2.

We continue and include in the next step also the spin–other-orbit, and in a third calculation the spin–spin interaction. This calculation is outlined in table 7.3.

Finally we perform active set, GBT calculations for $n \leqslant 4$. Before moving on to the BREIT + CI part, we must remember that GBT was satisfied in the variational calculation of MCHF. When we add the Breit operators to the Hamiltonian the GBT is not valid any longer. The earlier deleted CSFs therefore have to be reintroduced. All CSFs with $n \leqslant 4$ are stored in n4.c, and copied to cfg.inp, while the orbitals obtained in our active set GBT calculations are found in n4.w. The resulting fine structure is represented in figure 7.2.

The spin–orbit interaction gives a very predictable pattern, since it follows the Landé interval rule. This is a direct consequence of (7.27). In this case it implies that the distance between the $J = 2$ and 1 levels is twice as big as

Table 7.3. Full Breit and CI calculation for the fine structure of the $1s2p$ ^3P of He I.

```
>Breit
 Indicate the type of calculation
 0 => non-relativistic Hamiltonian only;
 1 => one or more relativistic operators only;
 2 => non-relativistic operators and selected relativistic:
>2
 Is full print-out requested? (Y/N)
>y
 All relativistic operators ? (Y/N)
>y
 THE TYPE OF CALCULATION IS DEFINED BY THE FOLLOWING PARAMETERS -
      BREIT-PAULI OPERATORS             IREL   = 2
      PHASE CONVENTION PARAMETER        ICSTAS = 1
 STATE  (WITH  1 CONFIGURATIONS):
 -------------------------------
 THERE ARE  2 ORBITALS AS FOLLOWS:
      1s  2p
 CONFIGURATION  1 ( OCCUPIED ORBITALS= 2 ):  1s( 1)  2p( 1)
                          COUPLING SCHEME:     2S1    2P1
                                                        3P0
 All Interactions? (Y/N):
>y
                    MULTIPLYING FACTOR     TYPE OF INTEGRAL
 (CONFIG   1/HO /CONFIG   1)
                         1.00000000     I  ( 1s, 1s)
                         1.00000000     I  ( 2p, 2p)
 (CONFIG   1/SO /CONFIG   1
                         6.00000000     S-O( 2p, 2p)
 (CONFIG   1/Rij/CONFIG   1)
                         1.00000000     F 0( 1s, 2p)
                        -0.33333333     G 1( 1s, 2p)
 (CONFIG   1/SOO/CONFIG   1)
                       -18.00000000     N 0( 2p, 1s/ 2p, 1s)
                       -12.00000000     N-1( 2p, 1s/ 1s, 2p)
                        -6.00000000     V 0( 1s, 2p/ 2p, 1s)
                         6.00000000     V 0( 2p, 1s/ 1s, 2p)
                         6.00000000     N 1( 1s, 2p/ 2p, 1s)
 (CONFIG   1/SS /CONFIG   1)
                        12.00000000     N 0( 2p, 1s/ 2p, 1s)
 MATRIX ELEMENTS CONSTRUCTED USING THE SPHERICAL HARMONIC PHASE
 CONVENTION OF CONDON AND SHORTLEY, THEORY OF ATOMIC STRUCTURE
```

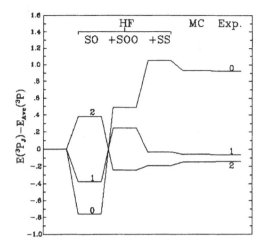

Figure 7.2. The fine structure (in cm^{-1}) of $1s2p$ ^3P of He I in different approximations. HF = Hartree–Fock, SO = including only spin–orbit, SOO = including spin–other-orbit, SS = including spin–spin, MC = multiconfiguration approach, Exp. = experimental value from Martin (1987).

between the $J = 1$ and 0. The spin–other-orbit contributions give an inverted fine structure, but the Landé interval rule still applies (see equation (7.28)). Spin–spin, however, has a different symmetry according to (7.29) and the fine structure becomes more irregular. The effect of correlation is not negligible, but the interaction with the ^1P is, in this special case.

7.7 The Blume–Watson approach

It is apparent from table 7.3 that there are a large number of integrals involved in a full Breit–Pauli calculation. Fortunately, it is not necessary to compute all possible interactions in a many-electron ion, since the core contributions can be dealt with in one of two different ways. The first is to replace the nuclear spin–orbit operator in (7.9) of the valence electrons by a more general, potential form (Cowan 1981, p 93)

$$\mathcal{H}^p_{SO} = \frac{\alpha^2}{2} \frac{1}{r} \frac{\partial V}{\partial r} \sum_{i=1}^{N}{}' l_i \cdot s_i, \qquad (7.33)$$

where $V(r)$ is the total field from the nucleus and the core electrons and where the prime indicates summation over outer (or valence) electrons. This seemingly simple formula leads to a number of complications, the most important being the decision on how to define the derivative of the non-local, exchange potential.

An alternative to this was derived by Blume and Watson (1962). Their theory is based on two observations concerning the total fine-structure Hamiltonian in (7.8). First, the contributions from closed shells are zero for the $\mathcal{H}_{SO}$ and $\mathcal{H}_{SS}$ parts. Second, the contributions from spin–other-orbit interaction between a valence and a core electron could be represented by an effective spin–orbit operator (the contributions from $\mathcal{H}_{SOO}$ within the core is trivially zero). The Blume–Watson, fine-structure Hamiltonian is therefore

$$\mathcal{H}_{FS}^{BW} = \mathcal{H}_{SO}^{BW} + \mathcal{H}_{SOO}' + \mathcal{H}_{SS}', \tag{7.34}$$

where again the primes indicate that the sums only run over outer, or valence, electrons. The $\mathcal{H}_{SO}^{BW}$ is given by

$$\mathcal{H}_{SO}^{BW} = \sum_i{}' \zeta_c(l_i) l_i \cdot s_i. \tag{7.35}$$

In this formula, the $\zeta_c(l_i)$ is an effective spin–orbit parameter, which contains the nuclear spin–orbit and all the core–valence spin–other-orbit interactions.

The Blume–Watson approach is implemented in the BREIT program. It is important to remember that this implies that it is inconsistent to include the spin–orbit effect, without including the spin–other-orbit, in any system with a core.

7.8 Systems with two valence electrons

We found that the two-body, spin-dependent effects were very important for helium and affected the fine structure substantially. If we do a similar investigation of heavier but homologous elements, we arrive at the results shown in the first columns of table 7.4. In this table, we use the same CSF expansion as

Table 7.4. Energy of $nsnp\ ^3P_J$, relative to the $nsnp\ ^3P_0$ level in some neutral, two-electron systems.

Element	J	Hartree–Fock		Valence correlation	Core valence	Exp.
		$E_{SO} + E_{SOO}$	$+ E_{SS}$			
He	1	−0.25	−1.08	−0.99	—	−0.996
	2	−0.74	−1.24	−1.08	—	−1.074
Be	1	0.92	0.38	0.54	0.63	0.64
	2	2.75	2.43	2.65	2.98	2.99
Mg	1	14.64	14.39	14.26	21.45	20.059
	2	43.91	43.76	43.11	61.77	60.773
Ca	1	34.79	34.68	34.30	57.74	52.162
	2	104.37	104.30	102.38	159.63	158.042

Table 7.5. Energy of $3s3p$ ^3P_J, relative to the $3s3p$ $^3P_0$ level in neutral magnesium from a fixed-core MCHF calculation. Core orbitals were obtained from Mg^{+2}.

Model	Step	N_{CSF}	$E(^3P_1)$	$E(^3P_2)$	LF
HF		1	14.39	43.76	2.04
	Active space approach				
VC	AS($n = 3$)	2	14.67	44.52	2.03
	AS($n = 4$)	10	14.15	42.86	2.03
	AS($n = 5$)	26	14.26	43.11	2.02
CV	AS($n = 3$)	23	16.19	48.96	2.02
	AS($n = 4$)	310	19.75	57.77	1.93
	AS($n = 5$)	786	21.45	61.77	1.88
	Selected single and doubles				
VC	AS($n = 4$)	10	14.15	42.86	2.03
CV	$2p \rightarrow 3p, 4p$	20	16.35	46.97	1.87
	$2p3s \rightarrow npmd$	72	17.47	51.88	1.97
	$2p3p \rightarrow 4p^2, ndmd$	93	18.50	54.63	1.95
	$2p3s \rightarrow nsmp$	106	19.26	56.44	1.93
	$2p \rightarrow 5p$	115	20.52	59.00	1.88
^1P			19.69	59.00	2.00
Expt			20.06	60.77	2.03

in the previous chapter. It is clear that while the total spin–orbit effect (which is a correction to the nuclear potential) increases rapidly with the size of the system, the spin–spin is almost constant and actually decreases slightly. The latter is due to the fact that spin–spin is a correction to the electron–electron interaction. Since the increase in atomic radius going from beryllium to calcium increases the separation of the two valence electrons, there is a decrease in their mutual interaction. Relatively speaking the two-body interactions are mainly important for small systems.

7.9 A limited model for core–valence correlation

The calculations referred to in table 7.4, which are based on the active space approach as described in the previous chapter, increase rapidly in size, especially for the core–valence calculation. Since the codes supplied are mainly designed for more limited calculation, we will describe an alternative to the active set calculations, concentrating on the $3s3p$ ^3P term in neutral magnesium. In the calculations above the active set was increased stepwise from a maximum principal quantum number of $n = 3$ to 5. In the outer correlation calculations all CSFs with closed $1s, 2s$ and $2p$ subshells were included, and the GBT applied, while for the core–valence calculations we used all CSFs of the form $1s^2 2s^2 2p^5 n l n' l' n'' l''$. In the $n = 5$ step this corresponds to 786 CSFs. The resulting fine structures from these active set calculations are given in table 7.5. We also give the results from a more limited calculation, where we only include some selected single and double replacements. We start with the active set calculation, $n = 4$ for outer correlation, with GBT applied. Then we add the CSF generated by the GENCL run shown in table 7.6. A listing of cfg.inp from table 7.5 is given in table 7.7. The set of orbitals $4s$, $4p$, $4d$ and $5p$ will be used as core–valence orbitals. The question on replacements will therefore be answered in table 7.8. Up to the second to last step, the results resemble the ones from the three times larger $AS(n = 4)$ calculation. It is also clear that the main improvement when adding the $n = 5$ orbitals to the active set is represented by the $2p \rightarrow 5p$ substitution in our limited calculation, a sort of spin-polarization.

The Landé factor

$$LF = \frac{E(^3P_2) - E(^3P_1)}{E(^3P_1) - E(^3P_0)} \tag{7.36}$$

is also given in table 7.5. It is interesting to see that the experimental value is very close to 2, which is reproduced by our outer-correlation calculations. When

Table 7.6. GENCL run for some single-replacement calculations for table 7.5.

```
>Gencl
                    Header   ?
>Mg I
         Closed Shells   ?
>  1s  2s
         Reference Set   ?
>2p(6)3s(1)3p(1)
                    2   ?
>
              Active Set   ?
>
           Replacements   ?
>2p=3p
                    2   ?
>2p=4p
                    3   ?
>
           Final Terms   ?
>3P
                    2   ?
>
```

Table 7.7. Listing of cfg.inp from table 7.5.

```
Mg I
 1s   2s
 2p( 6)  3s( 1)  3p( 1)
         1S0      2S1      2P1      2S0      3P0
 2p( 5)  3s( 1)  3p( 2)
         2P1      2S1      1S0      3P0      3P0
 2p( 5)  3s( 1)  3p( 2)
         2P1      2S1      1D2      3P0      3P0
 2p( 5)  3s( 1)  3p( 2)
         2P1      2S1      3P2      1P0      3P0
 2p( 5)  3s( 1)  3p( 2)
         2P1      2S1      3P2      3P0      3P0
 2p( 5)  3s( 1)  3p( 1)  4p( 1)
         2P1      2S1      2P1      2P1      1P0      2S0      3P0
 2p( 5)  3s( 1)  3p( 1)  4p( 1)
         2P1      2S1      2P1      2P1      1P0      2P0      3P0
 2p( 5)  3s( 1)  3p( 1)  4p( 1)
         2P1      2S1      2P1      2P1      1P0      2D0      3P0
 2p( 5)  3s( 1)  3p( 1)  4p( 1)
         2P1      2S1      2P1      2P1      3P0      2S0      3P0
 2p( 5)  3s( 1)  3p( 1)  4p( 1)
         2P1      2S1      2P1      2P1      3P0      2P0      3P0
 2p( 5)  3s( 1)  3p( 1)  4p( 1)
         2P1      2S1      2P1      2P1      3P0      2D0      3P0
 2p( 5)  3s( 1)  3p( 1)  4p( 1)
         2P1      2S1      2P1      2P1      3P0      4S0      3P0
 2p( 5)  3s( 1)  3p( 1)  4p( 1)
         2P1      2S1      2P1      2P1      3P0      4P0      3P0
 2p( 5)  3s( 1)  3p( 1)  4p( 1)
         2P1      2S1      2P1      2P1      3P0      4D0      3P0
```

we include the core–valence correlation, the size of the fine structure improves, but the LF is too small. The reason for this is that we have not included one important effect, the interaction between different LS terms. In this case, it is the influence of the $3s3p\ ^1P_1$ that has to be included. We do that by extending our limited core–valence model.

We will attempt to add one set of orbitals, $6s$, $6p$ and $6d$, to represent the difference in orbitals between the $3s3p\ ^3P$ and ^1P. We start by including all CSFs with a closed $2p$ subshell that can be generated from the active set

$$\{3s, 3p, 3d, 4s, 4p, 4d, 4f, 5p, 6s, 6p, 6d\}.$$

Table 7.8. The replacements used in the limited core–valence correlation calculation on $3s3p\ ^3P$ of neutral magnesium.

Step	Replacements
$2p \rightarrow 3p, 4p$	$2p = 3p$
	$2p = 4p$
$2p3s \rightarrow npmd$	$2p.3s = 3p.3d$
	$2p.3s = 3p.4d$
	$2p.3s = 3d.4p$
	$2p.3s = 4p.4d$
$2p3p \rightarrow 4p^2$	$2p.3p = 4p(2)$
$2p3p \rightarrow ndmd$	$2p.3p = 3d(2)$
	$2p.3p = 3d.4d$
	$2p.3p = 4d(2)$
$2p3s \rightarrow nsmp$	$2p.3s = 3p.4s$
	$2p.3s = 4s.4p$
$2p \rightarrow 5p$	$2p = 5p$

Table 7.9. Extra replacements for $3s3p\ ^1P$.

Step	Replacements
$2p \rightarrow 6p$	$2p = 6p$
$2p3s \rightarrow npmd$	$2p.3s = 3p.6d$
	$2p.3s = 3d.6p$
	$2p.3s = 4p.6d$
	$2p.3s = 4d.6p$
	$2p.3s = 6p.6d$
$2p3p \rightarrow 4p^2$	$2p.3p = 4p(2)$
$2p3p \rightarrow ndmd$	$2p.3p = 3d.6d$
	$2p.3p = 4d.6d$
	$2p.3p = 6d(2)$
$2p3s \rightarrow nsmp$	$2p.3s = 4s.6p$
	$2p.3s = 4p.6s$
	$2p.3s = 6s.6p$

In addition to the substitutions listed in table 7.8, we add the ones listed in table 7.9.

This shows that the agreement with the Landé interval rule is in the outer-correlation case a coincidence. The effect of core–valence correlation and the interaction with $^1P_1$ cancel each other.

7.10 Exploring complex spectra

Most of our examples so far have discussed simple spectra where quality results can be obtained by including various corrections. There are, however, situations where information is needed for a *spectrum*, where many results are required. Clearly such calculations cannot have the same accuracy as detailed calculations. In this section we will describe a preliminary calculation for levels in N II which point to the problems that need to be addressed.

In N II, the $2s^2 2p 3d$ ^3P and ^1P terms interact with similar terms of the lower-lying $2s2p^3$ and $2s^2 2p 3s$ states, and the higher $2s^2 2p 4s$ state. At the same time, an outer $3d$ often has considerable term dependence, and the many terms of $2s^2 2p 3d$ produce an environment where there may be significant spin–orbit (or Breit–Pauli) mixing even in this relatively light atom. The spin–orbit parameter for $3d$ will be small and so the spin–orbit parameter for $2p$ will play a significant role. Somewhat arbitrarily, we choose the spectrum up to $2s^2 2p 4s$ ^3P as the region to explore. In a later chapter, we will discuss the decay of these levels for which we also need to explore the $2s^2 2p^2$ and $2s^2 2p 3p$ levels.

Orthogonality constraints will play a role in the development of a strategy for such a study. The BREIT program assumes orbitals with different labels are orthogonal, thus non-orthogonal orbitals should be used with care (it is possible to combine the non-fine-structure int.1st from a NONH calculation with the fine-structure integrals from a BREIT calculation, though this needs to be done manually). Because only states with the same parity interact, we may divide the levels into odd and even categories. In theory, the two sets of calculations could proceed independently, but we view the problem as one where we have a $\{2s^2 2p, 2p^3, 2s2p^2\}$ core to which we couple an additional electron which might be another $2p$. Thus the first step is to compute the 'core' orbitals that are common to all states. This can be done by performing a calculation for $\{2s^2 2p, 2p^3\}$ ^2P with a common $1s^2$ core. The next step is to generate an orbital basis that will allow us to simultaneously represent the odd or even states of interest. MCHF does not have the ability to simultaneously optimize orbitals for a number of levels. Table 7.10 summarizes how a basis can be obtained.

Table 7.10. Strategy for obtaining the orbital basis for odd and even states from fixed-core MCHF calculations.

nl	odd	nl	even
$3s$	$\{2s^2 2p, 2p^3\}3s$ ^3P	$3p, 3d$	$\{2s^2 2p, 2p^3\}\{2p, 3p\}$
$4s$	$\{2s^2 2p, 2p^3\}4s$ ^3P		$\{2s^2, 2p^2\}\{3p^2, 3d^2\}$ ^3P
$3d_1$	$\{2s^2 2p, 2p^3\}3d_1$ ^3D	$4p, 4d$	$\{2s^2 2p, 2p^3\}\{2p, 3p, 4p\}$
$3p$	$\{2s2p^3, 2s2p^2 3p, 2s2p3p^2\}$ ^3P		$\{2s^2, 2p^2\}\{3p4p, 3d4d\}$ ^3D
$4d_1$	$\{2s^2 2p, 2p^3\}\{3d_1, 4d_1\}$ ^1D	$3s$	$2s^2 2p^2, 2p^4, 2s2p^2 3s$ ^3P

In this table, we use the set notation to indicate all possible combinations with members from one set combined with those of another, coupled to the designated term. Thus $\{2s^2, 2p^2\}\{3p^2, 3d^2\}$ refers to the four CSFs that can be formed by selecting a member from the first set and another from the second, coupled to ^3P. Calculations are performed in the order indicated with only the new orbitals varied. The radial functions from these calculations may be concatenated and store in a file as odd.w or even.w, respectively. Let us proceed next with the calculation for the odd states. The program GENCL may be used to generate all possible terms of the configurations listed in table 7.10, or in order to limit the number of CSFs, terms may be limited to $^{5,3}S$, $^{1,3}P$, $^{1,3}D$ and $^{1,3}F$. Running BREIT and then saving cfg.inp as odd.c, all information is ready for a CI run, where as many as ten eigenvalues need to be requested. Since nitrogen is a relatively light atom, the mass polarization correction was included. The result will be stored in a file odd.j which contains information for the different LSJ levels. A similar procedure can be applied to the even levels where the terms of interest are $^{1,3}S$, $^{1,3}P$ and $^{1,3}D$. The files odd.j and even.j may be concatenated and stored as atom.j, say. The program LEVELS will display the spectrum, with energies relative to the lowest. Table 7.11 displays a portion of the spectrum. The results in this table show a typical situation. Our computational model was appropriate for a Rydberg atom, where an outer electron is not strongly correlated with the core. Some configuration states, including the ground state, are not well represented in this model and their energies are too high. The difference between observed and calculated energy, relative to the lowest, is fairly constant within the levels of a term, but the difference is not constant for all terms. Thus there is a term dependence that our calculations have not captured. Though such a program is not available in the MCHF package, a common technique in atomic physics is to add LS-dependent constants to the diagonal elements of the interaction matrix so that the proper term energy separation is obtained.

Looking carefully at the $2s^2 2p3p$ terms we see a wide fluctuation in the needed corrections. For $2s^2 2p3d$ the corrections are almost constant with one exception. The $^{1,3}D$ terms are very close together and, when compared with observation, the order is reversed. The ^1D should be 346 cm^{-1} *lower* than the ^3D, whereas in our spectrum it is 180 cm^{-1} *higher*. The fine-structure splitting of the 'target' configuration $2s^2 2p$ is 167.7 cm^{-1} and so an appreciable Breit–Pauli mixing of the $J = 2$ levels can be expected. Table 7.12 shows the mixing from our present calculation. This information can be obtained from the COMP program. So we see some rather strong mixing, much larger than was the case for the intercombination lines investigated earlier. However, these results are not reliable, since they were based on an incorrect order for two closely spaced $J = 2$ levels. A conclusion of this exploratory study is that great care should be given to the correct determination of these $J = 2$ levels. At the same time, these results provide an explanation of the problem. It was mentioned in section 6.1.1.3 that an important strong interaction was the $p^2 \leftrightarrow sd$ ^1D interaction. Applied to our situation, it predicts a possible strong interaction between $2s2p^3$ and $2s^2 2p3d$

Table 7.11. A portion of the computed exploratory spectrum for N II.

ENERGY LEVELS:			Z = 7	6 electrons	
Configuration	Term	J	Total Energy (a.u.)	Energy Level (cm-1)	Obs-Calc.
2s(2).2p(2)	3P	.0	-53.9600344	.00	0.0
		1.0	-53.9598233	46.32	2.8
		2.0	-53.9594685	124.19	7.1
2s.2p(3)2D	1D	2.0	-53.8788440	17818.52	-4212
2s(2).2p_2P.3p	1P	1.0	-53.2244875	161427.55	3184
	3D	1.0	-53.2156764	163361.30	3161
		2.0	-53.2154135	163418.98	3164
		3.0	-53.2149987	163510.02	3169
	3S	1.0	-53.2069047	165286.38	3607
	3P	.0	-53.1976819	167310.47	3263
		1.0	-53.1975321	167343.34	3265
		2.0	-53.1972790	167398.91	3268
	1D	2.0	-53.1782849	171567.47	2645
	1S	.0	-53.1601183	175554.41	2720
2s(2).2p_2P.3d	3F	2.0	-53.1256001	183129.97	3382
		3.0	-53.1253469	183185.55	3386
		4.0	-53.1249831	183265.39	3388
	3D	1.0	-53.1220266	183914.23	3524
		2.0	-53.1219467	183931.78	3531
		3.0	-53.1217875	183966.72	3526
	1D	2.0	-53.1211248	184112.16	2980
	3P	2.0	-53.1151956	185413.41	3445
		1.0	-53.1149670	185463.59	3446
		.0	-53.1148424	185490.94	3447
	1F	3.0	-53.1132174	185847.56	3488
	1P	1.0	-53.1090947	186752.34	3369

certain terms. Looking at our spectrum we see that $2s2p^3\ ^1D$ is much too high relative to the observed. As a consequence, the denominator in (6.1) was too small and the effect of the interaction was to raise the energy of the upper level too much. In designing our strategy, not much attention was given to the $2s2p^3$ terms. For a better spectrum, more correlation needs to be included in these CSFs even if our goal was the prediction of the $2s^22p3d$ terms. Note that for many $2s^22p3p$ terms the difference between observed and calculated is similar to many for $2s^22p3d$, implying that the energy difference between such levels is in good agreement with observation.

Table 7.12. Breit-Pauli mixing for some of the $J = 2$ levels of $2s^22p3d$ from the exploratory calculation for N II.

```
2s(2).2p_2P.3d1   3D    2.0   -53.12194668
     .9618162   2s(2).2p_2P.3d1    3D
     .2043052   2p(3)2P1.3d1       3D
     .1393349   2s(2).2p_2P.3d1    1D
    -.0704853   2s.2p(3)2D3        3D
    -.0470481   2s(2).2p_2P.3d1    3P
     .0434475   2s(2).2p_2P.3d1    3F

2s(2).2p_2P.3d1   1D    2.0   -53.12112477
     .9402026   2s(2).2p_2P.3d1    1D
     .2033199   2s.2p(3)2D3        1D
     .2002931   2p(3)2P1.3d1       1D
    -.1415311   2s(2).2p_2P.3d1    3D
     .0785311   2s.2p(2)1D2_2D.3p  1D
    -.0580705   2s(2).2p_2P.3d1    3F
```

7.11   Z-dependence of relativistic effects

The relative importance of relativistic effects increases as a function of Z. This can be understood by looking at the variation, along an isoelectronic sequence, of the different interactions. The electrostatic effect is governed by F^k and G^k integrals, that are proportional to Z. The most important fine-structure effect is frequently the spin–orbit, which is given by (??). The corresponding expectation value is

$$E_{SO} \propto Z \int_0^\infty P(nl; r)\frac{1}{r^3} P(n'l'; r)\,\mathrm{d}r.$$

By using the substitution $r \to Zr$, we see that in a hydrogenic approach $E_{SO} \propto Z^4$. This is true in general for relativistic effects. It is clear that the relative

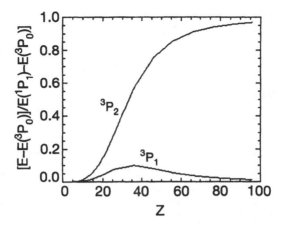

Figure 7.3. The level structure of the $2s2p$ configuration in the beryllium isoelectronic sequence.

contribution from relativistic effects increases rapidly with Z. We illustrate this with the $2s2p$ example in beryllium-like ions. In figure 7.3 we show the energy level structure of this configuration, normalized by the difference between the $^1P_1$ and $^3P_0$ level, as obtained from Breit–Pauli calculations including only the $2s2p$ ^3P and ^1P terms. It is clear that the LS approximation breaks down for high Z, and the structure changes from a triplet–singlet to two pairs. For the high-Z end the best description is not LS coupling any longer, but jj coupling.

When this transformation occurs, the LS composition of the eigenvectors changes. We can understand this change from a simple two-configuration model (we will return to this in chapter 9). If we assume that the two $J = 1$ levels of beryllium, for example, can be written

$$\Psi(^3P_1) = c_1\Phi(2s2p\ ^3P_1) + c_2\Phi(2s2p\ ^1P_1)$$

and

$$\Psi(^1P_1) = c_1\Phi(2s2p\ ^1P_1) - c_2\Phi(2s2p\ ^3P_1)$$

then the mixing coefficient to first order is given by

$$\left(\frac{c_2}{c_1}\right)^2 = \left(\frac{\langle 2s2p\ ^1P_1|\mathcal{H}_{FS}|2s2p\ ^3P_1\rangle}{E(^1P_1) - E(^3P_1)}\right)^2.$$

The numerator is given by a relativistic integral, whereas the denominator contains both relativistic and electrostatic integrals. The Z-dependence of the integrals is Z^4 and Z, respectively, which implies that the mixing strength is

$$\left(\frac{c_2}{c_1}\right)^2 \propto Z^6$$

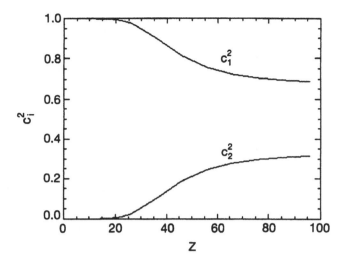

Figure 7.4. The mixing coefficients for $^3P_1$ and $^1P_1$ in $2s2p$ in beryllium-like ions.

for small Z where relativistic effects can be ignored in the denominator, but approaching a constant for large Z. This is illustrated in figure 7.4, which shows that the mixing of the two levels increases rapidly along the sequence.

7.12   Exercises

(i)   Calculate the fine structure of the ground term in first-row atoms, from boron to fluorine, by using a one-term approach. When is the fine structure inverted? What happens in nitrogen, where the $2p$ subshell is half filled?

(ii)   Investigate the structure of the $2s^2 2p^2$ configuration in carbon-like ions. Plot the level position as function of Z, as we did for beryllium-like ions in figure 7.3.

(iii)   Derive the jj-coupling designations for the $J = 1$ levels of $2s2p$. Express these in LS coupled functions. From this, interpret the asymptotic behaviour, for high Z, of the mixing coefficients in figure 7.4.

(iv)   Using the radial functions from an average-energy Hartree–Fock calculation, investigate the order of the $2s^2 2p3d$ levels in N II for all possible terms. Are terms in the correct order? Is a close degeneracy observed? How do results compare with the more elaborate study described in section 7.10?

Chapter 8

Isotope and Hyperfine Effects

8.1 The effects of the nucleus

The non-relativistic Hamiltonian†,

$$\mathcal{H}_0 = \sum_{i=1}^{N} \left(-\frac{\nabla_i^2}{2m} - \frac{Z}{r_i} \right) + \sum_{i<j}^{N} \frac{1}{r_{ij}}, \tag{8.1}$$

is valid under the assumption that the atomic nucleus can be treated as an infinitely heavy point charge. In reality, the nucleus is built from protons and neutrons and has both a finite mass and an extended charge distribution. These properties of the nucleus affect the energy level structure of an atomic system and, at least in theory, need to be included for an accurate determination of transition energies and other atomic properties. Frequently the effects of the nucleus are smaller than the uncertainty in the calculation of the correlation contribution. The corrections due to the finite nuclear mass dominate for light atoms whereas the extended charge (or finite volume) correction is important for heavy atoms. For systems where the non-relativistic approach with Breit–Pauli corrections is appropriate, these corrections are small and to a good approximation can be treated in first-order perturbation theory with wave functions for infinite mass as zero-order wave functions.

8.2 Mass shift

With a finite nuclear mass M, an N-electron atom should be considered as an $(N + 1)$-particle system. If we denote by R_0 the co-ordinates of the nucleus with respect to a fixed origin, and by R_i those of the electrons, the Hamiltonian

† In atomic units, the mass of the electron, $m = 1$, but it is included explicitly to avoid misunderstandings.

operator of this system can be written

$$\mathcal{H}_M = -\frac{\nabla^2_{R_0}}{2M} + \sum_{i=1}^{N}\left(-\frac{\nabla^2_{R_i}}{2m}\right) + V, \tag{8.2}$$

where V, the sum of the Coulomb interactions between the $(N+1)$ particles in the system, depends only on relative distances. A transformation of co-ordinates to the centre of mass

$$R = \frac{1}{M+Nm}(MR_0 + mR_1 + \ldots + mR_N), \tag{8.3}$$

and relative co-ordinates

$$r_i = R_i - R_0 \quad i = 1, \ldots, N, \tag{8.4}$$

yields the Hamiltonian

$$\mathcal{H}_M = -\frac{\nabla^2_R}{2M_{\text{tot}}} + \sum_{i=1}^{N}\left(-\frac{\nabla^2_{r_i}}{2\mu}\right) - \frac{1}{M}\sum_{i<j}^{N}\nabla_{r_i}\cdot\nabla_{r_j} + V, \tag{8.5}$$

where $M_{\text{tot}} = M + Nm$ is the total mass of the system and $\mu = Mm/(M+m)$ is the reduced mass of the electron. The first term represents the kinetic energy of the centre of mass, which can be neglected if R is not an important co-ordinate. Then, the equation to be solved (in terms of co-ordinates relative to the nucleus) is†

$$\left[\sum_{i=1}^{N}\left(-\frac{\nabla^2_{r_i}}{2\mu} - \frac{Z}{r_i}\right) + \sum_{i<j}^{N}\frac{1}{r_{ij}} - \frac{1}{M}\sum_{i<j}^{N}\nabla_{r_i}\cdot\nabla_{r_j}\right]\psi_M(r) = E_M\psi_M(r), \tag{8.6}$$

where the Coulomb interactions have been expressed explicitly. Multiplying by m/μ, and defining a new length variable, $\rho_i = (\mu/m)r_i$, we get the transformed equation

$$\left[\sum_{i=1}^{N}\left(-\frac{\nabla^2_{\rho_i}}{2m} - \frac{Z}{\rho_i}\right) + \sum_{i<j}^{N}\frac{1}{\rho_{ij}} - \frac{\mu}{Mm}\sum_{i<j}^{N}\nabla_{\rho_i}\cdot\nabla_{\rho_j}\right]\psi_M\left(\frac{m}{\mu}\rho\right)$$

$$= \frac{m}{\mu}E_M\psi_M\left(\frac{m}{\mu}\rho\right). \tag{8.7}$$

If, for a moment, we exclude the *specific mass shift* term (also referred to in the literature as the *mass polarization* term)

$$\mathcal{H}_M^{sms} = -\frac{\mu}{Mm}\sum_{i<j}^{N}\nabla_{\rho_i}\cdot\nabla_{\rho_j}, \tag{8.8}$$

† The spin co-ordinates are suppressed for brevity.

we have

$$\left[\sum_{i=1}^{N} \left(-\frac{\nabla_{\rho_i}^2}{2m} - \frac{Z}{\rho_i} \right) + \sum_{i<j}^{N} \frac{1}{\rho_{ij}} \right] \psi_M \left(\frac{m}{\mu} \rho \right) = \frac{m}{\mu} E_M \psi_M \left(\frac{m}{\mu} \rho \right), \qquad (8.9)$$

where now

$$E_M = \frac{\mu}{m} E_0 = \left(\frac{M}{M+m} \right) E_0 \qquad (8.10)$$

and

$$\psi_M \left(\frac{m}{\mu} \rho \right) = \psi_0(\rho), \qquad (8.11)$$

with E_0 and ψ_0 being, respectively, the eigenvalue and eigenfunction for infinite mass. Thus, the wave function is more extended for an atom with a light nucleus as compared to an atom with an infinitely heavy one. The energy shift resulting from the scaling of the wave function is known as the *normal mass shift* and is given by

$$E_M^{nms} = E_M - E_0 = \left(\frac{M}{M+m} - 1 \right) E_0 = -\frac{m}{M+m} E_0. \qquad (8.12)$$

The normal mass shift correction is often accounted for through the introduction of a finite mass Rydberg constant,

$$R_M = \frac{M}{M+m} R_\infty, \qquad (8.13)$$

to be used when converting from atomic units to cm^{-1}.

Treating the specific mass shift operator as a perturbation with the eigenfunctions $\psi_M(m\rho/\mu) = \psi_0(\rho)$ as zero-order functions, we get an additional shift known as the *specific mass shift* (or *mass-polarization correction*)

$$E_M^{sms} = \langle \psi_0 | -\frac{\mu}{Mm} \sum_{i<j}^{N} \nabla_i \cdot \nabla_j | \psi_0 \rangle. \qquad (8.14)$$

For later use in determining the difference of this contribution between isotopes, it is convenient to introduce the specific mass shift parameter, S, which is independent of the nuclear mass and depends only on the electronic wave function

$$S = \langle \psi_0 | -\sum_{i<j}^{N} \nabla_i \cdot \nabla_j | \psi_0 \rangle. \qquad (8.15)$$

The specific mass shift is then given as $E_M^{sms} = \mu S/Mm$. Adding the normal and specific mass shift corrections we finally obtain

$$E_M = E_0 + E_M^{nms} + E_M^{sms}. \qquad (8.16)$$

The above shows the corrections to the energy, but because of the spatial expansion of the wave function, the finite nuclear mass will also affect other properties. For example, Breit–Pauli operators which scale like $1/r^3$, must be corrected by a factor of $(\mu/m)^3$. However, in first-order perturbation theory it is customary to assume $\mu/m = 1$ except for the definition of R_M. The MCHF programs compute all quantities in the infinite mass unit of length but uses R_M for conversion to cm^{-1}. The specific mass shift operator may be included in a CI calculation thereby affecting the mixing coefficients of the CSFs in the wave function expansion.

8.2.1   Specific mass shift in the gradient form

To evaluate the specific mass shift when the eigenfunctions ψ_0 are approximated by configuration expansions

$$\Psi(\gamma LS) = \sum_i c_i \Phi(\gamma_i LS), \tag{8.17}$$

the specific mass shift operator must be put into tensorial form. Since the reduced one-electron matrix elements of the gradient operator can be factored into an angular and a radial part according to

$$\langle nl \| \nabla \| n'l' \rangle = \langle l \| \mathbf{C}^{(1)} \| l' \rangle \langle nl | \frac{\partial}{\partial r} - \frac{l(l+1) - 2 - l'(l'+1)}{2r} | n'l' \rangle, \tag{8.18}$$

where $l = l' + 1$ or $l = l' - 1$ (Brink and Satchler 1968) the specific mass shift operator can be written

$$\mathcal{H}_M^{sms} = -\frac{\mu}{Mm} \sum_{i<j}^N \nabla_i \cdot \nabla_j = -\frac{\mu}{Mm} \sum_{i<j}^N \nabla_{r_i} \nabla_{r_j} \mathbf{C}^{(1)}(i) \cdot \mathbf{C}^{(1)}(j). \tag{8.19}$$

with

$$\nabla_r = \frac{\mathrm{d}}{\mathrm{d}r} - \frac{l(l+1) - 2 - l'(l'+1)}{2r}. \tag{8.20}$$

This tensorial form is the same as that for the $k = 1$ term in the expression for the electrostatic interaction between the electrons

$$\sum_{i<j}^N \frac{1}{r_{ij}} = \sum_{i<j}^N \sum_k \frac{r_<^k}{r_>^{k+1}} \mathbf{C}^{(k)}(i) \cdot \mathbf{C}^{(k)}(j). \tag{8.21}$$

Thus, the computational apparatus set up for the calculation of the electrostatic interaction matrix elements can, with small modifications of the radial part, be used also for the specific mass shift. The modification of the radial part that has to be done amounts to replacing the Slater integrals with the so-called *Vinti* integrals $J(nl, n'l')$ given by

$$
\begin{aligned}
J(nl, n'l') &= \int_0^\infty r^2 R(nl; r) \nabla_r R(n'l'; r)\, dr \\
&= \int_0^\infty r^2 R(nl; r) \left[\frac{d}{dr} - \frac{l(l+1) - 2 - l'(l'+1)}{2r} \right] R(n'l'; r)\, dr
\end{aligned}
$$

(8.22)

where $|l - l'| = 1$.

As an example, let us evaluate the specific mass shift for a state described by the configuration expansion

$$
\Psi(1s2p\ ^3P) = c_1 \Phi(1s2p\ ^3P) + c_2 \Phi(2p3d\ ^3P).
$$

(8.23)

Running the NONH program for the wave function expansion above, the expression for the $k = 1$ term of the electrostatic interaction can be extracted from the int.1st file and is given by

$$
\langle \Psi(1s2p\ ^3P)| \sum_{i<j}^2 \frac{r_<}{r_>^2} \mathbf{C}^{(1)}(i) \cdot \mathbf{C}^{(1)}(j) |\Psi(1s2p\ ^3P)\rangle
$$

$$
= c_1^2 \frac{(-1)}{3} G^1(1s, 2p) + c_2^2 \frac{(-1)}{15} G^1(2p, 3d) + 2c_1 c_2 \frac{(-\sqrt{2})}{3} R^1(1s2p, 2p3d).
$$

(8.24)

Since the specific mass shift and the $k = 1$ term of the electrostatic interaction have the same tensorial form, the former can be calculated using the same expression, but now the radial integrals must be changed according to

$$
G^1(1s, 2p) \rightarrow -\frac{\mu}{Mm} J(1s, 2p) J(2p, 1s)
$$

(8.25)

$$
G^1(2p, 3d) \rightarrow -\frac{\mu}{Mm} J(2p, 3d) J(3d, 2p)
$$

(8.26)

and

$$
R^1(1s2p, 2p3d) \rightarrow -\frac{\mu}{Mm} J(1s, 2p) J(2p, 3d).
$$

(8.27)

8.2.2   Specific mass shift in the Slater form

As shown by Vinti (1940), there exists another expression for the specific mass shift operator known as the Slater form

$$
\mathcal{H}_M^{sms} = \frac{\mu}{2Mm} \sum_{i<j}^N \left[\sum_k \frac{r_<^k}{r_>^{k+1}} \mathbf{C}^{(k)}(i) \cdot \mathbf{C}^{(k)}(j) \right]
$$

$$
+ \frac{\mu}{2Mm} \sum_{i<j}^N \left[Z r_i r_j \mathbf{C}^{(1)}(i) \cdot \mathbf{C}^{(1)}(j) \left(\frac{1}{r_i^3} + \frac{1}{r_j^3} \right) \right].
$$

(8.28)

In this form, the mass shift is determined as a sum over Slater integrals, where those with $k = 1$ are modified with extra terms. As for the gradient form, the expression of the specific mass shift in the Slater form has the same angular symmetry as the electrostatic interaction between electrons. The only difference would be, as for the gradient form, that the radial integrals must be changed

$$R^k(ab, cd) \rightarrow \frac{\mu}{2Mm} R^k(ab, cd)$$

$$+ \frac{\mu}{2Mm} Z\delta_{1k} \left(\langle a|r|c\rangle\langle b|r^{-2}|d\rangle + \langle b|r|d\rangle\langle a|r^{-2}|c\rangle \right).$$

$$(8.29)$$

The two forms of the specific mass shift operator give the same result for exact non-relativistic wave functions. For approximate wave functions, however, they will differ and this can be used as a measure of the accuracy of the calculated values. In many cases there are large cancellations between the contributions from the two terms of the Slater operator, and this may lead to inaccurate values of the total shift. Generally, the gradient form gives more reliable values of the shift and is the preferred form of the operator.

8.2.3 Physical interpretation of the specific mass shift

The mass shift is the sum of the normal and specific shifts and can be interpreted as the kinetic energy of the nuclear motion relative to the centre-of-mass (Bethe and Salpeter 1957, section 37). The normal mass shift affects all levels in a similar manner, raising the energy by $\frac{-m}{M+m} E_0$. The specific mass shift (or mass polarization correction) on the other hand, may be either positive or negative, depending on the electronic state. If the electrons in the atom move completely independently of each other, then the specific mass shift vanishes. If, instead, the electrons move predominantly in the same direction, then the nucleus moves around to balance the electron motion. Individual electrons affect each other by the electrostatic interaction. In addition, the antisymmetry requirement for the wave function implies, as was discussed in section 1.2.2, some correlation in the motion between the electrons. We will look at the effect of the latter for the $1s2p\ ^{1,3}P$ terms of helium. The antisymmetry condition gives an exchange term in the expression for the electrostatic interaction. For $1s2p\ ^{1,3}P$ we have (see equation (2.37))

$$\langle 1s2p\ ^{1,3}P|\frac{1}{r_{12}}|1s2p\ ^{1,3}P\rangle = F^0(1s, 2p) \pm \frac{1}{3}G^1(1s, 2p) \qquad (8.30)$$

where the positive sign is for the singlet ($S = 0$) term. To obtain the expression for the specific mass shift we make the substitution

$$G^1(1s, 2p) \rightarrow -\frac{\mu}{Mm} J(1s, 2p)J(2p, 1s) \qquad (8.31)$$

giving

$$\langle 1s2p \ ^{1,3}P|\mathcal{H}_M^{sms}|1s2p \ ^{1,3}P \rangle = \pm\frac{\mu}{3Mm}|J(1s,2p)|^2. \tag{8.32}$$

Here we have used the permutation symmetry $J(1s,2p) = -J(2p,1s)$ of the Vinti integral. Physically, this means that in the singlet state the two electrons move mostly in the same direction and in the triplet state more frequently in opposite directions.

8.3 Field shift

Due to the finite size of the nucleus, the potential deviates from the Coulomb potential of a point charge Z. Since s electrons have a finite probability to be within the nuclear volume, the potential deviation will lead to an energy shift. If we let $V(r)$ denote the potential from the extended nuclear charge distribution of an isotope with mass M, then the energy correction for electrons moving in this potential as compared to the ones moving in the Coulomb potential can be written

$$E_M^{fs} = -\int_{R^3} \left(V(r) - \frac{Z}{r}\right) \rho_e(r) \, \mathrm{d}^3r, \tag{8.33}$$

where $\rho_e(r)$ is the electron charge distribution. The integration is formally over the entire space, but the difference in potential $(V(r) - Z/r)$ is only different from zero within the nuclear volume. Since the electron charge distribution is almost constant over this volume, $\rho_e(r)$ can be replaced with the value at $r = 0$ and taken outside the integral. Using the identity $\nabla^2 r^2 = 6$, the energy shift can be expressed

$$\begin{aligned} E_M^{fs} &= -\frac{1}{6}\rho_e(0) \int_{R^3} \left(V(r) - \frac{Z}{r}\right) \nabla^2 r^2 \, \mathrm{d}^3r \\ &= -\frac{1}{6}\rho_e(0) \int_{R^3} r^2 \, \nabla^2 \left(V(r) - \frac{Z}{r}\right) \mathrm{d}^3r, \end{aligned} \tag{8.34}$$

where the latter follows from integration by parts and the zero boundary condition at infinity. Using the relations

$$\nabla^2 \left(\frac{1}{r}\right) = -4\pi \delta(r), \tag{8.35}$$

and Poisson's equation

$$\nabla^2 V(r) = -4\pi \rho_n(r), \tag{8.36}$$

where $\rho_n(r)$ is the nuclear charge distribution, it is seen that

$$E_M^{fs} = \frac{2\pi}{3}\rho_e(0) \int_{R^3} r^2 \rho_n(r) \, \mathrm{d}^3r. \tag{8.37}$$

Finally, using the definition of the so-called mean-square radius of the nucleus

$$\langle r_M^2 \rangle = \frac{\int r^2 \rho_n(r)\, d^3 r}{\int \rho_n(r)\, d^3 r} \tag{8.38}$$

we obtain the expression

$$E_M^{fs} = \tfrac{2}{3}\pi Z \rho_e(0) \langle r_M^2 \rangle. \tag{8.39}$$

Often the charge distribution is not known, and instead a simple model is used in which the nuclear charge is uniformly distributed in a sphere of radius R. Inserting this distribution into (8.38) we get

$$R^2 = \tfrac{5}{3} \langle r_M^2 \rangle. \tag{8.40}$$

The radius R is often referred to as the equivalent radius (denoted Req in our program). For isotopes with mass number $A > 16$ the equivalent radius (in fm) can be obtained from the expression

$$R = 1.115A^{1/3} + 2.151A^{-1/3} - 1.742A^{-1}. \tag{8.41}$$

The theory and experimental results underlying this approximate formula are discussed in Elton (1968). It should, however, be stressed that for many isotopes the experimental radius differs substantially from the values predicted by this simple formula. Specifically, this formula may give large errors for the difference in radius between two isotopes of the same element.

8.3.1 Computational aspects

The electron density at the origin is given by the expectation value

$$\rho_e(0) = \langle \psi_0 | \sum_{i=1}^{N} \delta(r_i) | \psi_0 \rangle = \langle \psi_0 | \sum_{i=1}^{N} \frac{1}{4\pi} r_i^{-2} \delta(r_i) | \psi_0 \rangle, \tag{8.42}$$

where, for the last expression, the relation $4\pi r^2 \delta(r) = \delta(r)$ between the three- and one-dimensional delta functions has been used. Since the delta function is rotationally invariant, the operator

$$\sum_{i=1}^{N} \frac{1}{4\pi} r_i^{-2} \delta(r_i) \tag{8.43}$$

is a scalar and has the same angular properties as the one-electron part of the Hamiltonian,

$$\sum_{i=1}^{N} \left(-\frac{1}{2}\nabla_i^2 - \frac{Z}{r_i} \right). \tag{8.44}$$

Thus, to calculate the electron density for a state described by a configuration expansion (8.23), the expression for the one-electron part of the interaction, obtained from the int.1st file, can be used. The change that needs to be made amounts to the replacement

$$I(nl, n'l') \rightarrow \frac{1}{4\pi} \int_0^\infty P(nl; r) r^{-2} \delta(r) P(n'l'; r) \, dr$$

$$= \frac{1}{4\pi} \frac{P(nl; r)}{r} \frac{P(n'l'; r)}{r} \bigg|_{r \rightarrow 0} = \frac{1}{4\pi} \delta_{l,0} \delta_{l',0} AZ(nl) AZ(n'l').$$

(8.45)

In the expression above $AZ(nl)$ is the radial slope at the origin and is printed as AZ(nl) in the summary file, for example, of an MCHF calculation.

The major contribution to the electron density at the origin comes from the inner s electrons. For heavy atoms, these electrons are appreciably influenced by relativistic effects leading to higher densities. To account for these effects, the non-relativistic density should be multiplied with a correction factor, frequently denoted by $f(Z)$. Numerical values of the correction factor have been tabulated (Aufmuth *et al* 1987).

8.4 Level isotope shift

When the effects of the finite nuclear mass and extended charge distribution are taken into account, the isotopes of an element, characterized by having the same number of protons but different number of neutrons, will all have slightly different energy levels. The shift between the energy levels (for the same quantum state) of two different isotopes is called the *level isotope shift*. For an isotope with mass M and nuclear radius $\langle r_M^2 \rangle$, the mass and field shift corrected energy is given by

$$E_M = E_0 + E_M^{nms} + E_M^{sms} + E_M^{fs}$$

$$= E_0 - \frac{m}{M+m} E_0 + \frac{\mu}{Mm} S + \frac{2\pi}{3} Z \langle r_M^2 \rangle \rho_e(0),$$

(8.46)

where E_0 is the energy of the Hamiltonian. The level isotope shift, $E_{M'M}$, between two isotopes M' and M is then obtained as

$$E_{M'M} = E_{M'} - E_M = E_0 \left(\frac{m}{M+m} - \frac{m}{M'+m} \right)$$

$$+ S \left(\frac{\mu}{M'm} - \frac{\mu}{Mm} \right) + \frac{2\pi}{3} Z \rho_e(0) \left(\langle r_{M'}^2 \rangle - \langle r_M^2 \rangle \right).$$

(8.47)

Values of the differences in mean-square radii can be found in the tabulation by Aufmuth *et al* (1987).

8.5   Transition isotope shift

The transition isotope shift, or more commonly the isotope shift, is the difference
in transition energy between two isotopes. For an isotope with mass M and
nuclear radius $\langle r_M^2 \rangle$, the theoretical transition energy, ΔE_M, between an upper
level (2) and a lower level (1) can be written as

$$
\Delta E_M = E_M^{(2)} - E_M^{(1)}
$$

$$
= E_0^{(2)} - \frac{m}{M+m} E_0^{(2)} + \frac{\mu}{Mm} S^{(2)} + \frac{2\pi}{3} Z\rho_e^{(2)}(0)\langle r_M^2 \rangle
$$

$$
- \left[E_0^{(1)} - \frac{m}{M+m} E_0^{(1)} + \frac{\mu}{Mm} S^{(1)} + \frac{2\pi}{3} Z\rho_e^{(1)}(0)\langle r_M^2 \rangle \right].
$$

(8.48)

The transition isotope shift, $\Delta E_{M'M}$, between two isotopes M' and M is then
obtained as

$$
\Delta E_{M'M} = \Delta E_{M'} - \Delta E_{M'}
$$

$$
= E_0^{(2)} \left(\frac{m}{M+m} - \frac{m}{M'+m} \right) + S^{(2)} \left(\frac{\mu}{M'm} - \frac{\mu}{Mm} \right)
$$

$$
+ \frac{2\pi}{3} Z\rho_e^{(2)}(0) \left(\langle r_{M'}^2 \rangle - \langle r_M^2 \rangle \right) - \left[E_0^{(1)} \left(\frac{m}{M+m} - \frac{m}{M'+m} \right) \right.
$$

$$
\left. + S^{(1)} \left(\frac{\mu}{M'm} - \frac{\mu}{Mm} \right) + \frac{2\pi}{3} Z\rho_e^{(1)}(0) \left(\langle r_{M'}^2 \rangle - \langle r_M^2 \rangle \right) \right].
$$

Comparing with expression (8.47), it is seen that the transition isotope shift is
merely the difference in level isotope shift between the upper and lower levels.

In many cases the transition isotope shift can be accurately measured with
laser techniques or obtained from Fourier transform spectra. If the electronic
quantities S and $\rho_e(0)$ can be calculated accurately for both the upper and the
lower state the difference in nuclear radii can be deduced (King 1984).

8.5.1   Transition isotope shift for $1s2s\ ^3S-1s2p\ ^3P$ in Li II

The first calculation of the isotope shift in a many-electron atom was done by
Hughes and Eckart in 1930. Using linear combinations of products of hydrogen-
like wave functions, they calculated the isotope shift between ^{7}Li and ^{6}Li in the
$1s2s\ ^3S-1s2p\ ^3P$ transition of Li II. The calculated value of 1.09 cm^{-1} was in
good agreement with an early measurement of Schüler in 1927 giving 1.1 cm^{-1}.
The isotope shift has since then been remeasured with high accuracy using laser
techniques giving the value 1.1597(1) cm^{-1}. As an example of isotope shift
calculations, we will determine the isotope shift in the above transition using

the reduced two electron expansions of chapter 5 for the $1s2s\ ^3S$ and $1s2p\ ^3P$ states. Table 8.1 displays the use of ISO for a familiar expansion for $1s2p\ ^3P$ involving orbitals with $n \leqslant 3$. It assumes the results of the MCHF calculations have been stored in 3P.c and 3P.w, that the int.1st produced by NONH is still available. Notice that the program can use expansion information from three sources—MCHF, in which case the expansion coefficients are part of name.c, or CI where expansion coefficients are stored in either name.1 (non-fine structure) or name.j (with fine-structure splitting). In this example, there is no core and so all contributions involving a core are zero. For the integrals outside the core, the program has given contributions to the specific mass shift parameter (8.15) in the gradient form and in the Slater form. For the latter there are two parts, PART1 and PART2, corresponding to the two terms of the operator (8.28). It is the sum of the total contributions from these two parts that must be compared to the total specific mass shift parameter in the gradient form. Notice the extensive cancellation that occurs in the calculation of the sum in the Slater integral form. For brevity, full print-out has not been requested, but it is possible to display the contribution from every integral in the energy expression. Then the program has printed the specific and normal mass shift corrections for an isotope with mass, $M = 7$. Finally, the level isotope shift has been determined between the two masses as shown. To obtain the isotope shift for the $1s2s\ ^3S - 1s2p\ ^3P$ transition a similar calculation must be performed for the ^3S state. The nuclear masses for the two isotopes have been taken from the atomic mass table of Wapstra and Audi (1985).

The specific mass shift contribution (in the gradient form) to the level isotope shift of the $1s2p\ ^3P$ state is 0.868 776 99 cm^{-1} whereas the normal mass shift contribution is $-14.353\ 110\ 81$ cm^{-1}. The total level isotope shift is then $(0.868\ 776\ 99 - 14.353\ 110\ 81)$ cm^{-1} $= -13.484\ 333\ 82$ cm^{-1}. In the same way the level isotope shift of the $1s2s\ ^3S$ state is given by $(-0.050\ 282\ 71 - 14.590\ 553\ 67)$ cm^{-1} $= -14.640\ 836\ 38$ cm^{-1}. The transition isotope shift is the difference between the level isotope shifts of the two states $(-13.484\ 333\ 82 - (-14.640\ 836\ 38))$ cm^{-1} ≈ 1.1565 cm^{-1}, and compares very well with the experimental value of 1.1597 cm^{-1}. The field shift contribution to the transition isotope shift is very small and has been neglected. For the $1s2s\ ^3S$ state there is a large cancellation between the contributions to the specific mass shift parameter from the two terms of the operator in the Slater form. In the subtraction significant digits are lost, and the final value is less accurate than the value obtained from the gradient form of the operator.

8.5.2 Cancellation effects for specific mass shifts

A number of studies have shown that the Hartree–Fock model yields unreliable values of the specific mass shift which are more of qualitative than quantitative interest. The failure of the HF model is best understood from an example. Let us consider the $3s^2\ ^1S - 3s3p\ ^1P$ transition in Mg. In table 8.2 the different con-

Table 8.1. Isotope shift calculation of the $1s2p\ ^3P$ state in Li with a wave function expansion over the the $\{1s2p_1,\ 2s3p_1,\ 2p_23d\}$ set of configuration states.

```
>Iso
 Name of State
>3P
 Select type of the input file:
    1  name.c
    2  name.l
    3  name.j
>1
 Intermediate printing (y/n)
>n
 Mass shift (1), field shift (2) or both (3) ?
>1
 Default Rydberg constant (y/n)
>y
 Contribution to the isotope mass shift (y/n)
>y
 Enter two masses
>7.0160030,6.0151214
                       ----------------------
                          MASS CONTRIBUTION
                       ----------------------
                       CONTRIBUTIONS TO SPECIFIC MASS SHIFT
                       Gradient Form    Slater Integral Form
                       -------------    --------------------
                           TOTAL        PART1        PART2
    Total for core        0.00000000   0.00000000   0.00000000
    Total for integrals  -0.30425347   0.25457059  -0.58449005
    Total for core-outer  0.00000000   0.00000000   0.00000000
                       -------------------------------------------
    Total contribution                 0.25457059  -0.58449005
       in atomic units:  -0.30425347          -0.32991945
  1  SPECIFIC MASS SHIFT CORRECTION FOR MASS      7.00
         Gradient Form                     -5.23313246  cm-1
         Slater Integral Form              -5.67458500  cm-1
         Rydberg constant was             109728.7161   cm-1
  2  NORMAL MASS SHIFT CORRECTION          86.46472880  cm-1
  3  ISOTOPE SHIFT FOR MASSES               7.02 -   6.02
       a)  SPECIFIC MASS CONTRIBUTION
         Gradient Form                      0.86877699  cm-1
         Slater Integral Form               0.94206461  cm-1
       b)  NORMAL MASS CONTRIBUTION        -14.35311081  cm-1
```

Table 8.2. Contributions to the specific mass shift parameters S for $3s^2$ ^1S and $3s3p$ ^1P in Mg.

Contribution	$S(^1S)$	$S(^1P)$	$S(^1P) - S(^1S)$
Core	−27.38096	−27.42015	−0.0392
Core–valence	−0.22971	−0.18089	0.0488
Valence	0.00000	0.04449	0.0445
Total	−27.61067	−27.55653	0.0541
Exp[a]			0.0375

[a]Beverini *et al* (1990).

tributions to the specific mass shift parameter S are shown for both states. From the table it is clear that there are very large, but cancelling, contributions from the core. Thus relatively small changes of the individual core contributions due to different correlation effects may have a dramatic influence on the difference. In general, to obtain accurate values of the specific mass shift in a transition, very accurate calculations including core–core correlation need to be performed for both states. In addition the correlation effects must be well balanced.

8.6   Field shift correction for $3d^8(^3F)4p$ $^4D_{5/2}$ in Ni II

As an example of the finite volume correction and level field shift calculation, we use a Breit–Pauli wave function allowing for the mixing of all $3d^84p$ terms, as J values allow. Radial functions were determined from a HF calculation for $3d^8(^3F)4p$ ^4G and stored as Ni.w. GENCL was used to generate all possible terms that can couple to $J = 5/2$; BREIT generated the int.1st; after moving cfg.inp to Ni.c; CI was used to determine selected eigenvectors in Ni.j. Table 8.3 shows some of the output for the lowest eigenvalue, namely $3d^8(^3F)4p$ $^4D_{5/2}$ in Ni II. In the output, first are listed some of the CSFs that were included in the Ni.c file. Then we have the Breit–Pauli energy for the state under consideration. The electronic contributions to the field shift are the contributions to the so-called modified charge density $4\pi\rho_e(0)$ of (8.45). From these, a finite volume correction is computed for $M = 58$. An equivalent radius (R) was used of 4.84168120 fm. Finally, the level field shift contribution is computed using the factor $f(Z) = 1635.65$ MHz fm^{-2} as a relativistic correction for the density at the nucleus, and with $(\langle r_{M'}^2 \rangle - \langle r_M^2 \rangle) = 0.293$ fm^2, corresponding to the change in mean radii for the masses 58 and 60. The numerical values for $f(Z)$ and the change in nuclear radii were taken from Aufmuth *et al* (1987). Notice that, in absolute terms, the finite volume correction is fairly large but the effect on a spectrum will be small when the electron density at the nucleus changes very little from one term to the next as would be the case for this configuration.

Table 8.3. Finite volume and field shift corrections for $3d^8(^3F)4p\ ^4D_{5/2}$ of Ni II.

```
>Iso
 Name of State
>Ni
  Select type of the input file:
      1  name.c
      2  name.l
      3  name.j
>3
  Intermediate printing (y/n)
>n
  Mass shift (1), field shift (2) or both (3) ?
>2
  Maximum and minimum value of 2*J
>5,5
  Default Rydberg constant (y/n)
>y
  Contribution to the isotope field shift (y/n)
>y
 Enter f(Z) (MHz/fm^2) and delta <r^2> (fm^2)
>1635.65,0.293
        CONFIGURATIONS
     1   3d(8)1D2.4p_2D
     2   3d(8)1D2.4p_2F
     3   3d(8)1G2.4p_2F
        ...
     9   3d(8)3F2.4p_4D
    10   3d(8)3F2.4p_4F
    11   3d(8)3F2.4p_4G

  2*J-Value     Energy        State
      5     -1518.24050535  3d(8)3F2.4p_4D
            --------------------------------------------
             ELECTRONIC CONTRIBUTION TO THE FIELD SHIFT
            --------------------------------------------
  i)   Core
     Total for core                       187641.38887252
 ii)   Outer Electrons
     Total for outer electrons                 0.00000000
     Total contribution (au)             187641.38887252
  1  FINITE VOLUME CORRECTION FOR MASS           58.00
        Volume correction                  1284.67084080  cm-1
           Req was                            4.84168120  fm
  2  FIELD SHIFT CONTRIBUTION                   26.78230131  cm-1
```

8.7 Hyperfine structure

At the end of the last century it was shown experimentally that the fine-structure levels of atoms in many cases were not single levels, but consisted of a series of closely spaced levels, the so-called hyperfine structure (hfs). In 1924 Pauli suggested that the hfs could be due to an interaction between the magnetic field generated by the electrons and the magnetic moment of the nucleus. Later it was realized that the magnetic interaction could not fully explain the hyperfine structure, and it was shown that part of the splitting is due to an electric interaction between the electrons and the non-spherical nuclear charge density.

8.8 Hyperfine interaction

The hyperfine structure of the atomic energy levels is caused by the interaction between the electrons and the electromagnetic multipole moments of the nucleus. The contribution to the Hamiltonian can be represented by an expansion in multipoles of order K

$$\mathcal{H}_{hfs} = \sum_{K \geqslant 1} \mathbf{T}^{(K)} \cdot \mathbf{M}^{(K)}, \qquad (8.49)$$

where $\mathbf{T}^{(K)}$ and $\mathbf{M}^{(K)}$ are spherical tensor operators of rank K in the space respectively spanned by the electronic $|\gamma J M_J\rangle$ and nuclear $|\nu I M_I\rangle$ wave functions (Lindgren and Rosén 1974). The $K = 1$ term represents the magnetic dipole interaction and the $K = 2$ term the electric quadrupole interaction. Higher-order terms are much smaller and can often be neglected.

For an N-electron atom the electronic tensor operators are, in atomic units,

$$\mathbf{T}^{(1)} = \frac{\alpha^2}{2} \sum_{i=1}^{N} \left[2 \, l^{(1)}(i) r_i^{-3} - g_s \sqrt{10} \left[\mathbf{C}^{(2)}(i) \times s^{(1)}(i) \right]^{(1)} r_i^{-3} \right.$$
$$\left. + g_s \frac{8}{3} \pi \delta(r_i) s^{(1)}(i) \right] \qquad (8.50)$$

and

$$\mathbf{T}^{(2)} = - \sum_{i=1}^{N} \mathbf{C}^{(2)}(i) r_i^{-3}, \qquad (8.51)$$

where $g_s = 2.00232$ is the electronic g-factor and $\delta(r)$ the three-dimensional delta-function.

The magnetic dipole operator (8.50) represents the magnetic field due to the electrons at the site of the nucleus. The first term of the operator represents the field caused by orbital motion of the electrons and is called the *orbital* term. The second term represents the dipole field due to the spin motion of the electron and is called the *spin-dipole* term. The last term, known as the *Fermi contact* term, represents the contact interaction between the nucleus and the electrons and is proportional to the electron *spin density* at the nucleus. Since only s electrons has a finite probability of being at the site of the nucleus the spin density is the difference in density, at the nucleus, of s electrons with spin up and spin down. The electric quadrupole operator (8.51) represents the electric field gradient at the site of the nucleus.

The nuclear operators are

$$\mathbf{M}^{(1)} = \mu_N \sum_{i=1}^{nuc} \nabla \left(r_i \mathbf{C}^{(1)}(i) \right) \cdot \left[g_l l^{(1)}(i) + g_s s^{(1)}(i) \right] \tag{8.52}$$

and

$$\mathbf{M}^{(2)} = \sum_{i=1}^{prot} r_i^2 \mathbf{C}^{(2)}(i) = \sum_{i=1}^{prot} (3z_i^2 - r_i^2), \tag{8.53}$$

where g_l and g_s are, respectively, the orbital and spin g-factors of the nucleons. $\mu_N = \mu_B/M$ is the nuclear magneton. For the nuclear magnetic dipole operator the sum is over the nucleons (protons and neutrons) whereas for the nuclear electric quadrupole operator the sum is over the protons alone.

The conventional nuclear magnetic dipole moment μ_I and electric quadrupole moment Q are defined as the expectation values of the operators $\mathbf{M}^{(1)}$ and $\mathbf{M}^{(2)}$ in the nuclear state $|\nu I I\rangle$ with the maximum component of the nuclear spin, $M_I = I$

$$\begin{aligned} \langle \nu I I \mid M_0^{(1)} \mid \nu I I \rangle &= \mu_I \\ \langle \nu I I \mid M_0^{(2)} \mid \nu I I \rangle &= \tfrac{1}{2} Q. \end{aligned} \tag{8.54}$$

The quadrupole moment has a simple interpretation and provides a measure of the charge density distribution of the nucleus. Thus, if the charge density has complete spherical symmetry the electric quadrupole moment vanishes. If, on the other hand, the nuclear charge distribution is deformed, the electric quadrupole moment may have a large value (Shalit and Feschbach 1974, chapter 1).

The nuclear moments μ_I can be experimentally determined and may be thought of as known quantities. For a recent tabulation see Raghavan (1989). The Q on the other hand is often determined semi-empirically (Sundholm and Olsen 1992).

8.9   Angular properties of the hyperfine states

When the hyperfine contributions $\mathcal{H}_{hfs}$ are added to the normal electronic Hamiltonian $\mathcal{H}_0$, the total Hamiltonian

$$\mathcal{H} = \mathcal{H}_0 + \mathcal{H}_{hfs} \tag{8.55}$$

does not commute with the electronic J or nuclear I angular momentum operators. Instead we have

$$[\mathcal{H}, F^2] = [\mathcal{H}, F_z] = 0, \tag{8.56}$$

where $F = I + J$ is the total angular momentum of the electrons and the nucleus (compare the breakdown of LS coupling when the relativistic operators are added).

This implies that the eigenfunctions of $\mathcal{H}$ describing hyperfine states can be characterized with the angular quantum numbers F and M_F where

$$|I - J| \leqslant F \leqslant I + J \tag{8.57}$$

and

$$M_F = -F, -F + 1, \ldots, F - 1, F. \tag{8.58}$$

8.10 First-order hyperfine energies

The hyperfine interaction is very weak and can to a good approximation be treated in perturbation theory with the eigenfunctions of $\mathcal{H}_0$ as zero-order functions. These functions may be taken as products of the electronic $|\gamma J M_J\rangle$ and nuclear $|\nu I M_I\rangle$ eigenfunctions and we have

$$\mathcal{H}_0|\gamma J M_J\rangle|\nu I M_I\rangle = E_{\gamma J}|\gamma J M_J\rangle|\nu I M_I\rangle. \tag{8.59}$$

Unlike the correct functions describing hyperfine states, the product functions are not eigenfunctions to F^2 and F_z. However, by applying the Clebsch–Gordan expansion, the product functions may be transformed into an equivalent set of functions with the correct symmetry,

$$|\gamma \nu J I F M_F\rangle = \sum_{M_J M_I} \langle J I M_J M_I | J I F M_F\rangle|\gamma J M_J\rangle|\nu I M_I\rangle. \tag{8.60}$$

Now, according to first-order perturbation theory, if we only consider the magnetic dipole and electric quadrupole terms of the hyperfine interaction, the hyperfine correction to the electronic energy $E_{\gamma J}$ can be written as the sum

$$E_{hfs}(JIF) = E_{M1}(JIF) + E_{E2}(JIF) \tag{8.61}$$

where

$$E_{M1}(JIF) = \langle \gamma \nu J I F M_F \mid \mathbf{T}^{(1)} \cdot \mathbf{M}^{(1)} \mid \gamma \nu J I F M_F\rangle \tag{8.62}$$

and

$$E_{E2}(JIF) = \langle \gamma \nu J I F M_F \mid \mathbf{T}^{(2)} \cdot \mathbf{M}^{(2)} \mid \gamma \nu J I F M_F\rangle. \tag{8.63}$$

Using the tensor algebra (A.56), these corrections can also be written as

$$E_{M1}(JIF) = (-1)^{I+J+F} \left\{ \begin{matrix} I & J & F \\ J & I & 1 \end{matrix} \right\} \langle \gamma J \| \mathbf{T}^{(1)} \| \gamma J \rangle \langle \nu I \| \mathbf{M}^{(1)} \| \nu I \rangle$$

(8.64)

$$E_{E2}(JIF) = (-1)^{I+J+F} \left\{ \begin{matrix} I & J & F \\ J & I & 2 \end{matrix} \right\} \langle \gamma J \| \mathbf{T}^{(2)} \| \gamma J \rangle \langle \nu I \| \mathbf{M}^{(2)} \| \nu I \rangle.$$

(8.65)

Normally, the dependence on the F quantum numbers are factored out and the energies are expressed in terms of the hyperfine interaction constants (A and B factors),

$$A(J, J) = \frac{\mu_I}{I} \frac{1}{[J(J+1)(2J+1)]^{\frac{1}{2}}} \langle \gamma J \| \mathbf{T}^{(1)} \| \gamma J \rangle,$$

$$B(J, J) = 2Q \left[\frac{J(2J-1)}{(J+1)(2J+1)(2J+3)} \right]^{\frac{1}{2}} \langle \gamma J \| \mathbf{T}^{(2)} \| \gamma J \rangle.$$

(8.66)

The energy corrections are then given by

$$E_{M1}(JIF) = \tfrac{1}{2} A(J, J) C$$

(8.67)

and

$$E_{E2}(JIF) = B(J, J) \frac{\tfrac{3}{4} C(C+1) - I(I+1)J(J+1)}{2I(2I-1)J(2J-1)},$$

(8.68)

where $C = F(F+1) - J(J+1) - I(I+1)$.

The hyperfine components of a fine-structure level are said to form a *hyperfine-structure multiplet*. Normally, the magnetic dipole interaction dominates and the energy difference between two neighbouring hyperfine levels F and $F-1$ is approximately

$$\Delta E(JIF) = E_{M1}(JIF) - E_{M1}(JIF-1) = A(J, J)F.$$

(8.69)

This is the Landé interval rule for magnetic hyperfine structure.

8.11   First-order wave functions

The wave functions describing the hyperfine states are, to first order in the hyperfine interaction, given by

$$|\gamma \nu J I F M_F\rangle^{(1)} = |\gamma \nu J I F M_F\rangle + \sum_{\gamma' J'} c_{\gamma' J'} |\gamma' \nu J' I F M_F\rangle$$

(8.70)

where

$$c_{\gamma' J'} = \frac{\langle \gamma \nu J I F M_F | \mathcal{H}_{hfs} | \gamma' \nu J' I F M_F \rangle}{(E_{\gamma J} - E_{\gamma' J'})}.$$

(8.71)

The sum in the perturbation expansion is over all the electronic states $\gamma' J'$ such that $E_{\gamma' J'} \neq E_{\gamma J}$. Formally, this sum should also be over excited nuclear states, but, since the nuclear excitation energies are very large compared to the electronic energies, we may completely neglect the influence of these states. In numerical calculations the sum over electronic states must of course also be truncated. In most cases the sum over electronic states can be restricted to the fine-structure levels of the same LS term, other states being unimportant due to the large value of $E_{\gamma J} - E_{\gamma' J'}$ as compared to the hyperfine-structure matrix elements.

If we only consider the magnetic dipole and electric quadrupole terms of the hyperfine interaction, the matrix elements entering the expansion (8.70) can be written as the sum of the two terms

$$E_{M1}(JIF, J'IF) = \langle \gamma \nu JIFM_F \mid \mathbf{T}^{(1)} \cdot \mathbf{M}^{(1)} \mid \gamma \nu J'IFM_F \rangle \quad (8.72)$$

and

$$E_{E2}(JIF, J'IF) = \langle \gamma \nu JIFM_F \mid \mathbf{T}^{(2)} \cdot \mathbf{M}^{(2)} \mid \gamma \nu J'IFM_F \rangle \quad (8.73)$$

where for the two terms $J' = J \pm 1$ and $J' = J \pm 1, J \pm 2$, respectively. Defining the hyperfine-interaction constants off-diagonal in J as

$$A(J, J - 1) = \frac{\mu_I}{I} \frac{1}{[J(J+1)(2J+1)]^{1/2}} \langle \gamma J \| \mathbf{T}^{(1)} \| \gamma (J-1) \rangle$$

$$B(J, J - 1) = \frac{Q}{2} \left[\frac{J(J-1)}{(J+1)(2J-1)(2J+1)} \right]^{1/2} \langle \gamma J \| \mathbf{T}^{(2)} \| \gamma (J-1) \rangle$$

$$B(J, J - 2) = \frac{Q}{4} \left[\frac{J(J-1)(2J-1)}{(2J-3)(2J+1)} \right]^{1/2} \langle \gamma J \| \mathbf{T}^{(1)} \| \gamma (J-2) \rangle,$$

$$(8.74)$$

the off-diagonal matrix elements can be written

$$E_{M1}(JIF, (J-1)IF)$$
$$= \tfrac{1}{2} A(J, J-1)[(K+1)(K-2F)(K-2I)(K-2J+1)]^{1/2}$$
$$(8.75)$$

$$E_{E2}(JIF, (J-1)IF) = B(J, J-1)[(F+I+1)(F-I) - J^2 + 1]$$
$$\times \frac{[3(K+1)(K-2F)(K-2I)(K-2J+1)]^{1/2}}{2I(2I-1)J(J-1)}$$
$$(8.76)$$

$$E_{E2}(JIF, (J-2)IF) = B(J, J-2)[6K(K+1)(K-2I-1)]^{1/2}$$
$$\times \frac{[(K-2F-1)(K-2F)(K-2I)(K-2J+1)(K-2J+2)]^{1/2}}{2I(2I-1)J(J-1)(2J-1)}.$$
$$(8.77)$$

where $K = I + J + F$.

8.11.1   Magnetic hyperfine parameters

In light atoms where the relativistic effects can be neglected the diagonal and off-diagonal A and B factors can be expressed in terms of the J independent hyperfine parameters a_l, a_{sd}, a_c and b_q given by

$$
\begin{aligned}
a_l &= \langle \gamma LSM_LM_S \mid \sum_{i=1}^{N} l_0^{(1)}(i)r_i^{-3} \mid \gamma LSM_LM_S \rangle, \\
a_{sd} &= \langle \gamma LSM_LM_S \mid \sum_{i=1}^{N} 2C_0^{(2)}(i)s_0^{(1)}(i)r_i^{-3} \mid \gamma LSM_LM_S \rangle, \\
a_c &= \langle \gamma LSM_LM_S \mid \sum_{i=1}^{N} 2s_0^{(1)}(i)r_i^{-2}\delta(r_i) \mid \gamma LSM_LM_S \rangle, \\
b_q &= \langle \gamma LSM_LM_S \mid \sum_{i=1}^{N} 2C_0^{(2)}(i)r_i^{-3} \mid \gamma LSM_LM_S \rangle,
\end{aligned}
\tag{8.78}
$$

where $M_L = L$ and $M_S = S$. Explicit relations between these parameters and the A and B factors can be found in a review article by Hibbert (1975). The hyperfine parameters are interesting from a theoretical point of view since they are sensitive to different types of correlation effect. Frequently, the a_c parameter corresponding to the Fermi contact term is the most difficult parameter to calculate.

8.12   Computational aspects

In cases where the electronic wave functions of the fine-structure levels are given by configuration state expansions†

$$
\Psi(\gamma J) = \sum_j c_j \Phi(\gamma_j L_j S_j J) \tag{8.79}
$$

the electronic matrix elements can be written,

$$
\langle \gamma J \| \mathbf{T}^{(1)} \| \gamma' J' \rangle = \sum_{j,k} c_j c_k' \langle \gamma_j L_j S_j J \| \mathbf{T}^{(1)} \| \gamma_k L_k S_k J' \rangle \tag{8.80}
$$

$$
\langle \gamma J \| \mathbf{T}^{(2)} \| \gamma J \rangle = \sum_{j,k} c_j c_k' \langle \gamma_j L_j S_j J \| \mathbf{T}^{(2)} \| \gamma_k L_k S_k J' \rangle, \tag{8.81}
$$

† It should be noted that the wave functions (8.79) need not result from Breit–Pauli calculations. Instead, we may formally couple the L and S quantum numbers from the CSFs of an ordinary non-relativistic calculation to yield a J quantum number in which case we have

$$
\Psi(\gamma LSJ) = \sum_j c_j \Phi(\gamma_j LSJ) \,.
$$

and the evaluation reduces to the evaluation of reduced matrix elements between LSJ coupled CSFs. Using the 'uncoupling' formulae (A.56), (A.57) and (A.58) for the J quantum number the orbital, spin-dipole, Fermi contact and quadrupole terms of the matrix elements can be written in terms of matrix elements between LS coupled CSFs which in turn, using the program TENSOR (Robb 1973), can be reduced to a sum of one-particle matrix elements.

8.13 Configuration expansions for hyperfine structure

The hyperfine interaction is described by one-particle operators and, in the first approximation, only configurations generated by single orbital replacements from the configurations in the complex need to be included in the wave function expansion. In a many-electron atom orbital replacements may be made either from closed subshells defining the core or from valence subshells. CSFs generated by single replacements from the closed subshells represent a distortion, or *polarization*, of the core due to the electrostatic interaction with the valence electrons. In the next sections we will discuss the effects of the core polarization on the hyperfine structure.

8.14 Polarization effects

In the central-field approximation the charge distribution within a closed subshell is spherically symmetric. No electric or magnetic field is generated at the nucleus, and the contributions to the magnetic and electric hfs are zero. The contribution from a closed s-subshell is also zero, since the spin-densities from the two electrons are equal and the spin directions opposite. Polarization of the closed shells in the core due to the Coulomb interaction with open subshells can have a large effect on the hyperfine structure. The Coulombic exchange interaction reduces the repulsion between core and valence electrons with the same spin orientation, pulling the core electron towards the valence subshell. Especially, polarization of s-subshells is important. If the two s-electrons in the same shell have different spin-densities at the nucleus, a contact interaction is induced. Since inner s-electrons have high densities at the nucleus, a very small imbalance is sufficient to cause a net interaction which is comparable to that of an open valence subshell. To clarify this, we consider the $1s^2 2p\ ^2P$ state and look at an approximate wave function in terms of a single CSF with the magnetic quantum numbers $M_L = 1$ and $M_S = 1/2$. This wave function can be written as a Slater determinant

$$\Phi(1s^2 2p\ ^2P) = \frac{1}{\sqrt{6}} \begin{vmatrix} \phi(\alpha_1; q_1) & \phi(\alpha_1; q_2) & \phi(\alpha_1; q_3) \\ \phi(\alpha_2; q_1) & \phi(\alpha_2; q_2) & \phi(\alpha_2; q_3) \\ \phi(\alpha_3; q_1) & \phi(\alpha_3; q_2) & \phi(\alpha_3; q_3) \end{vmatrix}, \quad (8.82)$$

where $\alpha_1 = \{1s\ m_l = 0\ m_s = 1/2\}$, $\alpha_2 = \{1s\ m_l = 0\ m_s = -1/2\}$ and $\alpha_3 = \{2p\ m_l = 1\ m_s = 1/2\}$. Performing some algebra, the electrostatic energy may be written as a sum over direct and exchange contributions. However, exchange contributions are non-zero only if the orbitals have the same spin, namely α_1 and α_3. Thus, there is an asymmetry in the energy expression for the two s-orbitals. At the cost of increasing complexity, this asymmetry can be taken into account in the unrestricted Hartree–Fock (UHF) method, where radial functions associated with the two spin directions of the s-orbitals are allowed to be different. Solving the unrestricted Hartree–Fock equations, it is seen that the s orbital with $m_s = 1/2$ is pulled towards the valence shell leading to a net contact interaction for the closed $1s$ shell. However, it should be pointed out that when the two $1s$ orbitals are not equivalent the single determinant no longer is an eigenfunction of the total spin and the wave function is said to be spin contaminated.

The spin-polarization can also be described in the MCHF approach by adding a CSF of the form $\Phi(1s2s(^3S)2p\ ^2P)$ to yield an expansion

$$\Psi(1s^22p\ ^2P) = c_1\Phi(1s^22p\ ^2P) + c_2\Phi(1s2s(^3S)2p\ ^2P). \tag{8.83}$$

The asymmetry now enters through the expansion coefficient c_2 and the shape of the orbital $2s$. Note that the configuration $\Phi(1s2s(^1S)2p\ ^2P)$ may be omitted since, according to Brillouin's theorem, it is already included to first order. Although the configuration $\Phi(1s2s(^3S)2p\ ^2P)$ has a very small weight ($c_2 = -0.0033546$) and decreases the energy by only 0.000 024 54 au. the effect on the A factor, as can be seen from table 8.4, is very large.

In addition to the spin-polarization there is an *orbital polarization*, which is due to the interaction of the core with the non-spherical charge distribution of the valence orbital. This interaction distorts the spherical symmetry of the core leading to additional contributions also to the orbital, spin-dipolar and quadrupole terms. In the MCHF approach the orbital polarization can be

Table 8.4. Polarization contributions to the hyperfine interaction in the $1s^22p\ ^2P_{1/2,3/2}$ states of ^{7}Li (in MHz). In the polarization calculations the $1s$ and $2s$ orbitals were frozen and taken from the HF calculation.

Method	$A(^2P_{1/2})$	$A(^2P_{3/2})$	$B(^2P_{3/2})/Q$ (MHz barn^{-1})
HF	32.37	6.47	5.50
Spin polarization	9.15	−9.15	
Orbital polarization	1.13	−0.11	−0.65
Total	42.65	−2.79	4.85
UHF[a]	43.1	−4.2	
Experiment[b]	45.914(25)	−3.055(14)	

[a]Larsson (1970).
[b]Orth *et al* (1975).

described by including in the wave function expansion CSFs obtained single orbital substitutions from the core followed by a change in angular coupling. In the example of $1s^22p\ ^2P$ the CSFs $\{1s3d_1(^3D)2p\ ^2P, 1s3d_2(^1D)2p\ ^2P\}$ describe the orbital polarization and should be included in the expansion. Note that the $3d_1$ and $3d_2$ orbitals are non-orthogonal. Looking at table 8.4 it can be seen that the orbital polarization has an appreciable effect on the electric quadrupole constant. The effect on the magnetic dipole constant is, however, small compared to the spin-polarization contribution.

8.14.1 Hyperfine structure of the alkali atoms

As an example of hyperfine-structure calculations we will look at the spin and orbital polarization contributions for the $3p\ ^2P$ states in ^{23}Na. As the first step we perform a HF calculation for $3p\ ^2P$. The radial orbitals from the HF calculation are then used as the input for the polarization calculation. For the latter 14 CSFs are included. These CSFs are shown in table 8.5. Note that $1s1s'(^1S)2s^22p^63p\ ^2P$, $1s^22s2s'(^1S)2p^63p\ ^2P$ and $1s^22s^22p^52p'(^1S)3p\ ^2P$ are omitted from the expansion according to the GBT theorem. Further it should be noted that in polarization calculations non-orthogonal orbitals can be used extensively. In the present calculation one radial orbital is introduced for each coupling. In the MCHF calculation the HF orbitals are kept frozen, and only the polarization orbitals are optimized. The subsequent use of HFS to obtain the diagonal hyperfine interaction constants $A(J, J)$ and $B(J, J)$ is displayed in table 8.6. This run assumes that the results of the MCHF calculations have been stored in 2P.c and 2P.w. The output file 2P.h is shown in table 8.7. The output file displays the contributions (larger than the tolerance 1 MHz) to the hyperfine-interaction constants for all pairs of CSFs. At the end of the file the total interaction constants are shown together with the hyperfine parameters a_l, a_{sd}, a_c and b_q. Looking at the contribution from pairs of CSFs it is seen that the spin-polarization of the $1s$ subshell gives† 2×1.21876 MHz for $J = 1/2$ (the sign is reversed for the $J = 3/2$ contribution). This is to a large part counteracted by the spin-polarization contribution 2×-1.890566 MHz of the $2s$ subshell leading to a small total value for spin-polarization. For $3p\ ^2P$ the important effect is instead the orbital-polarization of the $2p$ subshell. Here $1s^22s^22p^52p_1(^1P)3p\ ^2P$, $1s^22s^22p^52p_2(^1D)3p\ ^2P$ and $1s^22s^22p^52p_5(^3D)3p\ ^2P$ give the major contributions to, respectively, the orbital term, the quadrupole term and the spin-dipole term.

In an HFS run with a print-out tolerance of 0 MHz, contributions from all pairs of CSFs will appear. In general the contributions from matrix elements between CSFs describing polarization of two different core subshells is negligible. In table 8.5 these would be of the type $\langle 1snl2s^22p^63p\ ^2P|O|1s^22sn'l'2p^63p\ ^2P\rangle$. This makes it possible to sum all the

---

† Only matrix elements $\langle i|O|j\rangle$, $i \geqslant j$, where O is the relevant hyperfine operator, are written to the hfs output file. Since $\langle i|O|j\rangle = \langle j|O|i\rangle$, an off-diagonal contribution should be counted twice.

Table 8.5. CSF expansion for spin- and orbital-polarization calculation for $3p\ ^2P$ in Na.

```
1s( 2)   2s( 2)   2p( 6)   3p( 1)
  1S0      1S0      1S0      2P1      1S0    1S0    2P0
1s( 1) 1s1( 1)   2s( 2)   2p( 6)   3p( 1)
  2S1      2S1      1S0      1S0      2P1    3S0    3S0    3S0    2P0
1s( 1) 3d1( 1)   2s( 2)   2p( 6)   3p( 1)
  2S1      2D1      1S0      1S0      2P1    1D0    1D0    1D0    2P0
1s( 1) 3d2( 1)   2s( 2)   2p( 6)   3p( 1)
  2S1      2D1      1S0      1S0      2P1    3D0    3D0    3D0    2P0
1s( 2)   2s( 1) 2s2( 1)   2p( 6)   3p( 1)
  1S0      2S1      2S1      1S0      2P1    2S0    3S0    3S0    2P0
1s( 2)   2s( 1) 3d3( 1)   2p( 6)   3p( 1)
  1S0      2S1      2D1      1S0      2P1    2S0    1D0    1D0    2P0
1s( 2)   2s( 1) 3d4( 1)   2p( 6)   3p( 1)
  1S0      2S1      2D1      1S0      2P1    2S0    3D0    3D0    2P0
1s( 2)   2s( 2)   2p( 5) 2p1( 1)   3p( 1)
  1S0      1S0      2P1      2P1      2P1    1S0    2P0    1P0    2P0
1s( 2)   2s( 2)   2p( 5) 2p2( 1)   3p( 1)
  1S0      1S0      2P1      2P1      2P1    1S0    2P0    1D0    2P0
1s( 2)   2s( 2)   2p( 5) 2p3( 1)   3p( 1)
  1S0      1S0      2P1      2P1      2P1    1S0    2P0    3S0    2P0
1s( 2)   2s( 2)   2p( 5) 2p4( 1)   3p( 1)
  1S0      1S0      2P1      2P1      2P1    1S0    2P0    3P0    2P0
1s( 2)   2s( 2)   2p( 5) 2p5( 1)   3p( 1)
  1S0      1S0      2P1      2P1      2P1    1S0    2P0    3D0    2P0
1s( 2)   2s( 2)   2p( 5) 4f1( 1)   3p( 1)
  1S0      1S0      2P1      2F1      2P1    1S0    2P0    1D0    2P0
1s( 2)   2s( 2)   2p( 5) 4f2( 1)   3p( 1)
  1S0      1S0      2P1      2F1      2P1    1S0    2P0    3D0    2P0
```

Table 8.6. Hyperfine-structure calculation of the $3p\ ^2P_{1/2,3/2}$ states in ^{23}Na.

```
>Hfs
 Name of state
>2P
 Indicate the type of calculation
 0 => diagonal A and B factors only;
 1 => diagonal and off-diagonal A and B factors;
 2 => coefficients of the radial matrix elements only:
>0
 Hyperfine parameters al,ad,ac and bq ? (Y/N)
>y
 Input from an MCHF (M) or CI (C) calculation ?
>m
 Full print-out ? (Y/N)
>y
 Tolerance for printing (in MHz)
>1
 Give 2*I and nuclear dipole and quadrupole moments
 (in n.m. and barn)
>3,2.217656,1
```

Table 8.7. Listing of output file 2P.h.

```
                    Hyperfine structure calculation
2P.c
Tolerance for printing      1.000000 MHz
Nuclear dipole moment       2.217656 n.m.
Nuclear quadrupole moment   1.000000 barns
2*I=   3
  Na    2P      -161.7864854
(config   1|hfs|config   1 )
Orbital term
matrix element      < 3p|R**(-3)| 3p>
  J    J'   weight   coeff       matrix element values  Aorb(MHz)
 1/2  1/2  .999966 2.666667     .1692                  31.830763
 3/2  3/2  .999966 1.333333     .1692                  15.915382
Spin-dipole term
matrix element      < 3p|R**(-3)| 3p>
  J    J'   weight   coeff       matrix element values  Adip(MHz)
 1/2  1/2  .999966 2.669759     .1692                  31.867671
 3/2  3/2  .999966 -.266976     .1692                  -3.186767
Quadrupole term
matrix element      < 3p|R**(-3)| 3p>
  J    J'   weight   coeff       matrix element values   B(MHz)
 3/2  3/2  .999966 .400000      .1692                  15.906368
(config   2|hfs|config   1 )
Fermi-contact term
matrix element      AZ(1s1)AZ( 1s)
  J    J'   weight   coeff    matrix element values Acont(MHz)
 1/2  1/2  -.000017 -.363308 2831.2052                1.218760
 3/2  3/2  -.000017  .363308 2831.2052               -1.218760
    ....
                            **
                    A factors in MHz
  J      J'       Orbital        Spin-dip       Cont        Total
 1/2    1/2      40.06185       41.71371     -1.35356    80.42200
 3/2    3/2      20.03093       -4.17137      1.35356    17.21311
                    B factors in MHz
  J      J'       Quadrupole
 1/2    1/2        .00000
 3/2    3/2      23.78529
                            **
                    Hyperfine parameters in a.u.
                    al            ad            ac          bq
                  .21301       -.04431       .04313     -.10123
                            **
```

Table 8.8. The effect of spin- and orbital-polarization for the $2s\ ^2S_{1/2}$ and $2p\ ^2P_{1/2,3/2}$ states in ^{7}Li (in MHz).

Method	$A(^2S_{1/2})$	$A(^2P_{1/2})$	$A(^2P_{3/2})$	$B(^2P_{3/2})/Q$
HF	289.55	32.37	6.47	5.50
$1s$	89.55	10.28	−9.26	−0.65
Total	379.10	42.65	−2.79	4.85
MCHF full correlation[a]	401.71	45.94	−3.098	5.30
Exp[b]	401.75	45.91	−3.055	

[a]Carlsson *et al* (1992a).
[b]Beckmann *et al* (1974), Orth *et al* (1975).

Table 8.9. The effect of spin- and orbital-polarization for the $3s\ ^2S_{1/2}$ and $3p\ ^2P_{1/2,3/2}$ states in ^{23}Na (in MHz).

Method	$A(^2S_{1/2})$	$A(^2P_{3/2})$	$A(^2P_{3/2})$	$B(^2P_{3/2})/Q$
HF	626.64	63.70	12.73	15.91
$1s$	40.32	3.19	−2.51	−0.29
$2s$	94.07	−3.66	3.78	−0.31
$2p$		17.19	3.19	8.46
Total	761.03	80.42	17.21	23.77
MCHF full correlation[a]	882.2	94.04	18.80	25.79
Exp[b]	885.81	94.42	18.69	

[a]Jönsson *et al* (1996).
[b]Beckman *et al* (1974), Carlsson *et al* (1992b), Carlsson (1988).

Table 8.10. The effect of spin- and orbital-polarization for the $4s\ ^2S_{1/2}$ and $4p\ ^2P_{1/2,3/2}$ states in ^{39}K (in MHz).

Method	$A(^2S_{1/2})$	$A(^2P_{3/2})$	$A(^2P_{3/2})$	$B(^2P_{3/2})/Q$
HF	145.01	16.42	3.28	23.22
$1s$	4.86	0.39	−0.29	−0.15
$2s$	7.50	−0.15	0.15	−0.18
$2p$		1.06	0.20	2.12
$3s$	20.16	−1.02	1.02	−0.16
$3p$		3.65	0.68	12.29
Total	177.53	20.35	5.04	37.14
Exp[a]	230.86	28.85	6.09	

[a]Arimondo *et al* (1977).

different polarization contributions from each subshell in the core. In tables 8.8, 8.9 and 8.10 the contributions from the individual core shells are shown for the lowest ^2S and ^2P states in ^{7}Li, ^{23}Na and ^{39}K. Due to the restrictions on the number of open subshells in the MCHF package, two calculations had to be done for K. In the first calculation the $2p$, $3s$ and $3p$ subshells were closed, and the polarization contributions were determined for the $1s$ and $2s$ subshells. In the second calculation the $1s$ and $2s$ subshells were closed, and contributions were determined from the $2p$, $3s$ and $3p$ subshells. From the tables it is seen that these small polarization calculations give substantial improvements compared to the HF values.

8.15 Exercises

(i) Show that

$$J(nl, n'(l-1)) = -J(n'(l-1), nl)$$
$$= \int r^2 R(nl; r) \left(\frac{\mathrm{d}}{\mathrm{d}r} - \frac{l-1}{r} \right) R(n'l'; r) \, \mathrm{d}r.$$
(8.84)

(ii) Run the NONH program for the expansion $\Psi = c_1 \Phi(1s2p\ ^3P) + c_2 \Phi(2p3d\ ^3P)$, requesting a full printout, and verify that the $k = 1$ term of the electrostatic interaction is given by (8.24).

(iii) Assume that the nuclear charge Z is uniformly distributed in a sphere with radius R. Show that the generated potential is given by

$$V(r) = \begin{cases} \dfrac{Z}{2R} \left(\dfrac{r^2}{R^2} - 3 \right) & r \leqslant R \\[2mm] -\dfrac{Z}{r} & r > R. \end{cases}$$
(8.85)

Treat the difference between this potential and the Coulomb potential as a small perturbation and show that the energy difference for a single s electron ψ is given by

$$\Delta E \simeq \tfrac{2}{5} \pi Z R^2 |\psi(0)|^2.$$
(8.86)

(iv) Perform active set valence correlation calculations for $2s^2\ ^1S$ and $2s2p\ ^1P$ in B II. Monitor the convergence of $S(2s^2\ ^1S)$, $S(2s2p\ ^1P)$ and $S(2s2p\ ^1P) - S(2s^2\ ^1S)$ as the active set is increased. Compare the obtained values with values from a full correlation calculation (Jönsson *et al* 1994). Finally calculate the ^{11,10}B transition isotope shift in $2s^2\ ^1S - 2s2p\ ^1P$.

(v) Run the hyperfine code for $1s2p\ ^3P_{J=1}$ and $1s2p\ ^1P_1$ states of Li II. Which has the larger hyperfine interaction constant?

(vi) Perform HF and spin-polarization calculations of the hyperfine interactions constants in $2s\ ^2S$, $3s\ ^2S$ and $4s\ ^2S$ states of ^{7}Li. How does the relative importance of the spin-polarization change when going from $2s\ ^2S$ to $4s\ ^2S$?

(vii) Perform HF and polarization calculations of the diagonal and off-diagonal hyperfine interactions constants in the $2s2p$ ^3P states of ^{9}Be. Use $I = 3/2$, $\mu_I = -1.177492$ and $Q = 1$. The experimental constants are $A(^3P_1) = -139.373(12)$ MHz, $A(^3P_2) = -124.5368(17)$ MHz, $A(^3P_2,^3P_1) = 16.44(40)$ MHz. How large are the errors for the two calculations. The experimental value for the electric-quadrupole interaction constant is $B(^3P_2) = 1.429(8)$ MHz. From the result of the spin-polarization calculation determine a preliminary value of Q.

(viii) Perform HF and polarization calculations of the diagonal hyperfine interactions constants in the $3s3p$ ^3P states of ^{24}Mg. Use $I = 3/2$, $\mu_I = -1.177492$ and $Q = 1$. Determine the polarization contributions from each core shell. Which shell gives the largest contribution for the different interaction constants?

(ix) Which of the two states $3s4p$ ^3P and $4s3p$ ^3P has the largest hyperfine interaction constants? Check numerically for ^{24}Mg.

(x) Perform HF and polarization calculations of the hyperfine interactions constant in the $2p^3$ $^4S_{3/2}$ state of ^{14}N. Determine the polarization contributions from the $1s$ and $2s$ subshells and explain why the hyperfine interaction constant is so difficult to evaluate for this state.

Chapter 9

Allowed and Forbidden Transitions

9.1 Introduction

So far we have treated the energy levels of an atomic system as stable states with infinite lifetimes. The presence of an electromagnetic field changes this picture. Absorption of photons will excite atoms and ions to higher levels, while stimulated emission will make excited ones decay. Einstein (1917) showed, using purely statistical arguments, that an excited ion will decay through *spontaneous emission* in the absence of an electromagnetic field. This process can also be derived by using a rigorous quantum electrodynamic treatment, where the field is quantized. In this chapter we start by discussing transitions between atomic levels in a general sense, postulating the operators responsible for the decay. After the first few sections we will concentrate on the most important few multipoles or terms in the expansion of the electromagnetic field. We will introduce a number of different properties describing the transition, and eventually arrive at the measurable transition rate and, directly linked to it, the lifetime of an excited state. We will start, however, with the most basic property, the component strength. Our aim is to arrive at computationally useful expressions as soon as possible. For a more thorough treatment of the theory of radiative transitions we refer to the book by Sobelman (1972). For a brief review of the many factors affecting spectra of atoms, see Martin and Wiese (1996).

9.1.1 Operators

An electromagnetic transition between two states is characterised by the angular momentum and parity of the corresponding photon. If the emitted or absorbed photon has angular momentum k and parity $\pi = (-1)^k$ then, following the terminology of classical electrodynamics, the transition is an electric multipole (Ek) transition. If, instead, the photon has parity $\pi = (-1)^{k+1}$ the transition is a magnetic multipole (Mk) transition. Each multipole is described by a transition operator $O_q^{\pi(k)}$ which is a spherical tensor operator of rank k and parity π. For electric and magnetic transitions, respectively, it has the form

$$E_q^{(k)} = \sum_{i=1}^{N} r^k(i) C_q^{(k)}(i), \tag{9.1}$$

or

$$M_q^{(k)} = \alpha\sqrt{k(2k-1)} \left[\frac{1}{k+1} MA_q^{(k)} + \frac{1}{2} g_s MB_q^{(k)} \right], \tag{9.2}$$

where

$$MA_q^{(k)} = \sum_{i=1}^{N} r^{k-1}(i) \left[\mathbf{C}^{(k-1)}(i) \times \mathbf{l}^{(1)}(i) \right]_q^{(k)} \tag{9.3}$$

and

$$MB_q^{(k)} = \sum_{i=1}^{N} r^{k-1}(i) \left[\mathbf{C}^{(k-1)}(i) \times \mathbf{s}^{(1)}(i) \right]_q^{(k)}. \tag{9.4}$$

9.1.2 Transition properties

To describe a transition that occurs between an upper state, $\gamma'J'M'$ and a lower state γJM we start with the most fundamental property, the *transition integral*

$$I_q^{\pi k}(\gamma JM, \gamma'J'M') = \langle \gamma JM | O_q^{\pi(k)} | \gamma'J'M' \rangle \tag{9.5}$$

and we define the *component strength* as

$$s^{\pi k}(\gamma JM, \gamma'J'M') = \sum_{q} \left| I_q^{\pi k}(\gamma JM, \gamma'J'M') \right|^2. \tag{9.6}$$

We will deal here only with systems where the energy levels are degenerate in the M quantum numbers, and the observable quantity is therefore the sum of the component strengths over these quantum numbers. The resulting property is the *line strength*:

$$S^{\pi k}(\gamma J, \gamma'J') = \sum_{M,M',q} \left| \langle \gamma JM | O_q^{\pi(k)} | \gamma'J'M' \rangle \right|^2. \tag{9.7}$$

From the Wigner–Eckart theorem (A.45) and the orthogonality relations (A.19) it is seen that the line strength is the square of the reduced transition matrix element, that is,

$$S^{\pi k}(\gamma'J', \gamma J) = \left| \langle \gamma J \| \mathbf{O}^{\pi(k)} \| \gamma'J' \rangle \right|^2. \tag{9.8}$$

The *transition rate* (or probability) for emission from the upper level to the lower level is given by

$$A^{\pi k}(\gamma'J', \gamma J) = 2C_k \left[\alpha \left(E_{\gamma'J'} - E_{\gamma J} \right) \right]^{2k+1} \frac{S^{\pi k}(\gamma'J', \gamma J)}{g_{J'}} \tag{9.9}$$

where $g_{J'}$ is the statistical weight of the upper level, namely

$$g_{J'} = 2J' + 1 \tag{9.10}$$

and

$$C_k = \frac{(2k + 1)(k + 1)}{k \left((2k + 1)!!\right)^2}. \tag{9.11}$$

The *oscillator strength* may refer to transition either in absorption or emission. The unmodified phrase implies the *absorption* oscillator strength where an atom in the lower state absorbs a photon and is excited to an upper state. In this case,

$$f^{\pi k}(\gamma J, \gamma' J') = \frac{1}{\alpha} C_k \left[\alpha \left(E_{\gamma' J'} - E_{\gamma J}\right)\right]^{2k-1} \frac{S^{\pi k}(\gamma J, \gamma' J')}{g_J}, \tag{9.12}$$

A similar expression applies to the *emission* oscillator strength where $\gamma' J'$ and γJ are interchanged, making the emission oscillator strength negative.

Of special importance is the weighted oscillator strength, or the *gf-value*. This property is, like the line strength, completely symmetrical (except for sign) between the two levels. It is given by

$$gf^{\pi k}(\gamma J, \gamma' J') = g_J f^{\pi k}(\gamma J, \gamma' J'). \tag{9.13}$$

The intensity of the spectral line, produced by the $\gamma' J' \rightarrow \gamma J$ transition, is proportional to the line strength and thus also to the gf-value.

Neither of these 'single line' properties are easy to measure. Most experiments yield the lifetime of the upper level. In this case we have to sum over multipole transitions to all lower lying levels. The lifetime, $\tau_{\gamma' J'}$, of level $\gamma' J'$ is

$$\tau_{\gamma' J'} = \frac{1}{\sum_{\pi k, \gamma J} A^{\pi k}(\gamma' J', \gamma J)}. \tag{9.14}$$

The relative intensity of lines originating from the same upper level, $\gamma' J'$, can be derived from the *branching ratio*, Q. For a transition from $\gamma' J'$ to γJ, this is defined as

$$Q = \tau_{\gamma' J'} \sum_{\pi k} A^{\pi k}(\gamma' J', \gamma J). \tag{9.15}$$

By using (9.9) and (9.2), we get

$$\begin{aligned} A^{Ek} &\propto \alpha^{2k+1} \\ A^{Mk} &\propto \alpha^{2k+3} \end{aligned} \tag{9.16}$$

which gives the relative size of the rate for the different multipole transitions. It is clear therefore that the largest transition rate will, in general, be for electric dipole (E1) radiation, dominating by at least a factor $1/\alpha^2$ over other types of transitions. For this reason E1 transitions are labelled *allowed* whereas higher order electric and all magnetic transitions are referred to as *forbidden*.

9.2   Matrix elements for transition operators

In the multiconfiguration approximation the wave functions are given by expansions

$$\Psi(\gamma J) = \sum_i c_i \Phi(\gamma_i L_i S_i J), \tag{9.17}$$

and the general reduced transition matrix element can be written

$$\langle \gamma J || \mathbf{O}^{\pi(k)} || \gamma' J' \rangle = \sum_{ij} c_i c_j' \langle \gamma_i L_i S_i J || \mathbf{O}^{\pi(k)} || \gamma_j' L_j' S_j' J' \rangle, \tag{9.18}$$

that is, as a sum of reduced matrix elements between CSFs. The line strength itself is the square of this sum of terms, some of which may be positive whereas others are negative, making possible both line strength enhancement and cancellation.

To evaluate the matrix elements between the LSJ coupled CSFs we start by recoupling J. The transition operators have two types of angular dependence (Robb 1973, 1975, Froese Fischer *et al* 1991). The entire electric multipole operator, $\mathbf{E}^{(k)}$, and the first term of the magnetic multipole operator, $\mathbf{MA}^{(k)}$, are spin independent and reduce in the same way†

$$\langle \gamma L S J || \mathbf{O}^{\pi(k)} || \gamma' L' S' J' \rangle$$

$$= \delta_{SS'} (-1)^{L+S+J'+k} \sqrt{[J, J']} \left\{ \begin{matrix} L & S & J \\ J' & k & L' \end{matrix} \right\} \langle \gamma L S || \mathbf{O}^{\pi(k)} || \gamma' L' S' \rangle. \tag{9.19}$$

The second term of the magnetic multipole operator, $\mathbf{MB}^{(k)}$, reduces according to

$$\langle \gamma L S J) || \mathbf{O}^{\pi(k)} || \gamma' L' S' J' \rangle$$

$$= \sqrt{[J, J', k]} \left\{ \begin{matrix} L & S & J \\ L' & S' & J' \\ k-1 & 1 & k \end{matrix} \right\} \langle \gamma L S || \mathbf{O}^{\pi(k)} || \gamma' L' S' \rangle \tag{9.20}$$

where in the $9j$-symbol $k-1$, 1 and k indicate the rank of the $\mathbf{MB}^{(k)}$ tensor in, respectively, spatial, spin and total space.

The transition operators are all one-electron operators,

$$\mathbf{O}^{\pi(k)} = \sum_{i=1}^{N} \mathbf{o}^{\pi(k)}(i), \tag{9.21}$$

† The indices i and j are suppressed for brevity.

and thus the LS reduced matrix elements can be written as sums over one-electron matrix elements. Two cases need to be considered. If the CSFs are identical then

$$\langle \gamma LS || \mathbf{O}^{\pi(k)} || \gamma LS \rangle = \sum_{nl} v(nl, nl) \langle nl || \mathbf{o}^{\pi(k)} || nl \rangle. \tag{9.22}$$

If instead the two CSFs differ by one orbital pair $(nl, n'l')$ then

$$\langle \gamma LS || \mathbf{O}^{\pi(k)} || \gamma' L'S' \rangle = v(nl, n'l') \langle nl || \mathbf{o}^{\pi(k)} || n'l' \rangle. \tag{9.23}$$

Here $v(nl, n'l')$ are angular coefficients and $\langle nl || \mathbf{o}^{\pi(k)} || n'l' \rangle$ one-electron reduced matrix elements. The latter are also frequently referred to as *transition integrals*.

The formulae above assume a single orthonormal orbital basis for both states. However, when wave functions are optimized independently, orbitals of the initial state will be non-orthogonal to those of the final state. The present codes allow for a limited amount of non-orthogonality (Hibbert *et al* 1988). The expressions for the matrix elements are quite similar to the orthonormal case, but overlap integrals may be present. We will discuss the restrictions later, but basically the transition codes can handle all two-electron systems with completely independent orbitals for valence correlation but a common core, and most three-electron systems with a common core. We will come back to this restriction after the next section.

9.3 Selection rules for radiative transitions

We can now deduce the selection rules for matrix elements of different multipole operators. We need to remember that the matrix element between two atomic state functions is given by the sum in (9.18), and that the selection rule for the sum can be different from the selection rules of the individual matrix elements between the CSFs. Let us therefore distinguish between *rigorous selection rules* that are valid for the whole CSF expansion, and *approximate selection rules* that are valid in the presence of negligible configuration interaction.

9.3.1 Rigorous selection rules

Rigorous selection rules apply to *all* CSF pairs. We know that all CSFs in an expansion for a given ASF have the same parity. So it is not surprising that the first rule relates to the parity of the transition operators. For the electric operators, parity is given by $(-1)^k$, while for the magnetic operators we have $(-1)^{k-1}$. If we denote the parity of the two states π and π' and consider $\frac{\pi'}{\pi}$, then

$$
\begin{aligned}
\mathbf{E}^{(k)}: & \quad \frac{\pi'}{\pi} = (-1)^k \\
\mathbf{M}^{(k)}: & \quad \frac{\pi'}{\pi} = (-1)^{k-1}.
\end{aligned}
\tag{9.24}
$$

For example, the electric dipole operator, $\mathbf{E}^{(1)}$, combines states of different parity, while the magnetic dipole, $\mathbf{M}^{(1)}$, and electric quadrupole, $\mathbf{E}^{(2)}$, combine states of equal parity.

The only other property that is common for all CSFs in an expansion for a given ASF is the total J-value. From the $6j$-symbol of (9.19) and the $9j$-symbol of (9.20) it is clear that all multipole operators give the selection rule

$$\Delta J = J - J' = 0, \pm 1, \ldots, \pm k \quad \text{and} \quad J + J' \geqslant k, \tag{9.25}$$

with the restriction that

$$J = J' = 0 \quad \text{is not allowed.} \tag{9.26}$$

9.3.2 Approximate selection rules

Other selection rules follow from a consideration of the J-independent reduced matrix elements of (9.22) and (9.23). Let us consider the rules on LS which lead to non-zero values of $v(nl, n'l')$ (or $v(nl, nl)$).

We can classify different angular momenta of a CSF as active or passive, depending on whether they 'participate' in the transition. Passive momenta will not change, while the active ones will obey a rule as in (9.25). The first rule to consider depends on the rank of the different operators, with respect to spatial and spin space. If we start with spin it is clear that the $\mathbf{E}^{(k)}$ operator is spin independent, and spins are always passive for electric multipole transitions. Hence the selection rule may be summarized as

$$\mathbf{E}^{(k)}: \quad \Delta S = 0. \tag{9.27}$$

The same selection rule is valid for the $\mathbf{MA}^{(k)}$ operator. The $\mathbf{MB}^{(k)}$ operator, however, involves a spin-operator of rank 1, and magnetic multipole transitions therefore can link CSFs that differ in spin by 1.

To deduce the selection rules for space angular momenta, we note that the corresponding tensor in the $\mathbf{E}^{(k)}$ operator is of rank k so that the selection rules are

$$\mathbf{E}^{(k)}: \quad \Delta L = 0, \pm 1, \ldots, \pm k; \qquad L + L' \geqslant k. \tag{9.28}$$

For the magnetic transitions we have to exercise a bit more care, since the space tensor is of rank k in $\mathbf{MA}^{(k)}$ and rank $k - 1$ in $\mathbf{MB}^{(k)}$. The selection rules for spin and space momenta are therefore linked

$$\mathbf{MA}^{(k)}: \quad \Delta S = 0 \qquad \Delta L = 0, \pm 1, \ldots, \pm k; \qquad L + L' \geqslant k$$
$$\mathbf{MB}^{(k)}: \quad \Delta S = 0, \pm 1 \quad \Delta L = 0, \pm 1, \ldots, \pm(k-1); \quad L + L' \geqslant k - 1. \tag{9.29}$$

Since the transition operator is a one-electron operator, it is clear from (9.23) that only one electron can have active angular momenta, that is only one of all the l quantum numbers obeys rules as in (9.29), while the rest are passive and unchanged. All angular momenta that are built from the angular momenta of this 'jumping electron' are also active, while all other intermediate ones are passive. All this is best understood from an example.

Let us look at transitions in nitrogen-like ions, where the ground term is $2p^3\ ^3P$. The first excited configurations are of the form $2p^2(\tilde{L}\tilde{S})nl$, where $\tilde{L}\tilde{S}$ denotes an intermediate coupling, which in this case is either ^3P, ^1D or ^1S. An electric dipole transition of the form

$$2p^2(\tilde{L}\tilde{S})nl\ LSJ - 2p^2(\tilde{L}'\tilde{S}')n'l'\ L'S'J'$$

makes the angular momenta l, l', L, L' and J, J' active, while $\tilde{L}, \tilde{L}', \tilde{S}, \tilde{S}', S, S'$ and all other one-electron momenta are passive. The selection rules therefore state that the parent term, represented by the characters with a tilde, will be conserved.

The discussion so far has been quite general, and the selection rules in (9.24), (9.25), (9.27), (9.28) and (9.29) are correct for all multipoles. The remaining selection rules pertain to the one-electron reduced matrix elements which depend on the type of transition and the 'jumping electron'.

(i) Electric multipole

The one-electron reduced matrix elements for electric multipole radiation are

$$\langle nl||r^k\mathbf{C}^{(k)}||n'l'\rangle = \langle l||\mathbf{C}^{(k)}||l'\rangle \int_0^\infty r^k P(nl; r) P(n'l'; r)\, \mathrm{d}r. \tag{9.30}$$

The reduced matrix element $\langle l||\mathbf{C}^{(k)}||l'\rangle$ introduces the triangle rule for l, l' and k. Thus for $k = 1$ we have $\Delta l = \pm 1$ whereas for $k = 2$ we have $\Delta l = 0, \pm 2$, with $l + l' \geqslant k$. These are no rules on Δn. These one-electron rules are the same as those for the total angular momenta of (9.28).

(ii) Magnetic multipole

The general rules in this case are the same as those for the total angular momenta. However, the magnetic dipole, $\mathbf{M}^{(1)}$, is of special interest. This transition operator simplifies to

$$\mathbf{M}^{(1)} = \sum_{i=1}^N \frac{\alpha}{2}\left[\mathbf{l}^{(1)}(i) + g_s\mathbf{s}^{(1)}(i)\right], \tag{9.31}$$

which gives

$$\mathbf{M}^{(1)} = \frac{\alpha}{2}\left[\mathbf{J}^{(1)} + (g_s - 1)\,\mathbf{S}^{(1)}\right]. \tag{9.32}$$

The reduced matrix elements on the right sides of (9.19) and (9.20) then reduce to simple expressions which give the selection rules

$$\Delta S = \Delta L = \Delta \gamma = 0. \tag{9.33}$$

To first order, magnetic dipole transitions only occur between fine-structure levels of the same term. The total reduced matrix element is

$$\langle \gamma L S J || \mathbf{M}^{(1)} || \gamma L S J' \rangle$$

$$= \frac{\alpha}{2} \left[\delta_{JJ'} \left(J'(J'+1)(2J'+1) \right)^{1/2} + (-1)^{L+S+J+1}[J, J']^{1/2} \right.$$

$$\left. \times \left\{ \begin{array}{ccc} L & S & J \\ 1 & J' & S \end{array} \right\} (S(S+1)(2S+1))^{1/2} \right].$$

(9.34)

9.3.3 Summary of selection rules

Thus the selection rules are determined by:

(i)   Rigorous rules that apply to the atomic state functions as a whole.
(ii) Approximate rules that apply to pairs of CSFs. These arise in two ways:
  (a) The expressions for $v(nl, n'l')$.
  (b) The one-electron reduced matrix elements $\langle nl || o^{\pi(k)} || n'l' \rangle$.

For the three most frequently occurring transitions, these rules are summarized in table 9.1.

Table 9.1. Selection rules for some allowed and forbidden transition.

E1 (allowed)	M1 (forbidden)	E2 (forbidden)
	Rigorous rules	
$\Delta J = 0, \pm 1$	$\Delta J = 0, \pm 1$	$\Delta J = 0, \pm 1, \pm 2$
(except $0 \leftrightarrow 0$)	(except $0 \leftrightarrow 0$)	(except $0 \leftrightarrow 0$, $1/2 \leftrightarrow 1/2, 0 \leftrightarrow 1$)
Parity change	No parity change	No parity change
	Approximate rules	
One-electron jumping	No change in	No change in configuration *or*
($\Delta l = 1$)	configuration	one-electron jumping
		($\Delta l = 0, \pm 2$)
$\Delta S = 0$	$\Delta S = 0$	$\Delta S = 0$
$\Delta L = 0, \pm 1$	$\Delta L = 0$	$\Delta L = 0, \pm 1, \pm 2$

Correlation and Breit–Pauli mixing are two reasons why some of the above rules are obeyed only approximately. Consider the transition from the $3s^2$ ^1S ground state of Mg. The 'one-electron jumping' rule would imply that the transition to $3p3d$ ^1P would not be allowed. However, since the ground state in fact has a large $3p^2$ ^1S component in its wave function expansion, a transition is allowed from the correlation component to the $3p3d$ ^1P final state. Similarly, a Breit–Pauli interaction may mix terms of different spins, such as $3s3p$ $^3P_1$ and $3s3p$ $^1P_1$. As a result of such mixing, the $3s^3$ $^1S_0 - 3s3p$ $^3P_1$ transition is 'allowed', but it will generally be much weaker than the transition to $3s3p$ ^1P. Hence such transitions are called *spin-forbidden* transitions or *intercombination lines*.

9.4 Computational aspects

It has already been mentioned that when non-orthogonal orbitals are present, extra terms may be present in (9.23), along with overlap integrals between orbitals of the two states. Clearly the transition integral involves the 'jumping electron' whereas the others are passive 'spectator electrons'. These spectators must have the same angular momentum. The codes assume that there will be at most two overlap integrals. Consequently, we could have the situation of $\langle n_1 l_1^{q_1} | n_1' l_1^{q_1} \rangle \langle n_2 l_2^{q_2} | n_2' l_2^{q_2} \rangle$, where $l_1 \neq l_2$. But when $l_1 = l_2$, the non-orthogonality becomes too difficult. Consider the transition $2p^3 3s$ $^1P - 2p^2 3p^2$ ^1D where we think of the jumping electron as being $3s - 3p$. Then the spectators are $2p^3$ on the left and $2p^2 3p$ on the right. One would expect overlap factors $\langle 2p|2p\rangle^2 \langle 2p|3p\rangle$, but in order to determine the proper coefficient, in general, the problem becomes difficult. The code assumes, that when $l_1 = l_2$, in either the initial or final state, that the subshells are singly occupied (see section 4.5).

Table 9.2 shows the evaluation of transition matrix elements with one CSF in the initial state and two in the final state, stored in files init.c and final.c, respectively. The core orbitals are assumed to be the same in both initial and final states but the program queries about the others. In this case we indicate that the initial and final state orbitals are *not* an orthonormal set and that there are *no* orbitals in common. For the evaluation of the $(1, 1)$ CSF pair, it is clear the jumping electron is $2p - 3s$ and the two spectator shells have different l-values. The matrix element is evaluated. For the $(1, 2)$ pair of CSFs, we could have $3s - 3p$ as a jumping electron but then all other electrons are spectators. On the right, these spectators come from two shells with the same l angular quantum number whereas on the left we have a triply occupied shell, a situation that the code cannot deal with. When orthogonality conditions are beyond the code, a message is included in the output, though the program will not stop. Thus it is important to check the MLTPOL output when extensive non-orthogonality is used, in order to make certain rules have not been violated.

Table 9.2. MLTPOL output showing the matrix elements between pairs of CSFs.

```
>Mltpol
 Name of Initial State
init
 Name of Final State
final
 THERE ARE  2 INITIAL STATE ORBITALS AS FOLLOWS:
   2p  3s
 THERE ARE  3 FINAL STATE ORBITALS AS FOLLOWS:
   2p  3s  3p
 Initial & final state orbitals an orthonormal set ? (Y/N)
>n
 List common orbitals, terminating with a blank orbital.
 Upper and lower case characters must match.
 Fixed format (18(1X,A3)) as inicated below:
 AAA AAA AAA AAA AAA AAA AAA .... etc (up to 18/line)
>
 THERE ARE  2 INITIAL STATE ORBITALS AS FOLLOWS:
   2p  3s
 THERE ARE  3 FINAL STATE ORBITALS AS FOLLOWS:
   2p  3s  3p
 INITIAL STATE CONFIGURATIONS:-
 ----------------------------
       1.    2p( 3)  3s( 1)
              2P1     2S1
                           1P0
 FINAL STATE CONFIGURATIONS:-
 ----------------------------
       1.    2p( 2)  3s( 2)
              1D2     1S0
                           1D0
       2.    2p( 2)  3p( 2)
              1S0     1D2
                           1D0
  Type of transition ? (E1, E2, M1, M2, .. or *)
>E1
1 ELECTRIC TRANSITION OF ORDER    1
 < INITIAL STATE ||E (1)|| FINAL STATE > =  SUM of
     COEFF*WEIGHT(INITIAL,I)*WEIGHT(FINAL,J)*< NL||E (1)||N'L'>
     COEFF    I  J < NL||E (1)||N'L'>
 -1.29099445  1  1 < 2p||E (1)|| 3s > < 2p| 2p>**2 < 3s| 3s>**1
 THE FOLLOWING SUBSHELLS HAVE A COMMON L-VALUE BUT CONTAIN TOO
 MANY ELECTRONS FOR THIS CODE       1  1  3  4
 THE MATRIX ELEMENT IS ( 1, 2)
  Type of transition ? (E1, E2, M1, M2, .. or *)
>*
```

9.5   Allowed transitions

The strongest transitions are the electric dipole (E1) transitions for which we have

$$S(\gamma J, \gamma' J') = \left| \langle \gamma J \| \sum_{i=1}^{N} r(i) \mathbf{C}^{(1)}(i) \| \gamma' J' \rangle \right|^2, \qquad (9.35)$$

$$A(\gamma' J', \gamma J) = \frac{4}{3} [\alpha \Delta E_J]^3 \frac{S(\gamma J, \gamma' J')}{g_{J'}}, \qquad (9.36)$$

$$gf(\gamma J, \gamma' J') = \tfrac{2}{3} \Delta E_J \, S(\gamma J, \gamma' J'), \qquad (9.37)$$

where

$$\Delta E_J = E_{\gamma' J'} - E_{\gamma J}. \qquad (9.38)$$

9.5.1   Multiplet values

The potentially many transitions from the fine-structure levels J' of the upper term $\gamma' L' S'$ to the fine-structure levels J of the lower term $\gamma L S$ are said to form a *multiplet*. Since the transitions will be closely spaced in wavelength, individual components may not be resolved in experiment. Furthermore, if the J-dependence of the lines is small, it is convenient to consider the ensemble as a single entity.

Let us define the *multiplet strength*† $S(\gamma L S, \gamma' L' S')$ as

$$S(\gamma L S, \gamma' L' S') = \sum_{J, J'} S(\gamma L S J, \gamma' L' S' J'). \qquad (9.39)$$

Similarly, we can define

$$f(\gamma L S, \gamma' L' S') = \Delta E_{LS} \frac{\sum_{J, J'} (2J + 1) f(\gamma L S J, \gamma' L' S' J') / \Delta E_J}{\sum_J (2J + 1)} \qquad (9.40)$$

where

$$\Delta E_{LS} = \frac{\sum_{J'} (2J' + 1) E_{\gamma' L' S' J'}}{\sum_{J'} (2J' + 1)} - \frac{\sum_J (2J + 1) E_{\gamma L S J}}{\sum_J (2J + 1)} \qquad (9.41)$$

is the difference of the weighted energies of the two terms. If we now define the statistical weight of the lower term

$$g_{LS} = \sum_J (2J + 1) = (2L + 1)(2S + 1), \qquad (9.42)$$

we get the formula,

$$gf(\gamma L S, \gamma' L' S') = g_{LS} f(\gamma L S, \gamma' L' S') = \tfrac{2}{3} \Delta E_{LS} S(\gamma L S, \gamma' L' S'), \qquad (9.43)$$

† The multiplet strength is sometimes also denoted S_M.

a formula of the same form as for a line, but a different statistical weight.

The above multiplet formulae are often presented in terms of wavelengths rather than transition energies, since wavelengths are the observed quantity. But for computational work, the present form is more convenient. Furthermore, for a non-relativistic LS calculation without any J-dependence, equation (9.43) is the formula of choice since the multiplet strength can be computed directly as

$$S(\gamma LS, \gamma'L'S') = \left|\langle\gamma LS||\mathbf{E}^{(1)}||\gamma'L'S'\rangle\right|^2.\tag{9.44}$$

This discussion assumed we had J-dependent data and wished to reduce the data to a single value for a multiplet, a situation that occurs in processing experimental data. In such cases, it is important to remember that wavelengths *in vacuo* should be used and not wavelengths in air†. In theoretical work, it may be desirable to use LS results to predict the gf-values or transition rates for the individual lines of a multiplet. Equation (9.19) may be used for this purpose.

9.5.2   Different gauges

In section 9.1 operators were defined for each type of transition, but in fact, all operators have alternative forms, corresponding to different gauges. These are most easily derived for the electric dipole operator. Using the commutator relation,

$$[\mathbf{r}, \mathcal{H}] = i\,\mathbf{p},\tag{9.45}$$

it is easy to show that there are two alternative forms of the electric dipole matrix element, namely the length form,

$$\langle\gamma JM|\sum_{i=1}^{N} r(i)C_q^{(1)}(i)|\gamma'J'M'\rangle,\tag{9.46}$$

and the velocity form,

$$\frac{1}{(E_{\gamma'J'} - E_{\gamma J})}\langle\gamma JM|\sum_{i=1}^{N}\nabla_q^{(1)}(i)|\gamma'J'M'\rangle.\tag{9.47}$$

The two forms are equivalent for exact solutions of the non-relativistic Hamiltonian. It is therefore customary to compute both forms, and use the agreement between the two results as one of the quality criteria for the calculation. It is important to remember that the agreement between length and velocity form is a necessary, not a sufficient, condition for accurate wave functions.

---

† It is customary to report wavelengths *in vacuo* below 2000 Å and in air above this point (Cowan 1981, page 7).

As a matter of fact, there is a third form of the transition operator that is sometimes used, the accelerator form,

$$\frac{1}{(E_{\gamma'J'} - E_{\gamma J})^2} \langle \gamma J M | \sum_{i=1}^{N} \nabla_q^{(1)}(i) V(i) | \gamma' J' M' \rangle, \qquad (9.48)$$

where $V(i) = -\frac{Z}{r_i}$ is the nuclear potential.

In the present transition codes, length and velocity values are reported only for LS transitions.

9.6 LS calculations for allowed transitions

To illustrate the use of the MCHF package, we will start by looking at an allowed transition. We choose the resonance line in beryllium-like nitrogen, $2s^2 \; ^1S - 2s2p \; ^1P$. We include only valence correlation and use an active set, together with a reduced form of the expansion for ^1S and use the GBT to eliminate $2s3p \; ^1P$ from the expansion for the upper term. The $1s$ core is obtained from a Hartree–Fock calculation for the excited state and kept fixed all through the calculations. The active set is successively increased to $n = 4$. We will illustrate the intermediate runs where $n = 3$. We assume information about the initial state is stored in n3_1S.c and n3_1S.w, and for the final state in n3_1P.c and n3_1P.w. These calculations have the same common $1s$ orbital for the $1s^2$ core. We assume the MLTPOL program has been run for these two states.

Table 9.3. The mltpol.1st from a MLTPOL run for the $n = 3$ resonance line in N IV.

```
1.41421356E 1( 2s   1, 2p   1)( 2s| 2s) 1
1.41421356E 1( 2s   1, 3p   2)( 2s| 3s) 1
0.81649658E 1( 2p   2, 2s   1)( 2p| 2p) 1
0.81649658E 1( 2p   2, 3s   2)( 2p| 3p) 1
0.81649658E 1( 2p   2, 3d   3)( 2p| 2p) 1
0.81649658E 1( 2p   2, 3d   4)( 2p| 3p) 1
1.41421356E 1( 3s   3, 2p   1)( 3s| 2s) 1
1.41421356E 1( 3s   3, 3p   2)( 3s| 3s) 1
0.81649658E 1( 3p   4, 2s   1)( 3p| 2p) 1
0.81649658E 1( 3p   4, 3s   2)( 3p| 3p) 1
0.81649658E 1( 3p   4, 3d   3)( 3p| 2p) 1
0.81649658E 1( 3p   4, 3d   4)( 3p| 3p) 1
0.63245553E 1( 3d   5, 2p   3)( 3d| 3d) 1
0.63245553E 1( 3d   5, 3p   4)( 3d| 3d) 1
              *
              *
```

The output of this calculation is stored in a file `mltpol.lst` and is displayed in table 9.3. Each line in the file contains a coefficient, the type of one-electron transition operator, the pair of orbitals for the transition integral and the pair of CSFs, followed by one overlap integral, in the present case.

In table 9.4 we show the corresponding LSTR run, which computes the transition integrals and gives the different transition properties for multiplet quantities. In the latter are listed the individual contributions to the transition matrix element, for both length and velocity form. The sums, given under the label SUM OF LENGTH FORM and VELOCITY FORM, give the sum from contributions of individual pairs of CSFs to the reduced matrix element. We

Table 9.4. LSTR run for the resonance line in N IV.

```
>Lstr
  Name of Initial State
>n3_1S
  Name of Final State
>n3_1P
  INTERMEDIATE PRINTING (Y OR N) ?
>y
  TOLERANCE FOR PRINTING ?
>0.0
  Initial State
  -------------
  N IV   1S         -51.1690485
  1      0.9649251  2s(2)_1S0
  2      0.2617578  2p(2)_1S0
  3     -0.0152045  3s(2)_1S0
  4      0.0031314  3p(2)_1S0
  5     -0.0127077  3d(2)_1S0
  Final   State
  -------------
  N IV   1P         -50.5633262
  1      0.9920397  2s.2p_1P
  2     -0.0317592  3s.3p_1P
  3     -0.1217692  2p.3d_1P
  4     -0.0045774  3p.3d_1P
 Default Rydberg constant (Y or N) ?
>y
  N IV    Z =  7.   Angular multiplicities =  1  3
                    Spin    multiplicities =  1  1
  Orbitals
 Initial:   6;  1s  2s  2p  3s  3p  3d
 Final  :   6;  1s  2s  2p  3s  3p  3d
```

Table 9.4. (continued)

```
            Transition Integrals        Length       Velocity
1-> 1:    2s-> 2p   < 2s| 2s>^ 1     1.4251637     0.9590277
1-> 2:    2s-> 3p   < 2s| 3s>^ 1     0.0008058     0.0034832
2-> 1:    2p-> 2s   < 2p| 2p>^ 1    -0.2234335     0.1557671
2-> 2:    2p-> 3s   < 2p| 3p>^ 1    -0.0000538    -0.0002767
2-> 3:    2p-> 3d   < 2p| 2p>^ 1     0.0405980     0.1405897
2-> 4:    2p-> 3d   < 2p| 3p>^ 1     0.0000387     0.0001340
3-> 1:    3s-> 2p   < 3s| 2s>^ 1    -0.0002949     0.0009609
3-> 2:    3s-> 3p   < 3s| 3s>^ 1     0.0009027     0.0002541
4-> 1:    3p-> 2s   < 3p| 2p>^ 1    -0.0000433     0.0000392
4-> 2:    3p-> 3s   < 3p| 3p>^ 1     0.0000861    -0.0002058
4-> 3:    3p-> 3d   < 3p| 2p>^ 1     0.0000107    -0.0000283
4-> 4:    3p-> 3d   < 3p| 3p>^ 1    -0.0000088     0.0000233
5-> 3:    3d-> 2p   < 3d| 3d>^ 1     0.0014438    -0.0049684
5-> 4:    3d-> 3p   < 3d| 3d>^ 1    -0.0000293    -0.0001325


  SUM OF LENGTH FORM            =     1.24518592
  SUM OF VELOCITY FORM          =     1.25466751
  E1 FINAL gf-VALUES: LENGTH  =    0.626110D+00
                      VELOCITY=    0.635682D+00
  ENERGY DIFFERENCE OF THE STATES:  1.32935469D+05 CM-1
                                    7.52244684D+02 ANGSTROMS(vac)
                                    6.05722300D-01 A. U.
  Rydberg constant for conversion:     109733.02
  TRANSITION PROBABILITY:(final->initial)=2.4600773D+01 10^8 s-1
```

Table 9.5. Results for gf-values of the resonance, $2s^2\ {}^1S - 2s2p\ {}^1P$ transition in N IV.

n_{max}	λ (Å)	gf_l	gf_v
	Valence correlation model		
HF	806.21	0.8417	0.4105
3	752.24	0.6261	0.6357
4	758.28	0.6224	0.6334
	Full correlation model[a]		
8	766.53	0.6086	0.6085
	Experiment[b]		
	765.15	0.619 ± 0.022	

[a]Fleming *et al* (1995).
[b]Engström *et al* (1979).

can see already at this point that the length and velocity form agree to within a few per cent, and that the contributions from the two forms do not always have the same sign. The results for the different steps, and from some large-scale calculations are given in table 9.5. It is clear that to get perfect agreement between length and velocity forms we need to include core effects.

9.7   Cancellations in the transition integral

In the previous example the computed oscillator strength was quite stable, and did not vary appreciably when the size of the calculation increased. In this section we will discuss some examples where this is not the case. The examples concern different types of cancellation in the transition integral in (9.18). These occur when the calculation is dominated by large contributions with opposite signs resulting in a small oscillator strength for the transition.

9.7.1   Cancellation induced by configuration interaction

The strong interaction between the $3s3d$ and $3p^2$ ^1D in magnesium-like ions gives us the first example of cancellation. Let us use a simple model with only two CSFs, for the upper state in Al II. According to this model the two ^1D states can be represented as

$$\Psi(3s3d\ ^1D) = c_1\Phi(3s3d\ ^1D) + c_2\Phi(3p^2\ ^1D)$$

and

$$\Psi(3p^2\ ^1D) = c_1\Phi(3p^2\ ^1D) - c_2\Phi(3s3d\ ^1D).$$

We optimize the orbitals in an MCHF calculation on the lower term, $3p^2$ ^1D. Then, with the same orbitals as input (wfn.inp), and with the proper phases added to cfg.inp, a second MCHF run is executed in order to get the new eigenvector and energy, but without any orbital optimization. For the lower atomic state we use $3s3p$ ^1P, obtained with a frozen core, and just the $3s$ and $3p$ orbitals are reoptimized. In tables 9.6 and 9.7 we give some output from the LSTR code for the $3s3p$ $^1P - 3p^2$ ^1D and $3s3p$ $^1P - 3s3d$ ^1D transitions, respectively. Neither of these calculations are accurate, but they illustrate the strong cancellation for the $3s3p - 3p^2$ case, and the transfer of oscillator strength to the $3s3p - 3s3d$ transition. For the latter the agreement between the length and velocity gf-values already in this very crude model is within 15%.

9.7.2   Radial cancellation

For electric dipole transitions the integral (9.23) is given by

$$\langle l||\mathbf{C}^{(1)}||l'\rangle \int_0^\infty r P(nl; r) P(n'l'; r)\, \mathrm{d}r. \tag{9.49}$$

Table 9.6. Results for the $3s3p$ $^1P - 3p^2$ ^1D transition.

```
   Initial State
   -------------
   Al II 1P        -241.4033937
   1    1.0000000   3s.3p_1P
   Final   State
   -------------
   Al II 1D        -241.3331079
   1    0.7375341   3p(2)_1D2
   2   -0.6753099   3s.3d_1D
   Al II   Z = 13.   Angular multiplicities =  3  5
                     Spin    multiplicities =  1  1
   Orbitals
 Initial:   5;  1s  2s  2p  3s  3p
 Final  :   6;  1s  2s  2p  3p  3s  3d
Transition Integrals                    Length      Velocity
1-> 1: 3s-> 3p < 3p| 3p>^ 1 < | >^ 0     3.1632199   10.4816142
1-> 2: 3p-> 3d < 3s| 3s>^ 1 < | >^ 0    -3.3943654   -8.7112119

            SUM OF LENGTH FORM              =      -0.2311470
            SUM OF VELOCITY FORM            =       1.7704023
        E1 FINAL gf-VALUES: LENGTH  =  0.250353D-02
                            VELOCITY=  0.146868D+00
ENERGY DIFFERENCE OF THE STATES: 1.54256470D+04 CM-1
                                 6.48271027D+03 ANGSTROMS(vac)
                                 7.02858000D-02 A. U.
Rydberg constant for conversion:        109735.16
TRANSITION PROBABILITY:(final->initial)= 7.9470933D-04 10^8 s-1
```

The second type of cancellation occurs in the radial part of this integral and is best illustrated by an example.

The $3s$ $^2S - 4p$ ^2P transition in sodium-like ions as a function of Z shows a minimum for Mg II. This can be understood by looking at the integrand of the radial integral for some of the ions in the sequence (see figure 9.1). There are two different contributions, one positive and one negative. In figure 9.2 we show the function

$$F(R) = \int_0^R r P(nl; r) P(n'l'; r) \, dr, \qquad (9.50)$$

for the same ions. From this figure the delicate balance for Mg II is obvious, where the final integral (when $R \to \infty$) is an order of magnitude smaller than the maximum value of this integral. In table 9.8 we give the final oscillator

Table 9.7. Results for the $3s3p\ ^1P - 3s3d\ ^1D$ transition.

```
    Initial State
    -------------
    Al II 1P        -241.4033937
    1     1.0000000    3s.3p_1P
    Final   State
    -------------
    Al II 1D        -241.1380669
    1     0.7375341    3s.3d_1D
    2     0.6753099    3p(2)_1D2
    Al II    Z = 13.   Angular multiplicities =  3  5
                       Spin    multiplicities =  1  1
    Orbitals
 Initial:   5;  1s  2s  2p  3s  3p
 Final  :   6;  1s  2s  2p  3p  3s  3d
 Transition Integrals                      Length      Velocity
1-> 1: 3s-> 3p < 3p| 3p>^ 1 < | >^ 0     2.8963473    2.5423453
1-> 2: 3p-> 3d < 3s| 3s>^ 1 < | >^ 0     3.7071259    2.5202431

             SUM OF LENGTH FORM              =     6.60347322
             SUM OF VELOCITY FORM            =     5.06258838
         E1 FINAL gf-VALUES: LENGTH  =   0.771319D+01
                             VELOCITY=   0.453351D+02
 ENERGY DIFFERENCE OF THE STATES: 5.82312947D+04 CM-1
                          1.71728965D+03 ANGSTROMS(vac)
                          2.65323500D-01 A. U.
 Rydberg constant for conversion:          109735.16
 TRANSITION PROBABILITY:(final->initial) = 3.4891204D+01 10^8 s-1
```

strengths, in length and velocity form, for some of the ions which show a clear minimum for Mg II. We have also performed a large-active-set ($n = 6$) calculation for this ion, including core–valence correlation. As a comparison we notice that core–valence correlation does not affect the gf-value much for Na I but leads to a drastic change for Mg II.

9.8   Core–valence effects on line strength

In chapter 6 we discussed core–valence correlation and in section 6.2.4 showed that the latter increases the binding energy, the percentage increase being related to the polarizability of the core. In this section we investigate the effect on the line strength.

For many years there was a long-standing discrepancy between theory and

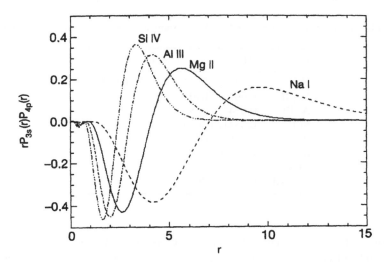

Figure 9.1. The transition integrand between $3s\ ^2S$ and $4p\ ^2P$ in Na I-like ions.

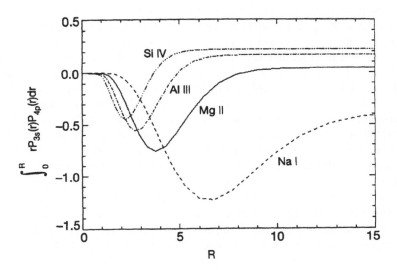

Figure 9.2. The integral $\int_0^R rP(nl;r)P(n'l';r)\,\mathrm{d}r$ as a function of R for some Na I-like ions. The cancellation between different parts of the transition integral ($R \to \infty$) is almost complete for Mg II, and persists to a smaller extent throughout the sequence.

Table 9.8. gf-values for the $3s\ ^2S - 4p\ ^2P$ transition in some sodium-like ions.

Ion	HF	+ CP
Na I	0.0256/0.0239	0.0243/0.0259
Mg II	0.0078/0.0103	0.00227/0.00157
Al III	0.0245/0.0251	
Si IV	0.0641/0.0640	
P V	0.108/0.107	
S VI	0.152/0.150	
Cl VII	0.194/0.192	
Ar VIII	0.234/0.230	

experiment for the resonance transitions of lithium and sodium (for a review, see Godefroid *et al* 1996). In the case of lithium, the earlier beam-foil results (Gaupp 1982) were revised by ultra-precise measurements analysing the long-range vibrational energies of the singly excited diatomic molecule (McAlexander *et al* 1996). When theory was corrected by small relativistic effects, they came into perfect agreement. For sodium, the situation was somewhat different. Earlier theoretical results had not included enough correlation effects and the experimental value needed to be revised (Volz *et al* 1996, Oates *et al* 1996).

Let us investigate the study of the sodium transition with the present code. Clearly, given the limited non-orthogonality allowed between the initial and final states, we will need a common core, and, as suggested in chapter 6, we choose Hartree–Fock orbitals for $1s^2 2s^2 2p^6\ ^1S$. The major core–valence effect arises from the interaction with $2p^6$ (more electrons, somewhat larger mean radius than $2s$) and the wave function expansion can easily be derived using GENCL, selecting and active set expansion to virtual orbitals† $2p, 3s, 3p, \ldots$ and Type 1 generation (one hole in $2p^6$). Table 9.9 shows results from this calculation and compares it with a more accurate theoretical calculation (Jönsson *et al* 1996) where orthogonality constraints are relaxed and large-scale expansions are permitted, using a parallel computing environment for some of the largest runs. The length form of the line strength is independent of the transition energy and decreases smoothly as the size of the active set is increased. The velocity form is less stable in this case. Comparing our length value with the large-scale calculations which include core–valence correlation with *all* shells, we see the line strength of the latter decreasing slightly more rapidly. When the present results have essentially converged, we have what appears to be an excellent agreement with the observed transition energy but this is somewhat accidental in that the relativistic shift will increase the value by about 35 cm^{-1}. Even

† This does not quite correspond to the formal definition, but was an easy way of implementing the desired rule.

Table 9.9. Core–valence effects on the line strength of the $3s$–$3p$ resonance transition in Na I.

| AS | CV with $2p^6$ | | CV with $1s^2 2s^2 2p^6$ [a] | |
	S_l/S_v	ΔE (cm^{-1})	S_l	ΔE (cm^{-1})
HF	41.00	15882	40.82	15913
3	38.34/41.72	16705	38.41	16768
4	37.43/37.61	16904	37.20	16971
5	37.26/37.65	16945	36.97	17028
6	37.20/37.74	16964	36.89	17048
			36.86	17060
Core–core SD			37.40	16885
Core–core SDT			37.35	16885
Relativistic core effects			37.26	
Experiment			37.26(5)[b]	16968

[a] Jönsson *et al* (1996).
[b] Volz *et al* (1996).

so, without the velocity form result, one might think the present line strength is accurate. The length and velocity discrepancy alerts us to the possibility of omitted effects. In the present case, the additional correlation, mainly from SD core–core correlation, cancels with a relativistic correction arising from a different relativistic core.

But the important factor to note is the reduction in line strength as the active set increases. Since both the $3s$ and $3p$ radial functions are oscillating functions, there could be an increase (or decrease) in cancellation, but it is easy to show that the decrease is due to a contraction of the orbitals: applying GBT to the $3s$ of $2p^6 3s$ and the $3p$ of $2p^6 3p$ so that there are no other $2p^6 nl$ CSFs, the mean radius of the fixed-core MCHF $3s$ decreases by a factor of 0.966 and $3p$ by a factor 0.971. The line strength has decreased more than the product of these two factors, so there has also been a change in the cancellation.

For the $2p^6$ ^1S–$2p^5 3s$ ^1P transition, an appropriate correlation model for the transition energy was investigated in table 6.10. There we treated $1s^2 2s^2 2p^5$ as the core, the outer electron being $2p'$ (non-orthogonal to $2p$) and $3s$, for the initial and final state, respectively. We included core–valence correlation and SD from $2p^5$. Though the transition energy is in good agreement with experiment, an LSTR run yields length and velocity forms of the line strength as 0.413 and 0.346, respectively. The discrepancy again points to some omitted effects, possibly contributions from the $2s^2$ subshell or triple excitations.

9.9   Spin-forbidden transitions

In chapter 7 we discussed different effects that arise from relativistic corrections in our calculations. Among these effects we mentioned the mixing of different *LS* terms. We will discuss this mixing in more detail, introduce a limited 'first-order' model and investigate its effect on transition probabilities for several examples.

9.9.1   A first-order model

Let us consider a two-electron system where the ground state is $ns^2\ {}^1S_0$ and the first two excited terms are $nsnp\ {}^3P$ and ${}^1P$. The spin-forbidden transition between the ${}^1S_0$ and ${}^3P_1$ can be understood from a first-order model equivalent to assuming that the wave functions for the atomic state dominated by the ${}^3P_1$ can be written as

$$|{}^3P_1\rangle = c_1|nsnp\ {}^3P_1\rangle + c_2|nsnp\ {}^1P_1\rangle$$

where, in the general case, each ket vector on the right side is a sum over CSFs. The line strength for the intercombination line is therefore

$$S({}^3P_1, {}^1S_0) = \left|\langle {}^3P_1\|\mathbf{E}^{(1)}\|{}^1S_0\rangle\right|^2$$
$$\approx c_2^2\left|\langle nsnp\ {}^1P_1\|\mathbf{E}^{(1)}\|ns^2\ {}^1S_0\rangle\right|^2$$

or

$$S({}^3P_1, {}^1S_0) \approx c_2^2\ S({}^1P_1, {}^1S_0). \tag{9.51}$$

Thus the intercombination line strength is proportional to the allowed ${}^1P_1 - {}^1S_0$ line strength. Furthermore, by using a two×two interaction matrix model for the odd levels, including only $nsnp\ {}^3P_1$ and ${}^1P_1$, we get

$$c_2^2 \approx \left|\frac{\langle nsnp\ {}^3P_1|\mathcal{H}_{BP}|nsnp\ {}^1P_1\rangle}{\Delta E_{\text{term}}(nsnp)}\right|^2, \tag{9.52}$$

where

$$\Delta E_{\text{term}}(nsnp) = E({}^1P_1) - E({}^3P_1). \tag{9.53}$$

Thus, in a first-order model, $S({}^3P_1, {}^1S_0) \propto S({}^1P_1, {}^1S_0)/|\Delta E_{\text{term}}(nsnp)|^2$.

There is no good model for the off-diagonal matrix element between the two terms. However, the same Breit–Pauli integrals usually appear as in the fine-structure separation of the term, ${}^3P$ in the present case. If we look at only the spin–orbit contribution we get

$$\langle nsnp\,{}^3P_1|\mathcal{H}_{SO}|nsnp\,{}^1P_1\rangle = -\sqrt{2}\langle np|\mathcal{H}_{SO}|np\rangle,$$

while the energy difference between the the ${}^3P_2$ and ${}^3P_0$ is

$$\Delta E_{fs}({}^3P) = E({}^3P_2) - E({}^3P_0) = -3\langle np|\mathcal{H}_{SO}|np\rangle.$$

For cases where the spin–orbit operator is the major contributor to the fine-structure splitting, it is reasonable to conclude that, approximately,

$$S(^3P_1, {}^1S_0) \propto \left| \frac{\Delta E_{fs}(^3P)}{\Delta E_{\text{term}}(nsnp)} \right|^2 S(^1P_1, {}^1S_0). \tag{9.54}$$

Whereas the transition energy and the term energy separation may be used to improve the predicted transition probabilities for intercombination lines, the role of the fine-structure splitting is more tenuous, in general.

9.9.2 The $2s^2\,{}^1S_0 - 2s2p\,{}^3P_1$ transition

Let us continue using transitions in beryllium-like N IV as an example. To compute the intercombination line between $2s^2\,{}^1S_0$ and $2s2p\,{}^3P_1$ we need to represent not only these two terms, but also the ones that induce the transition. As we already pointed out, the most important source is the interaction between $2s2p\,{}^3P_1$ and $2s2p\,{}^1P_1$. The BREIT code does not allow non-orthogonal orbitals, so we need to use a common orbital set for both terms as well as a common $1s^2$ core. Somewhat arbitrarily, the $1s$ will be an HF obrital for $1s^22s^2\,{}^1S$. At the same time the representation of the two terms has to be balanced, so that the different properties in (9.54) are well represented. There are different ways of approaching this, e.g. one can obtain one set of orbitals on ^3P, then a second on ^1P, then back to ^3P and so on. Such a *cross-wise optimization* will unfortunately destroy the possibility of a smooth behaviour of different properties.

A more systematic approach is what we will label the LS-dependent $(N, N + 1)$ method. In this we name one of the terms 'primary' (the ^1P in our example) and the other 'secondary' (the ^3P). For each step we now optimize orbitals with principal quantum number $n \leqslant N$ on the primary term. These are then used for the secondary term, where additional orbitals with $n = N + 1$ (but no increase in the maximum l) are added and optimized. As an example, in the $(2, 3)$ step we optimize $2s$ and $2p$ on $2s2p\,{}^1P$, and $3s$ and $3p$ on $2s2p\,{}^3P$. The $3s$ and $3p$ orbitals serve as corrections for the LS-dependence of the orbitals. Note that we do not include $3d$ for the ^3P, but keep the maximum l the same for the two terms. Even though this is now a spin-forbidden transition, the MLTPOL run is exactly the same as earlier E1 transition runs. Assuming this has been performed, the LSJTR run for this case is illustrated in table 9.10, where in the CI run producing the eigenvector information, we only asked for one even, $J = 0$ level and one odd, $J = 1$ level.

The procedure is repeated up to $N = 4$. The results for some of the important energies, according to (9.54), are given in table 9.11. We have also included the second most important contributor to the intercombination rate, the $2p^2\,{}^3P_0$, which mixes with $2s^2\,{}^1S_0$ and adds a contribution of the form $2p^2\,{}^3P - 2s2p\,{}^3P$ to the transition matrix element.

In table 9.12 we give the line strengths and transition probabilities from different approximations. Results are given both without and with the inclusion of the $2p^2\,^3P$ term.

When computing the transition probabilities, according to (9.36), we can introduce the experimental transition energy, by multiplying with the factor,

$$A_{\delta\lambda}(^3P_1, {}^1S_0) = \left(\frac{\Delta E_{\text{trans}}(exp)}{\Delta E_{\text{trans}}(ai)}\right)^3 A_{ai}(^3P_1, {}^1S_0), \qquad (9.55)$$

where the *ai* subscript denotes *ab initio* results. This correction can be applied for all transitions. It is correct to all orders, since it comes directly from the definition of *A*-values.

There also are other possible scalings. According to (9.54), when the first-order model applies, the line strength can be corrected with the term energy separation factor

$$\delta_{\text{term}} = \left(\frac{\Delta E_{\text{term}}(exp)}{\Delta E_{\text{term}}(ai)}\right),$$

Table 9.10. LSJTR run for (2, 3) calculation of the $2s^2\,^1S_0 - 2s2p\,^3P_1$ intercombination line in N IV.

```
>Lsjtr
  Name of Initial State
>even_n23
  Name of Final State
>odd_n23
  INTERMEDIATE PRINTING (Y OR N) ?
>y
  TOLERANCE FOR PRINTING ?
>0.00001
 Initial State
 -------------
   1    2s(2)_1S0
   2    2p(2)1S0_1S
   3    2p(2)_3P2
   4    2p.3p_3P
   5    3p(2)_3P2
 Final    State
 -------------
   1    2s.2p_1P
   2    2s.2p_3P
   3    2p.3s_3P
   4    2s.3p_3P
   5    3s.3p_3P
            N IV      Z =  7.    N =   4
  Default Rydberg constant (Y or N) ?
>y
```

Table 9.10. (continued)

```
    N IV  1S  -51.1672407  =>   N IV  1P  -50.5159806
ELECTRIC TRANSITION OF ORDER   1
  2*JI =  0 2*JF =  2
  --------------------

  Line  2s(2)_1S0 - 2s.2p_3P
  --------------------------

2s(2)_1S0      -> 2s.2p_1P  :  2s -> 2p   -0.0011595
2p(2)1S0_1S    -> 2s.2p_1P  :  2p -> 2s    0.0001868
2p(2)_3P2      -> 2s.2p_3P  :  2p -> 2s    0.0001380
ENERGY DIFFERENCE OF THE STATES:  6.8769918D+04 CM-1
                                  1.4541242D+03 ANGSTROMS
                                  3.1335110D-01 A. U.
  2s(2)_1S0  2J =   0
  TOTAL ENERGY =       -51.1948139
     0.963335  0.268301  0.000278 -0.000049  0.000002
  2s.2p_3P  2J =    2
  TOTAL ENERGY =       -50.8814628
    -0.001399  0.998545 -0.044139  0.030915 -0.001440
SUM OF LENGTH FORM                      =    2.0772214D-06
FINAL OSCILLATOR STRENGTH (GF)          =    4.3389072D-07
TRANSITION PROBABILITY IN EMISSION (Aki) =   4.5624336D+02
```

Listing of `tr.lsj` output:

```
        N IV    Z = 7.   N =   4
  0  -51.19481393  2s(2)_1S0
  2  -50.88146283  2s.2p_3P
  68769.92 CM-1       1454.12 ANGS(VAC)       1454.12 ANGS(AIR)
E1  S =  2.07722D-06   GF =  4.33891D-07   AKI =  4.56243D+02
*
```

so that

$$A_N({}^3P_1, {}^1S_0) = A_{\delta\lambda}({}^3P_1, {}^1S_0)/\delta_{\text{term}}^2 \tag{9.56}$$

is a corrected value value of better accuracy than the computed result. We refer to this as a 'normalized' result, correcting for the observed transition energy that is valid to all orders, and for the term-energy separation that is correct to first order. The latter, of course, assumes that a suitable model is used for the Breit–Pauli mixing. The fine-structure correction, is not on as firm a footing. If we define the ratio of the observed separation to the computed,

$$\delta_{fs} = \left(\frac{\Delta E_{fs}(exp)}{\Delta E_{fs}(ai)} \right),$$

Table 9.11. Excitation energies, term splitting and fine-structure energies (in cm^{-1}) for the $2s^2 - 2s2p$ transition in N IV, from different LS-dependent $(N, N+1)$ calculations.

Step	ΔE_{trans} $^3P_1^o-^1S_0$	ΔE_{Term} $^1P_1^o-^3P_1^o$	$^3P_0-^1S_0$	ΔE_{fs} $^3P^o$	^3P
		Valence correlation model			
(2, 3)	68770	74439	178618	202.2	189.3
(3, 4)	67723	65851	177759	200.3	191.8
(4, 5)	67722	64793	177671	202.0	194.1
		Full correlation model (Fleming *et al* 1995)			
(8, 9)	67284	63515	175542	206.3	198.3
		Experiment (Ölme 1970)			
	67272	63422	175535	207.1	197.5

Table 9.12. Line strengths (in 10^{-6} au) and transition probabilities (in s^{-1}) for the $2s^2 - 2s2p$ intercombination line in N IV. Results from *ab initio* calculations, without and with $^1S_0$ and $^3P_0$ mixing are tabulated. *Ab initio* A-values are corrected for the experimental wavelength ($A_{\delta\lambda}$), and also fine structure and term splitting (A_C).

Step	S_{ai} w/o ^3P	with ^3P	S_C	A_{ai} w/o ^3P	with ^3P	$A_{\delta\lambda}$	A_C
		Valence correlation model					
(2, 3)	2.84	2.08	3.00	623	456	427	617
(3, 4)	3.18	2.41	2.77	666	505	495	570
(4, 5)	3.39	2.62	2.88	711	550	539	591
		Full correlation model (Fleming *et al* 1995)					
(8, 9)		2.82	2.85		580	580	586

and use this to 'adjust' the transition rate so that

$$A_C(^3P_1, {}^1S_0) = \left(\frac{\delta_{fs}}{\delta_{term}}\right)^2 A_{\delta\lambda}(^3P_1, {}^1S_0). \tag{9.57}$$

The computed value generally is an improvement if the the computed fine-structure splitting was incorrect by a large amount (as is the often the case in a valence correlation calculation), but once an accurate calculation has been performed, the discrepancy in the splitting may be due to factors not related to the Breit–Pauli mixing and the adjusted value is not an improvement. However, it could be useful as an uncertainty estimate.

It is interesting to see how well, in this simple case, the corrected values reproduce the most accurate, full correlation results. In the general case we have to be careful, since we do not include the effect of core–valence correlation on the transition matrix element between different CSFs when we apply these scaling methods.

A more careful investigation can be made. In table 9.12 we used a scaling based only on the properties of the $2s2p$ configuration. If we look at table 9.10 we see that the main contributions to the total transition matrix element are of two sorts $\langle {}^1S_0 \| \mathbf{E}^{(1)} \| {}^1P_1^o \rangle$ and $\langle {}^3P_0 \| \mathbf{E}^{(1)} \| {}^3P_1^o \rangle$ where the first, in the $(2, 3)$ approach, is $-0.001\ 595 + 0.000\ 186\ 9 = -0.001\ 408$ and the second is $0.000\ 138\ 0$ according to table 9.10. Each of these transition matrix elements should, in a more accurate approach, be scaled separately. The first one with the factors discussed above, using the fine structure of $2s2p\ {}^3P^o$ and the term splitting in the $2s2p$ configuration. The second one, though, arises due to the mixing between the $2s^2\ {}^1S_0$ and the $2p^2\ {}^3P_0$. The term splitting to be used is the energy difference between these two, and the fine-structure splitting is the difference of ${}^3P_2$ and ${}^3P_0$ of $2p^2$. In reality the mixing between ${}^1S_0$ and ${}^3P_0$ is a two-step process, induced by the Breit–Pauli interaction between the $2p^2\ {}^3P$ and $2p^2\ {}^1S$, and the electrostatic interaction between the latter and the $2s^2\ {}^1S$, ground term. This illustrates that the scaling process used above has to be done with care, and could be used primarily to estimate the accuracy of a property rather than to predict the final answer. The large-scale accurate calculation is still needed to get a correct result.

9.9.3 Lines induced by accidental degeneracy of levels

The analysis of the $2s2p$ intercombination line was relatively simple, since it is a clear cut case from a perturbation theory point of view. The electrostatic interaction and the distance between different terms are much larger than the relativistic effects represented by the fine structures and the off-diagonal matrix elements between the terms. The intercombination rate is therefore much smaller than the rate for allowed transitions. This makes experimental determination of these rates a true challenge, that has only been met in the last few years by using ion-trap methods (Kwong *et al* 1993).

There are many cases where the situation is much more complicated, and the intercombination lines are enhanced due to different effects. One of the most interesting cases, the decay of $2s3p\ {}^3P_1$, was investigated during the end of the 1970s (Engström 1979, Hibbert 1979). The intercombination line to the ground state here is much stronger than for the corresponding $2s2p$ case. This can be predicted already by referring to (9.54). For N IV the fine structure of $2s2p\ {}^3P$ is 207 cm^{-1}, while the term splitting between ${}^1P$ and ${}^3P$ is about 63422 cm^{-1}.

According to equation (9.54) the ratio of the line strengths is

$$2s2p: \quad \frac{S(^3P_1, {}^1S_0)}{S(^1P_1, {}^1S_0)} \approx \left(\frac{207}{63422}\right)^2 \approx 1 \times 10^{-5}.$$

The computed value of this ratio from the full correlation study by Fleming *et al* (1995) is $2.82 \times 10^{-6}/1.54$. Already in a single-configuration Hartree–Fock calculation the $2s3p$ of N IV has a fine structure of 49 cm^{-1} and a term splitting of 5180 cm^{-1}, implying a larger ratio for the intercombination and allowed line by one order of a magnitude. The effect of the plunging configuration $2p3s$ makes the enhancement even more pronounced, since it pushes the $2s3p$ ^1P down, closer to and actually below the $2s3p$ ^3P term. The final term splitting is only -1465 cm^{-1}, and the approximate line strength ratio is

$$2s3p: \quad \frac{S(^3P_1, {}^1S_0)}{S(^1P_1, {}^1S_0)} \approx \left(\frac{49}{-1465}\right)^2 \approx 1 \times 10^{-3},$$

an enhancement of a further two orders of magnitude.

To complete this qualitative discussion, we show the results of active set, GBT, $(N, N+1)$ calculations in table 9.13 for N IV and O V. The core in these calculations is computed from the $1s^2$ ^1S ground term of N VI and O VII. The principal terms are $2s3p$ ^3P and $2s^2$ ^1S, while the secondary ones are $2s3p$ ^1P and $2p^2$ ^3P, respectively. We include only the outer valence correlation here, and compare with a more detailed calculation including core-polarization along with independently optimized orbitals (Froese Fischer, Gaigalas and Godefroid 1997). For the latter, the correction for transition energy is essentially negligible, with the major correction coming from the fine-structure factor. It is also clear that the corrections applied to the valence correlation calculation have over-estimated the adjustment needed, but would have yielded acceptable uncertainty estimates.

The measured quantity is the total lifetime of the $2s3p$ $^3P_1$ in comparison with the $2s3p$ $^3P_0$ and $2s3p$ $^3P_2$. The former may decay to both $2s^2$ $^1S_0$ and $2s3s$ $^1S_0$ whereas the latter may each decay to $2s3s$ $^3S_1$. Assuming decay rates to the $2s3s$ configurations are similar, the intercombination rate is

$$A(^3P_1, {}^1S_0) \approx \frac{1}{\tau(^3P_1)} - \frac{1}{\tau(^3P_2)}. \tag{9.58}$$

In real life the situation is more complicated than this, and the best approach would be to compute transition rates for all transitions from the different $2s3p$ ^3P fine-structure levels.

An even more dramatic enhancement of the intercombination line strength has been observed for the $3s4p$ ^3P lifetimes in magnesium-like ions. In this case the plunging configuration is $3p3d$, which changes position with $3s4p$ in the iso-electronic sequence between P IV and Cl VI as we discussed in section 6.1.4. In

Table 9.13. Transition energies, term splitting, fine-structure energies (in cm^{-1}) and transition probabilities (in 10^6 s^{-1}) for the $2s3p$ $^3P_1-2s^2$ $^1S_0$ intercombination line in N IV and O V.

Source	Step	ΔE_{trans}	ΔE_{term}	ΔE_{fs}	A_{ai}	$A_{\delta\lambda}$	A_C
			N IV				
Present:							
	HF	388698	+5180	48.9	0.29	0.32	4.46
	(3, 4)	405908	−1422	50.1	2.87	2.87	2.83
	(4, 5)	405801	−1381	50.0	3.46	3.46	3.23
	(5, 6)	405814	−1355	50.2	3.43	3.44	3.06
Other theories:							
	CIV3[a]						3.07
	MCHF[b]	406131	−1451	51.6	3.27	3.27	3.20
	MCDF[c]	406040	−1190		1.40		
Experiments:							
	Moore[d]	405988	−1465	51.2			
	Beam-foil[e]						3.3 ± 2.7
			O V				
Present:							
	HF	563211	+6272	110.2	2.4	2.6	27.1
	(3, 4)	582785	−2074	111.9	16.0	16.0	17.5
	(4, 5)	582682	−1916	111.9	20.1	20.1	18.8
	(5, 6)	582682	−1916	111.9	20.2	20.2	18.9
Other theories:							
	CIV3[a] *ab initio*	581444	−1050		66.8		18.3
	MCHF[b]	583041	−2002	114.3	19.85	19.83	19.52
	MCDF[c]	582770	−1780		16.4		
Experiments:							
	Moore[d]	582843	−2018	113.9			
	Beam-foil[e]						22 ± 6

[a]Hibbert (1979).
[b]Also including core-polarization (Froese Fischer, Gaigalas and Godefroid 1997).
[c]Valence correlation only (Fritzsche and Grant 1994).
[d]Moore (1949).
[e]Engström *et al* (1979).

S V this results in a close degeneracy of $3s4p$ ^3P and ^1P, with a term splitting of only 373 cm^{-1}, compared to a fine structure of 356 cm^{-1}. The intercombination line, e.g., the $3s^2$ $^1S_0$ − $3s4p$ $^3P_1$, contributes over 40% to the sum of all transition probabilities from the $^3P_1$ level, as can be seen in table 9.14.

Table 9.14. Transition probabilities[a] (in 10^9 s^{-1}) for lines from the $3s4p$ ^3P fine structure levels in S V together with lifetimes (in ns) of the three levels.

Lower level	$^3P_0$	$^3P_1$	$^3P_2$
$3s^2$ $^1S_0$		0.4561	
$3s3d$ $^3D_1$	1.2627	0.2608	0.0093
$^3D_2$		0.7545	0.1633
$^3D_3$			1.1260
$^1D_2$		0.1305	0.0000
$3p^2$ $^3P_0$		0.0014	
$^3P_1$	0.0205	0.0031	0.0003
$^3P_2$		0.0022	0.0239
$^1D_2$		0.3903	0.0000
$^1S_0$		0.0223	
$3s4s$ $^3S_1$	0.2970	0.2396	0.3037
$^1S_0$		0.0208	
τ(computed)[a]	0.634	0.437	0.615
τ(exp.)[b]		0.37 ± 0.04	0.72 ± 0.07

[a]Brage and Hibbert (1989).
[b]Reistad *et al* (1984).

9.10 Branching ratios in complex spectra

Earlier, in section 7.10, we discussed ways of exploring complex spectra. Our example was the spectrum in N II up to $2s^22p4s$ ^3P. We found that, because of the small exchange between $2p$ and $3d$ and LS-dependent correlation, the separation of some of the terms of $2p3d$ was not much larger than the fine-structure splitting of the $2p$, and Breit–Pauli mixing could be appreciable, even in such a light atom. By creating a file for the final state, where the odd.j file contains only one eigenvector, we can obtain transition rates for E1 decay to the lower-lying even states, executing first MLTPOL, then LSJTR which produces the tr.lsj file and finally LINES to display the data.

Table 9.15 shows the results. Immediately evident is that the strongest line strength occurs for the $2p3p$ $^3S_1 - 2p3d$ $^3P_2$ transition whereas the largest transition rate from $2p3d$ $^3P_2$ is to $2p^2$ $^3P_2$. In fact, summing up the transition rates we get a lifetime if 0.304 ns for this transition. Thus, assuming these data are correct and other types of decay are negligible, the branching ratio for decay to this lower level would be 0.826. But also clear from this table is that a given state may decay to many lower levels and that for E1 transitions, the rate varies as ΔE^3 (E in the line list of table 9.15). Thus, if the objective were to accurately predict the lifetime of the upper level, the even state that needs to be computed accurately is $2s^22p^2$ ^3P. The most recent experimental value for

Table 9.15. Decay rates for the $2p3d\ ^3P_2$ level of N II.

```
Line List for N      ( Z =  7.) with  6 electrons
-------------------------------------------------------------
Transition Array
    Multiplet  Line   Type E(cm-1)   L(air)    S     gf    Aki
-------------------------------------------------------------
2s(2).2p(2)                 2s(2).2p_2P.3d
    3P  3P   1.0- 2.0  E1 185367.1    539.5   .186  .105  4.79E+08
             2.0- 2.0  E1 185289.2    539.7   .734  .413  1.89E+09
    1D  3P   2.0- 2.0  E1 167594.9    596.7   .003  .001  5.30E+06

2s(2).2p_2P.3p              2s(2).2p_2P.3d
    1P  3P   1.0- 2.0  E1  23985.9   4168.0   .020  .001  1.09E+05
    3D  3P   1.0- 2.0  E1  22052.1   4533.4   .010  .001  4.22E+04
             2.0- 2.0  E1  21994.4   4545.3   .212  .014  9.16E+05
             3.0- 2.0  E1  21903.4   4564.2  1.757  .117  7.48E+06
    3S  3P   1.0- 2.0  E1  20127.0   4967.1 23.695 1.449  7.83E+07
    3P  3P   1.0- 2.0  E1  18070.1   5532.5  2.384  .131  5.70E+06
             2.0- 2.0  E1  18014.5   5549.5 11.311  .619  2.68E+07
    1D  3P   2.0- 2.0  E1  13845.9   7220.3   .026  .001  2.75E+04
```

this lifetime is 0.457 ± 0.020 (Bastin *et al* 1992) and the most recent theoretical value is 0.409 (Bell *et al* 1992). Since our computational model for the even states concentrated on the $2s^22p3p$ configurations, it is not surprising that our lifetime is in error. Thus two conclusions can be drawn from this study:

(i) The interaction of $2s2p^3\ LS$ terms with $2s^22p3d$ is important for odd states.

(ii) For lifetimes of $2s^22p3d$ levels the $2s^22p^2\ LS$ terms are the most important even states.

Given the current Breit–Pauli constraints, such calculations are complex. Semi-empirical methods may be used to adjust the diagonals of the interaction matrix, as discussed in section 7.10, or large-scale calculations need to be performed.

9.11 Forbidden lines

The next topic to investigate is truly forbidden lines in different atomic systems, arising due to higher-order multipoles of the electromagnetic field. We will continue to use the N IV system as our example.

9.11.1   Forbidden lines in two-electron systems

In figure 9.3 the three lowest terms in two-electron systems are shown, along with the possible single-photon decay channels (multiphoton decay falls outside the scope of this book). Most of these transitions can be computed from the already existing files for N IV, using the $(2, 3)$ calculation described in section 9.6. For the intraterm lines in ^3P we use the same files for lower and upper states† as shown in table 9.16.

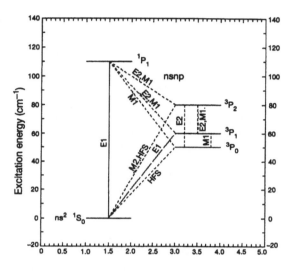

Figure 9.3. Transitions in two-electron systems with the ns^2 ground state. HFS implies hyperfine-induced transitions.

The LSJTR code is run in the same way as before, keeping in mind that we have the same upper and lower .c, .w and .j files. To get a list of the intraconfiguration lines we can use the LINES code, which is illustrated in table 9.17.

The final results, using the $(4, 5)$ calculation, are given in table 9.18, together with results from large-scale calculations. Only lines with a significant *branching ratio* will be of interest, that is lines with a rate not too much smaller than the line with the largest rate originating from the same upper level. These transitions are marked with an '*' in table 9.18. The M2 transition from $^3P_2$ was quite unexpected, when discovered in astrophysical, low-density plasmas. The competing process of transitions induced by hyperfine interaction was first assumed to be responsible for it. This type of transition will be discussed in the next section.

---

† Some computer systems do not allow the same files to be used both for the initial state and final state. In such a case, it is simplest to make two copies of the same files.

Table 9.16. The MLTPOL run for forbidden lines within $2s2p\ ^3P$ in N IV.

```
>Mltpol
 Name of Initial State
>odd_n23
 Name of Final State
>odd_n23
 THERE ARE  4 INITIAL STATE ORBITALS AS FOLLOWS:
   2s  2p  3s  3p
 THERE ARE  4 FINAL STATE ORBITALS AS FOLLOWS:
   2s  2p  3s  3p
 Initial & final state orbitals an orthonormal set ? (Y/N)
>n
 List common orbitals, terminating with a blank orbital.
 Upper and lower case characters must match.
 Fixed format (18(1X,A3)) as inicated below:
 AAA AAA AAA AAA AAA AAA AAA .... etc (up to 18/line)
>
 THERE ARE  4 INITIAL STATE ORBITALS AS FOLLOWS:
 ...
   Type of transition ? (E1, E2, M1, M2, .. or *)
>E2
1 ELECTRIC TRANSITION OF ORDER   2
     COEFF      I    J    < NL||E (2)||N'L'>
    1.00000000   1    1    < 2p||E (2)|| 2p >    < 2s| 2s>** 1
    1.00000000   2    2    < 2p||E (2)|| 2p >    < 2s| 2s>** 1
 ...
    1.00000000   5    4    < 3p||E (2)|| 3p >    < 3s| 2s>** 1
    1.00000000   5    5    < 3p||E (2)|| 3p >    < 3s| 3s>** 1
        NUMBER OF TERMS IN THE ABOVE SUMMATION = 17
   Type of transition ? (E1, E2, M1, M2, .. or *)
>M1
1 MAGNETIC TRANSITION OF ORDER   1
     COEFF      I    J    < NL||M (1)||N'L'>
    1.00000000   1    1    < 2p||MA(1)|| 2p >    < 2s| 2s>** 1
    1.00000000   2    2    < 2p||MA(1)|| 2p >    < 2s| 2s>** 1
 ...
    1.73205081   5    5    < 3s||MB(1)|| 3s >    < 3p| 3p>** 1
    1.00000000   5    5    < 3p||MB(1)|| 3p >    < 3s| 3s>** 1
        NUMBER OF TERMS IN THE ABOVE SUMMATION = 49
   Type of transition ? (E1, E2, M1, M2, .. or *)
>*
```

Table 9.17. Results of the LINES program for forbidden lines in N IV.

```
>Lines
 Enter tolerance on line strength
>0.0
 Number of transitions =          9
 Select the line list order:
 1:  Energy (cm-1)
 2:  Wavelength (Angstroms) in Vacuum
 3:  Wavelength (Angstroms) in Air
 4:  Line Strength
 5:  gf Value
 6:  Transition Probability
 Enter your selection:
>2
 Line List for N IV   ( Z =  7.) with  4 electrons
```

```
 Transition Array
 Multiplet  Line  Type  E(cm-1)      L(air)      S      gf      Aki
```

Multiplet	Line	Type	E(cm-1)	L(air)	S	gf	Aki
2s.2p			2s.2p				
3P 1P	0.0- 1.0	M1	74500.4	1342.3	0.000	0.000	1.46E-02
	1.0- 1.0	E2	74439.3	1343.4	0.000	0.000	7.63E-04
	1.0- 1.0	M1	74439.3	1343.4	0.000	0.000	1.09E-02
	2.0- 1.0	M1	74298.2	1345.9	0.000	0.000	1.80E-02
	2.0- 1.0	E2	74298.2	1345.9	0.000	0.000	2.20E-04
3P	0.0- 2.0	E2	202.2	494423.5	0.653	0.000	4.94E-12
	1.0- 2.0	E2	141.1	708264.3	1.469	0.000	1.84E-12
	1.0- 2.0	M1	141.1	708264.3	2.500	0.000	3.79E-05
	0.0- 1.0	M1	61.0	1637585.8	2.000	0.000	4.09E-06

9.12 Hyperfine-induced transition

As discussed in section 9.3, the $J' = 0 \to J = 0$ transition is forbidden by the rigorous selection rules. These rules apply to atoms with nuclear spin $I = 0$. For atoms with $I \neq 0$ the off-diagonal hyperfine interaction mixes states with different J values and induces the $J' = 0 \to J = 0$ transition.

As an example let us consider the fine-structure levels $1s^2\,{}^1S_0$ and $1s2p\,{}^3P_0$. In the presence of hyperfine interaction these levels are split into hyperfine components characterized by the F quantum numbers†. In the simplest model, including only two terms in the perturbation expression (8.70), the wave function

---

† Since $J = 0$ there is only one hyperfine component. For this component $F = I$.

Table 9.18. Transition rates for forbidden lines between and within the lowest three terms in N IV.

Lower level	Upper level	Type	Present (4, 5)		Obs[a]
			ΔE (cm^{-1})	A (s^{-1})	
$2s2p\ ^3P_0$	$2s2p\ ^1P_1$	M1	74500.4	1.46E − 02	
$2s2p\ ^3P_1$	$2s2p\ ^1P_1$	E2	74439.3	7.63E − 04	
$2s2p\ ^3P_1$	$2s2p\ ^1P_1$	M1	74439.3	1.09E − 02	
$2s2p\ ^3P_2$	$2s2p\ ^1P_1$	M1	74298.2	1.80E − 02	
$2s2p\ ^3P_2$	$2s2p\ ^1P_1$	E2	74298.2	2.20E − 04	
$2s2p\ ^3P_0$	$2s2p\ ^3P_2$	E2	202.2	4.94E − 12	
$2s2p\ ^3P_1$	$2s2p\ ^3P_2$	E2	141.1	1.84E − 12	
$2s2p\ ^3P_1$	$2s2p\ ^3P_2$	M1	141.1	3.79E − 05	
$2s2p\ ^3P_0$	$2s2p\ ^3P_1$	M1	61.0	4.09E − 06	
$2s^2\ ^1S_0$	$2s2p\ ^1P_1$	E1	132515.7	2.45E + 09	*
$2s^2\ ^1S_0$	$2s2p\ ^3P_2$	M2	67862.9	1.19E − 02	*
$2s^2\ ^1S_0$	$2s2p\ ^3P_1$	E1	67722.3	5.50E + 02	*

[a]Observable in emission.

corresponding to $1s2p\ ^3P_0\ IF$ can be written

$$|1s2p\ ^3P_0\ IF\rangle^{(1)} = c_1|1s2p\ ^3P_0\ IF\rangle + c_2|1s2p\ ^3P_1\ IF\rangle, \qquad (9.59)$$

where $|1s2p\ ^3P_0\ IF\rangle$ and $|1s2p\ ^3P_1\ IF\rangle$ are the JIF coupled functions defined in (8.60). The expansion coefficient c_2 is given by

$$c_2 = \frac{\langle 1s2p\ ^3P_1\ IF|\mathcal{H}_{hfs}|1s2p\ ^3P_0\ IF\rangle}{\Delta E(^3P_1 - {}^3P_0)}. \qquad (9.60)$$

Neglecting all terms in the hyperfine interaction except the magnetic dipole, the hyperfine interaction matrix element can, according to (8.75), be expressed in terms of the off-diagonal magnetic dipole constant $A(^3P_1, {}^3P_0)$

$$\langle 1s2p\ ^3P_1\ IF|\mathcal{H}_{hfs}|1s2p\ ^3P_0\ IF\rangle = \sqrt{I(I+1)}A(^3P_1, {}^3P_0). \qquad (9.61)$$

To calculate the rate for the $1s2p\ ^3P_0\ IF \to 1s^2\ ^1S_0\ IF$ transition we note that (9.9) is generally valid for transitions between any levels, and we have

$$A(^3P_0\ IF, {}^1S_0\ IF) = \tfrac{4}{3}[\alpha\Delta E(^3P_1\ IF - {}^1S_0\ IF)]^3$$

$$\times\ c_2^2 \frac{1}{g_I}|\langle 1s^2\ ^1S_0\ IF\|\mathbf{E}^{(1)}\|1s2p\ ^3P_1\ IF\rangle|^2, \qquad (9.62)$$

where $g_J = 2J + 1$ is the statistical weight of the upper level. The electric dipole operator acts only on the electronic space, and we may decouple the nuclear and electronic parts in the reduced matrix element. From (A.57) we get

$$\langle 1s^2 \, {}^1S_0 \, IF \| \mathbf{E}^{(1)} \| 1s2p \, {}^3P_1 \, IF \rangle$$

$$= (-1)^{2I+1}(2I + 1) \left\{ \begin{matrix} 0 & I & I \\ I & 1 & 1 \end{matrix} \right\} \langle 1s^2 \, {}^1S_0 \| \mathbf{E}^{(1)} \| 1s2p \, {}^3P_1 \rangle$$

$$= (-1)^{2I+1} \sqrt{\frac{2I + 1}{3}} \langle 1s^2 \, {}^1S_0 \| \mathbf{E}^{(1)} \| 1s2p \, {}^3P_1 \rangle. \tag{9.63}$$

Inserting the above expression into (9.62) and noting that

$$\Delta E({}^3P_1 \, IF - {}^1S_0 \, IF) \approx \Delta E({}^3P_1 - {}^1S_0), \tag{9.64}$$

we have

$$A({}^3P_0 \, IF, {}^1S_0 \, IF) \approx c_2^2 A({}^3P_1, {}^1S_0). \tag{9.65}$$

Thus, the rate for the hyperfine-induced transition is equal to the rate for the intercombination transition times a cofficient related to the off-diagonal hyperfine interaction matrix element.

As a numerical example we look at the $J' = 0 \to J = 0$ transition in ${}^{19}F^{+7}$ ($I = 1/2$ and $\mu_I = 2.628868$). As the first step HF calculations were performed for the $1s^2 \, {}^1S$ and $1s2p \, {}^3P$ terms. At the HF level, the off-diagonal magnetic dipole constant $A({}^3P_1, {}^3P_0) = 380665$ MHz. Converting from MHz to cm^{-1} (see appendix C) this corresponds to 12.7 cm^{-1}. Multiplying with $\sqrt{I(I + 1)} = \sqrt{3}/2$ and dividing by the experimental fine-structure splitting $\Delta E({}^3P_1 - {}^3P_0) = 149 \, cm^{-1}$ (Klein *et al* 1985) we obtain $c_2 = 0.0738$. As the second step Breit–Pauli calculations were performed for $1s^2 \, {}^1S_0$ and $1s2p \, {}^3P_1$, where for the latter state the expansion $\{1s2p \, {}^3P_1, 1s2p \, {}^1P_1\}$ was used. The Breit–Pauli wave functions give $A({}^3P_1, {}^1S_0) = 1.77 \, ns^{-1}$. Multiplying with c_2^2 we finally get $A({}^3P_0 \, IF, {}^1S_0 \, IF) = 0.00964 \, ns^{-1}$. This is in reasonable agreement with the rate 0.0142 ns^{-1} obtained in a full correlation calculation by Aboussaïd *et al* (1995).

9.13   Z-dependence of transition properties

We remind ourselves that the nth one-electron moment scales, in the hydrogenic picture, as $(Z - \sigma)^{-n}$ (6.4). From (9.7) it then follows that the line strengths for a transition varies either as

$$S^{Ek} \propto (Z - \sigma)^{-2k} \tag{9.66}$$

or

$$S^{Mk} \propto (Z - \sigma)^{-2k+2}. \tag{9.67}$$

According to (9.9) we then get that the transition rates themselves vary as

$$A^{Ek} \propto \Delta E^{2k+1}(Z - \sigma)^{-2k} \tag{9.68}$$

or

$$A^{Mk} \propto \Delta E^{2k+1}(Z - \sigma)^{-2k+2}, \tag{9.69}$$

where ΔE is the transition energy. As we have already seen in section 6.3, this property has different Z-dependence for $\Delta n = 0$ and $\Delta n \neq 0$ transitions. When discussing forbidden transitions, we also need to consider transitions within a term, that is between fine-structure levels. Since the energy difference is in this case proportional to a Breit–Pauli integral, we can write

$$\Delta E_{\text{fine structure}} \propto (Z - \sigma)^4 \tag{9.70}$$

according to the discussion in section 7.11. The Z-dependence for the different cases are given in table 9.19.

Table 9.19. Z-dependence of different transitions.

Multipole	Transition	Z-dependence
Ek	$\Delta n = 0$	$(Z - \sigma)$
Ek	$\Delta n \neq 0$	$(Z - \sigma)^{2k+2}$
Ek	Fine structure	$(Z - \sigma)^{6k+4}$
Mk	$\Delta n = 0$	$(Z - \sigma)^3$
Mk	$\Delta n \neq 0$	$(Z - \sigma)^{2k+4}$
Mk	Fine structure	$(Z - \sigma)^{6k+6}$

It is clear that the relative importance of forbidden lines, especially the ones between fine structure levels, increases rapidly along the isoelectronic sequence.

9.14 Exercises

(i) Include core–valence correlation for Mg II, in an active set calculation. To represent the $4p$ term do two different calculations.

   (a) In the first start with Hartree–Fock for the $3p$ ^2P and include this term in the expansion for the $4p$. Keep the $3p$ orbital fixed while increasing the orbital set.

   (b) In a second, separate approach, use the GBT theorem and delete the $2p^6 3p$ ^2P in the expansion for $4p$.

   Observe how the cancellation is radial or CI induced depending on the model used.

(ii) From a small ($n = 3$) calculation, determine the extent of the contribution from $2p^5 3d$ ^1P to the line strength for the $2p^6$ ^1S $-$ $2p^5 3s$ ^1P transition in Ne.

(iii) From a 2×2 matrix model for an $nsnp$ configuration, verify (9.52) and (9.53).

(iv) Derive the relationship between transition probabilities that corresponds to (9.51):

$$A(^3P_1, {}^1S_0) \propto \left(\frac{\lambda_a}{\lambda_{ic}}\right)^3 c_2^2 A(^1P_1, {}^1S_0) \tag{9.71}$$

where λ_a and λ_{ic} are the wavelengths for the allowed and intercombination transitions, respectively.

(v) Use the 2×2 model to show that an LS-breaking interaction can be detected, not only by observed intercombination lines, but also by deviations from the Landé interval rule.

(vi) Perform the $(3, 4)$ GBT calculation as described in the text for N IV (remember to put the $2s3p$ state first!). Using the obtained orbitals, investigate the relative position of the $2s3p$ ^1P and ^3P in three different models:

(a) single-configuration Hartree–Fock;
(b) including only $2s3p$ and $2p3s$;
(c) also including $2p3d$;
(d) also including $2p4s$.

Interpret the results.

(vii) Use the discussion in sections 9.1.1 and 9.13 to show that the transition rate for a spin-forbidden, electric dipole line has a Z-dependence of $(Z - \sigma)^7$ for $\Delta n = 0$, and $(Z - \sigma)^{10}$ for $\Delta n \neq 0$.

Chapter 10

MCHF Continuum Wave Functions

10.1 Continuum processes

So far, all the processes that have been considered involved only the bound states of the wave equation. The Hamiltonian, viewed as an operator, has both a discrete and a continuous spectrum. The former corresponds to regions where isolated energies are eigenvalues of the wave equation whereas in the latter, any energy has associated with it an eigenfunction for an atomic state. Regions of the continuous spectrum describe systems where one or more of the electrons are 'free' electrons, not bound to a localized charge distribution. Correlation is still an important factor, either in the 'target' (N-electron system) of the free electron, or through interactions with configuration states of localized ($N + 1$)-electron systems.

An example is shown in figure 10.1. In this figure, the free electron has

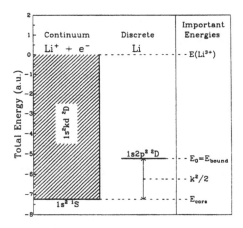

Figure 10.1. The $1s2p^2$ ^2D discrete perturber embedded in the $1s^2kd$ ^2D continuum in Li.

the orbital quantum numbers of a d orbital which couples to the two-electron ^1S target, the ground state of Li^+. In a simple model, this target or core could be represented by the Hartree–Fock wave function but clearly then some correlation has been omitted. The energy of the continuum state is $E_{target}+k^2/2$, where $k^2/2$ is the energy in hartrees of the free electron†. But embedded in this continuum is the energy of a localized charge distribution, labelled $1s2p^2\ ^2D$. This could be a Hartree–Fock wave function for a single configuration state but correlation effects may also be present. Most methods for determining a wave function in such cases partition the wave function into a discrete component whose radial functions all satisfy the bound state conditions at large r and the continuum component.

Many radiative and collision processes involve one or more continuum electrons. In this chapter we will only consider processes associated with photo-absorption. They often are depicted by reaction diagrams:

$$h\nu + A^*$$
$$\updownarrow$$
$$h\nu + A \leftrightarrow A^{**} \leftrightarrow e + A^{+*}$$
$$\updownarrow$$
$$e + A^+$$

In such diagrams, A is assumed to be a neutral atom with each * indicating an excited electron. Notice that each path has a reverse process associated with it: the absorption of a photon may leave the atom in a doubly excited state which may then decay, either to the initial state or another excited state. Or the excited state may undergo decay without emitting any photons, but releasing an electron. The latter is referred to as autoionization and is one of the process that determines the lifetime of A^{**}, often an unstable, doubly excited state.

In this chapter we will consider only simple continuum wave functions where there is only one possible continuum electron. This is the beginning of a bridge to the extensive literature associated with scattering theory and electron impact excitation and ionization.

10.2   Continuum functions

10.2.1   Simple one-electron systems

Consider the one-electron wave equation with $E = k^2/2$ for which the radial equation is

$$\left(\frac{d^2}{dr^2} + \frac{2[Z - Y(r)]}{r} - \frac{l(l+1)}{r^2} + k^2\right) P(kl; r) = 0, \qquad (10.1)$$

† In figure 10.1 the target is referred to as the core.

with $P_{kl}(0) = 0$. In this equation $Y(r)$ is a potential accounting for the screening of the nucleus by the target and might be the direct contribution as in a Hartree–Fock approximation. At large r, solutions of the differential equation are linear combinations of $\sin kr$ and $\cos kr$, the particular combination depending on the inner region. As in the case of bound state solutions, it is convenient to normalize the solutions. The usual convention is normalization per unit energy so that, in atomic units,

$$\int_0^\infty P(kl; r) P(k'l; r) \, dr = \delta(E - E'). \tag{10.2}$$

For the one-electron systems the solutions are the regular and irregular Coulomb functions, often denoted as $F(Z, k, l; r)$ and $G(Z, k, l; r)$ respectively, where the latter does not satisfy the boundary condition at the origin (Abramowitz and Stegun 1965). The asymptotic form for large r of the regular Coulomb function can be derived analytically as

$$F(Z, k, l; r) \rightarrow \left(\frac{2}{\pi k}\right)^{1/2} \sin\left(kr - \frac{1}{2}l\pi + \frac{Z}{k}\log 2kr + \sigma_l\right) \tag{10.3}$$

where σ_l, the Coulomb phase shift, is

$$\sigma_l = \arg \Gamma\left(l + 1 - i\frac{Z}{k}\right). \tag{10.4}$$

Equation (10.1) differs in that the function $Y(r) \rightarrow N$ at large r, where N is the number of electrons in the target. Thus, for large r, the effective nuclear charge is $Z_{\text{eff}} = Z - N$. At the same time, the modification of the potential in the inner region introduces another phase shift so that

$$P(kl; r) \rightarrow \left(\frac{2}{\pi k}\right)^{1/2} \sin\left(kr - \frac{1}{2}l\pi + \frac{Z_{\text{eff}}}{k}\log 2kr + \sigma_l + \delta_l\right) \tag{10.5}$$

where δ_l is the wave phase shift with respect to the Coulomb wave.

In the Wentzel–Kramer–Brillouin (Wentzel 1926, Kramers 1926, Brillouin 1926), or phase-amplitude, method, equation (10.1) is written as

$$\frac{d^2}{dr^2} P(kl; r) + W(r) P(kl; r) = 0, \quad W(r) = \frac{2[Z - Y(r)]}{r} - \frac{l(l+1)}{r^2} + k^2 \tag{10.6}$$

for which solutions have the form

$$P(kl; r) = \frac{A}{\zeta^{1/2}} \sin[\phi(r)] \tag{10.7}$$

$$\zeta^2 = W(r) + \zeta^{1/2} \frac{d^2}{dr^2} \zeta^{-1/2} \tag{10.8}$$

and

$$\phi(r) = \int_a^r \zeta(s) \, ds, \quad \text{or} \quad \phi'(r) = \zeta(r). \tag{10.9}$$

Many numerical methods are based on the latter form, at least in the asymptotic region (see Burgess 1963, for example).

10.2.2 Continuum Hartree–Fock equation

Let us assume that $\Phi(N; \gamma_t L_t S_t)$ is a configuration state function for an N-electron target state, constructed from bound state orbitals in that all approach zero exponentially for large r. Then the wave function for the Hartree–Fock continuum state $E = E_{\text{target}} + k^2/2$ is

$$\Psi(N + 1; E \gamma L S) = \mathcal{A} \left[\Phi(N; \gamma_t L_t S_t) \bullet |kl\rangle \right] \tag{10.10}$$

where $\bullet$ represents the coupling of the orbitals of the continuum electron with orbital quantum number l and radial factor $P(kl; r)$.

For continuum functions, the variational principle cannot be applied to the Rayleigh quotient, since the Hamiltonian $\mathcal{H}$ is no longer Hermitian. Instead, we can impose the Galerkin condition, $\langle \delta \Psi | \mathcal{H} - E | \Psi \rangle = 0$, for all allowed variations $\delta \Psi$. Introducing Lagrange multipliers for orthogonality constraints and integrating over all variables except the radial co-ordinate of the continuum electron we get the equation,

$$\left(\frac{d^2}{dr^2} + \frac{2}{r}[Z - Y(kl; r)] - \frac{l(l+1)}{r^2} + k^2 \right) P(kl; r)$$

$$= \frac{2}{r} X(kl; r) + \sum_n \delta_{l,l_n} \varepsilon_{kl,nl} P(nl; r), \tag{10.11}$$

which has exactly the same form as the Hartree–Fock equation for a singly occupied orbital of a bound state system, the only difference being the specified binding energy, $k^2/2$, and the boundary condition at infinity which refers only to amplitude.

10.2.3 Close-coupling equations

As in the case of bound state wave functions, the Hartree–Fock approximation omits the effect of correlation. A more accurate wave function is a linear combination of configuration states. The continuum problem differs from the general bound state problem in that the continuum orbital is always singly occupied and, under our present assumptions, there is only one such orbital. This simplifies the problem considerably and it is then possible to express the wave function as

$$\Psi(N + 1; E \gamma LS) = \Psi_b(N + 1; \gamma_b LS) + \Psi_k(N + 1; E \gamma_k LS)$$

$$\Psi_b(N + 1; \gamma_b LS) = \sum_{i=1}^{M_b} c_i \Phi(N + 1; \gamma_i LS) \tag{10.12}$$

$$\Psi_k(N + 1; E \gamma_k LS) = \mathcal{A} \left[\Psi_t(N; \gamma_t L_{\gamma_t} S_{\gamma_t}) \bullet |kl\rangle \right]$$

where

$$\Psi_t = \sum_{M_b+1}^{M} c_i \Phi(N; \gamma_i L_t S_t). \tag{10.13}$$

In the above, Ψ_b represents the bound or discrete perturber or, in some cases, the embedded Rydberg series, and Ψ_k represents the continuum orbital coupled to a correlated target wave function Ψ_t. For example, in figure 10.1, the perturber is $1s2p^2 \, {}^2D$ whereas the target of a kd continuum orbital is $1s^2 \, {}^1S$: both could be MCHF expansions. In this definition, it is assumed that Ψ_t is normalized. When the target is a single configuration state, the weight of the continuum configuration state is unity. The normalization of the wave function for the continuum state is implicitly determined by the normalization of the continuum orbital. Consequently, the expansion coefficients for the bound perturber may be larger than unity when the interaction with the continuum component is strong.

The equations for $\Psi(N + 1; E \gamma LS)$, referred to in the literature as close-coupling equations, are exactly the same as the MCHF equations for a multiconfiguration expansion (Burke 1996). It can be shown (Burke and Seaton 1971) that, in this case, $\delta[\langle\Psi|\mathcal{H} - E|\Psi\rangle - \tan\delta_l] = 0$, and that for solutions of the close-coupling equations, the error in $\tan\delta_l$ is of quadratic order in the error of the wave function.

Not all the 'bound' components of the wave function are perturbers in the sense that they introduce irregularities in the wave function; some arise from the orthogonality constraints. Consider $1sks \, {}^1S$, similar to the $1s2s \, {}^1S$ case studied earlier. The best approximation would be one where ks was not required to be orthogonal to the $1s$ target orbital. Let $\overline{ks}$ be non-orthogonal to $1s$. Then the $1s$ component may be projected out so that $|\overline{ks}\rangle = |ks\rangle + \alpha|1s\rangle$, and the single configuration state $1s\overline{ks} \, {}^1S$ becomes a linear combination of $1sks$ and $1s^2$. Note that, unlike the bound state problem, projecting out a bound component does not affect the normalization; $\overline{ks}$ and ks have the same asymptotic form.

Table 10.1 shows some of the output from the continuum program CMCHF. As before, the wave function expansion is given in the cfg.inp file. This usually is the output of an MCHF calculation for the target, so the target energy is given on the header card. The continuum orbital is coupled to the target ($1s$ in this case) and the bound configuration states inserted ahead of the continuum component. The queries are much the same as before with the addition of the request for information about the range of energies for which calculations are

Table 10.1. Continuum calculation for $1sks$ ^1S.

---

cfg.inp file

```
   H       1S          -0.5

   1s( 2)
      1S0
   1s( 1)  ks( 1)                        1.0000000
      2S1     2S1     1S0
```

CMCHF calculation

```
>Cmchf
 ATOM, TERM, Z in FORMAT(A,A,F) :
>H-,1S,1.
 There are   2 orbitals as follows:
    1s  ks
 Enter orbitals to be varied:
   (ALL,NONE,SOME,NIT=,comma delimited list)
>=1
 Default electron parameters ? (Y/N)
>y
 Default values (NO,REL,STRONG) ? (Y/N)
>y
 H-    1S         1.   220  2  1  2  F  T
 Default values for other parameters ? (Y/N)
>y
 Enter k^2: MIN, DELTA, MAX
>.2,.2,.2

 *** CALCULATIONS FOR k^2 =    0.2000
 Outer region reduced by  98 points in CSOLVE.
  End of range:  35.2512151145885397  au.
       < 1s| ks>= 1.2D+00
        ks       .5036787    -1.6381188    .6953687 c  1.00D+00
   1    8.603918
  ...

          ITERATION NUMBER  5
          ----------------
          CONVERGENCE CRITERIA = 2.9D-06
          E( ks 1s) =-3.30713E-01   E( 1s ks) = 1.00000E-05
          EL      ED/DELTA          AZ        NORM       DPM
       < 1s| ks>= 1.5D-05
          ks       1.1348311      -.6206052    .9999029 c  1.79D-06
    1    1.169144
```

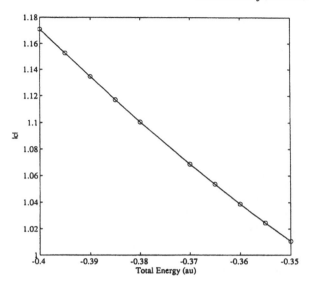

Figure 10.2. Absolute value of the $1s^2$ expansion coefficient in $1sks$ of H$^-$, as a function of the energy, in the vicinity of $E(1s^2) = -0.375$ au.

to be performed. Instead of diagonalization, the program needs to determine the weights of the discrete components, $1s^2$ in the present case. Suppose the interaction matrix is partitioned so that

$$\mathbf{H} = \begin{bmatrix} \mathbf{H}^{bb} & \mathbf{H}^{bk} \\ \mathbf{H}^{kb} & \mathbf{H}^{kk} \end{bmatrix}$$

and let $\mathbf{c}_b$ be the expansion coefficients of the bound configuration states and $\mathbf{c}_t$ the expansion coefficients for the target of the continuum component. Then the Galerkin conditions for the expansion coefficients are

$$(\mathbf{H}^{bb} - E)\mathbf{c}_b + \mathbf{H}^{bk}\mathbf{c}_t = 0. \tag{10.14}$$

When $(\mathbf{H}^{bb} - E)$ is singular, the present implementation is not able to obtain a solution. In the case of table 10.1, this singularity occurs at $E = -0.375$ au. Though the mixing coefficients change sign, figure 10.2 shows that the absolute value of the coefficient is totally regular. For this reason, states such as $1s^2$ are called 'pseudostates'. The calculation in table 10.1 is for $E = -0.4$ au; a solution for the Coulomb equation is obtained, omitting exchange which leads to a mixing coefficient of -0.026435 au. But as the continuum equation is solved by the SCF procedure including exchange, the final value becomes -1.171063 au.

The $1s2p^2$ ^2D perturber affects the wave function in quite a different manner. It is a configuration state which interacts with $1s^2kd$ and represents

Table 10.2. Wave function expansion for $1s2p^2\ {}^2D$ perturber imbedded in the $1s^2kd\ {}^2D$ continuum. The target energy is -7.2364152 au.

Li+	1S		−7.2364152		
1s(1)	2p(2)				
2S1	1D2	2D0			
1s(1)	3p(2)				
2S1	1D2	2D0			
1s(1)	3d2(2)				
2S1	1D2	2D0			
1s(1)	2s(1)	3d1(1)			
2S1	2S1	2D1	1S0	2D0	
1s(1)	2s(1)	3d3(1)			
2S1	2S1	2D1	3S0	2D0	
1s(2)	kd4(1)				1.0000000
1S0	2D1	2D0			

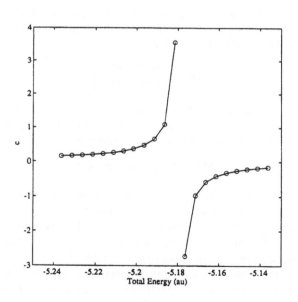

Figure 10.3. The $1s2p^2$ expansion coefficient in the $1s^2kd\ {}^2D$ continuum in Li as a function of the energy.

correlation but, at certain energies, becomes the dominant component of the wave function. Table 10.2 shows a cfg.inp file for this case. In order to capture as much correlation as possible, the d orbitals are all non-orthogonal. Figure 10.3 shows the mixing coefficient of the $1s2p^2$ component of the wave function as a function of energy. As the energy goes through the resonance, the mixing coefficient becomes extremely large and changes sign. The shape of this 'resonance' is related to its position E_r and width Γ. In a study of configuration interaction in the continuum, Fano (1961) showed that for an isolated resonance

$$c(E)^2 = \frac{1}{\pi} \frac{(1/2)\Gamma}{(E - E_r)^2 + (1/4)\Gamma^2}. \tag{10.15}$$

10.3 Photoionization or photodetachment

Suppose atoms are in an initial state with energy E_0 and that the first ionization limit is E_{γ_l}. The absorption of a photon of energy ω greater than $E_{\gamma_l} - E_0$ yields a photoelectron with kinetic energy $k^2/2$, so that the total energy of the final state is

$$E = E_0 + \omega = E_{\gamma_l} + k^2/2. \tag{10.16}$$

When the 'target' has positive charge, the process is photoionization, but in the case where the initial atom is a negative atom (also called an anion), the process is referred to as photodetachment.

Let the wave function for the initial state be Ψ_0 and the wave function for the continuum state be $\Psi_{\gamma E}$. Then the photoabsorption cross section is

$$\sigma_\gamma = \frac{4\pi^2}{3c} \omega \langle \Psi_0 || \sum_{j=1}^{N+1} r_j^{(1)} || \Psi_{\gamma E} \rangle^2 / (2L_0 + 1). \tag{10.17}$$

The factor $4\pi^2/3c = 2.68909$ MB, where MB stands for megabarn which equals 10^{-18} cm^2. In this formula ω is in atomic units.

At each energy E we may define a continuum oscillator strength

$$\frac{df}{dE} = \frac{2\omega}{3} \langle \Psi_0 || \sum_{j=1}^{N+1} r_j^{(1)} || \Psi_{\gamma E} \rangle^2 / (2L_0 + 1). \tag{10.18}$$

Combining this result with (10.17), we get

$$\sigma_\gamma = \frac{2\pi^2}{c} \frac{df}{dE} \tag{10.19}$$

$$= 4.03364 \quad \mathrm{MB_{au}} \frac{df}{dE} \tag{10.20}$$

where 'au' designates atomic units for energy-related quantities.

An interesting rule in connection with the oscillator strength distribution is the *Thomas–Reiche–Kuhn* sum rule (Thomas and Reiche 1925, Kuhn 1925)

$$S(0) = \sum_n f_n + \int_{E_n}^{\infty} \frac{\mathrm{d}f}{\mathrm{d}E} \, \mathrm{d}E \tag{10.21}$$

$$= N + 1 \tag{10.22}$$

where $\sum_n$ is the sum over the discrete states with oscillator strength f_n and $N+1$ is the number of electrons in the initial state. By representing each discrete f_n by an area, a histogram may be constructed such that each rectangle has height $f_n/\Delta E_n$ and width ΔE_n. The former represents the $\mathrm{d}f/\mathrm{d}E$ in the discrete spectrum and goes over smoothly across the ionization limit into the continuum. Figure 10.4 shows the oscillator strength distribution in hydrogen for transitions $1s \rightarrow np$ in the discrete region and $1s \rightarrow kp$ in the continuum. In order to determine the histogram, consider $E(n) = -1/(2n^2)$ as a continuous function of n and consider the tangents to the curve at each integer n. The intersections of these tangents define a set n^*, the ends of the rectangles and hence the width of each rectangle. As n increases $\Delta E_n \rightarrow \mathrm{d}E/\mathrm{d}n = 1/n^3$. The circles in figure 10.4 show the plot of $n^3 f_n$ as a function of $E(n)$. This also shows that, for a regular Rydberg series, $n^3 f_n$ should approach a constant as n increases.

In many situations, correlation in the initial state is extremely important but final state correlation should not be neglected. Let us consider photodetachment

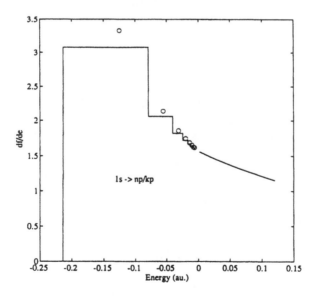

Figure 10.4. The $1s \rightarrow np/kp$ oscillator strength across the ionization threshold in hydrogen.

in H⁻. Using a natural orbital expansion, a wave function including orbitals up to $n = 6$ can be obtained by the methods discussed in chapter 5. For the continuum state, the easiest calculations are fixed-target types of calculation. Table 10.3 shows a possible wave function expansion. All the 9*l* orbitals are 'closed channels' in that they have a positive binding energy relative to their target, $E_{target} - E$. These are not spectroscopic orbitals so the only significance of the symbol '9' is in the determination of initial estimates. The target orbitals are all hydrogenic. By the MCHF conventions, all the orbitals will be orthogonal to 1*s* and 2*s*, but additional orthogonalities are required so that there is no overlap between the different channels. These orthogonality requirements are given after the configuration state list. In a CMCHF calculation, the only orbitals varied would be the 9*l* and kp_c orbitals. For certain energy regions some of the closed channels contribute very little causing convergence difficulties; calculations are simplified if these are omitted, but it is attractive to have the same expansion over the entire energy region of interest.

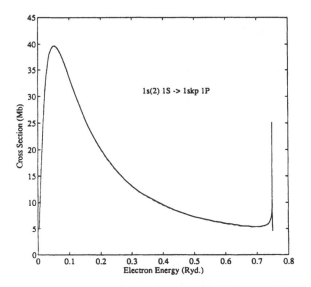

Figure 10.5. The $1s \rightarrow kp$ photodetachment cross section in H⁻ with variational correlation orbitals.

Table 10.4 shows an expansion using the non-orthogonal, natural orbital expansion for the representation of correlation. All orbitals are now orthogonal to 1*s*, and orbitals of the same symmetry and the same set index, are orthogonal; no additional orthogonalities need be specified. Since all orbitals are to be varied except 1*s*, it is convenient to have it appear first in the list of electrons produced by CMCHF; this is achieved by having the unoccupied 1*s* inserted in the first configuration state. Figure 10.5 shows the length and velocity form of the cross

Table 10.3. cfg.inp file describing a final state wave function for photodetachment of H⁻ with target energy of −0.5000000 au.

H-	1P	-0.5000000	
2s(1)	9p1(1)		
2S1	2P1	1P0	
2ph(1)	9s1(1)		
2P1	2S1	1P0	
2ph(1)	9d1(1)		
2P1	2D1	1P0	
3sh(1)	9p2(1)		
2S1	2P1	1P0	
3ph(1)	9s2(1)		
2P1	2S1	1P0	
3ph(1)	9d2(1)		
2P1	2D1	1P0	
3dh(1)	9p3(1)		
2D1	2P1	1P0	
3dh(1)	9f1(1)		
2D1	2F1	1P0	
1s(1)	kpc(1)		1.000000
2S1	2P1	1P0	
*****			
2ph 9p2			
2ph 9p3			
3sh 9s2			
3dh 9d2			

section, obtained by running first MLTPOL to generate the angular data and then PHOTO, similar to LSTR described in the earlier chapter. Notice that length and velocity are in excellent agreement and that a resonance occurs just before the $2s$ ionization limit where the electron energy would be 0.75 Ryd and is due to the $2s2p$ ^1P perturber.

Figure 10.6 shows a magnified portion of the peak near the threshold. The curves in excellent agreement are the results of the fully variational, natural orbital expansion (length and velocity barely distinguishable, appearing as a single solid curve) whereas the dotted and dot-dashed curves are length and velocity values for the fixed-target model. This figure shows the advantage of the variational approach for the correlation effects in the final state.

Table 10.4. cfg.inp file for a final state wave function for photodetachment in H⁻, using natural orbital expansion.

```
H-      1P        -0.5000000

1s( 0) 2s1( 1) 2p1( 1)
   1S0      2S1      2P1      2S0      1P0
3s1( 1) 4p1( 1)
   2S1      2P1      1P0
3p1( 1) 4s1( 1)
   2P1      2S1      1P0
3p2( 1) 4d2( 1)
   2P1      2D1      1P0
3d2( 1) 4p2( 1)
   2D1      2P1      1P0
3d1( 1) 9f1( 1)
   2D1      2F1      1P0
1s( 1) kpc( 1)                              1.000000
   2S1      2P1      1P0
```

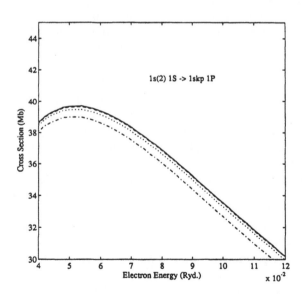

Figure 10.6. A comparison of fixed target length and velocity forms of the $1s \rightarrow kp$ photodetachment cross section in H⁻ with the cross sections from variational correlation orbitals.

10.4   Autoionization

Because of the interaction between a discrete component, Ψ_b with energy E_b, and the continuum, the former is spread over the adjacent continuum and results in a resonance line shape with centre $E_r = E_b + F(E)$, where $F(E)$ represents the shift of the resonance due to the interaction. Let $V_{E_b} = \langle \Psi_b | \mathcal{H} | \Psi_k \rangle$. Then, by the *golden rule* (Cowan 1981), the width at half height is

$$\Gamma = 2\pi |V_{E_b}|^2, \qquad (10.23)$$

and the decay rate for autoionization is given by

$$A = \frac{\Gamma}{\hbar} = \frac{2\pi V_{E_b}^2}{\hbar} = 2.5976 \times 10^{17} V_{E_b}^2 \quad \left[\frac{s^{-1}}{au} \right]. \qquad (10.24)$$

Finally, the lifetime, assuming we have only the one autoionizing decay process, is

$$\tau = 1/A. \qquad (10.25)$$

Notice that the golden rule only requires the bound and continuum components at a given energy and usually these are computed independently, omitting the interaction while determining the continuum orbital. Frequently it is simply a Hartree–Fock orbital. In some cases, like the $1sks\ ^1S$ considered earlier, it is desirable to compute the continuum including some interactions. In fact, it may happen that the wave function is not fully partitioned into disjoint bound and continuum components, that some overlap is present. Then the rule must be generalized to

$$V_{E_b} = \langle \Psi_b | \mathcal{H} - E_b | \Psi_k \rangle. \qquad (10.26)$$

Calculation of the autoionization rate using (10.24) can be performed by first using MCHF to obtain the position and composition of the perturber, then adding to the cfg.out file the continuum component which may either have a correlated target or be a single configuration state. Sufficient orthogonality must be maintained as required for the determination of the interaction matrix by NONH. The energy of the continuum electron is the difference of the energy of the target and the perturber. Ideally, both should be determined accurately, but the orthogonality constraints my require a compromise. In the calculation for Li ^2D, a wave function expansion as shown in table 10.6 was selected with $1s$ common to both the perturber and the continuum component. For the best possible perturber position, all orbitals were optimized on the latter. The results of the MCHF calculation (without the continuum CSF) were saved as Li2D.w and Li2D.c and the continuum component added. In the subsequent AUTO calculation the same $1s$ was taken for the target, but the energy of the target, when requested as input, was the Hartree–Fock energy that would have been obtained if the $1s$ could have been separately optimized. The AUTO calculation first determines a fixed-core Hartree–Fock continuum orbital and then evaluates the interaction between the continuum component and the perturber. Results are written to a file auto.dat which summarizes the relevant data for the calculation. The width of the ^2D is found to be 11.69 meV, somewhat larger than measured in an optical line-broadening experiment by Cederquist and Mannervik (1985) who obtained a value of 10.5 ± 0.3 meV. The experimental position is reported relative to the $1s^2 2s$ ^2S ground state of lithium. Adding the ionization potential of Li (0.19814 au) to the energy of the continuum electron, converting to eV, we get 60.05 meV, in good agreement with the experimentally observed value of 61.062 ± 0.006 (McIlrath and Lucatorto 1977). By totally ignoring correlation in the target, the present calculation has not dealt carefully enough with the resonance position. This was compensated by using a Hartree–Fock energy for the target, but such assumptions make results unreliable. A more careful study has been reported (Brage *et al* 1993) where results are compared with other theories and experiment. Even so, this example shows that ideally, one would like to independently optimize the orbitals describing the target and the orbitals for the perturber. This can be achieved through the use of non-orthogonal orbitals, without restrictions, as described by Zatsarinny (1996).

Many core-excited levels are metastable against autoionization. Consider the $2p^5 3s3p$ states of Na for which the only state with a lower energy is the Na$^+$ $2p^6$ ground state. Then the $2p^5 3s3p$ ^2L terms are perturbers in the $2p^6 kl$ continuum and, as long as the parity is the same, the $2p^5 3s3p$ ^2L terms may interact with a continuum and autoionize. A more careful analysis reveals that the decay may be to $2p^6 kd$ ^2D for $J = 5/2, 3/2$ and $2p^6 ks$ ^2S for $J = 1/2$, leaving the ion in the Na$^+$ $2p^6$ ^1S ground state. This suggests that only doublets can autoionize, but the spin–orbit interaction, or more correctly, the Breit–Pauli interaction, mixes the $2p^5 3s3p$ ^4L terms with the $2p^5 3s3p$ ^2L terms, thereby opening a radiationless decay channel also for the ^4L terms. For such cases, Breit–Pauli calculations must first be performed for all $2p^5 3s3p$ terms, both quartets and doublets, and correlation must be included†. Such calculations are not simple. Non-orthogonal orbitals may be used for the non-relativistic caclulations but the BREIT program assumes orbitals with different labels are orthogonal. It is possible, though not elegant, to merge two int.1st files, where the non-relativistic integrals are produced by NONH and only the relativistic integrals are generated by BREIT.

A simpler example is shown in tables 10.6–10.9. Here the non-relativistic example of table 10.5 has been extended to include also the quartet states, requiring a Breit–Pauli calculation. The GENCL run now generates all possible terms with J values overlaping those of $1s2p^2$. In the reference set, 1s(1)2p(1)3p(1) should be included: for the ^2D considered earlier, a reduced form was used, but for the Breit–Pauli calculation, where orbitals form a basis and are not varied these need to be included for better accuracy. The output from GENCL needs to be modified in the sense that the 3d3 of

---

† The complete output of such an example is included with the code.

Table 10.5. MCHf autoionization calculation for Li $1s2p^2$ ^2D using the expansion of table 10.2.

```
>Auto
 Name of State
>Li2D
 Select input file for discrete state(s):
    1  name.c
    2  name.l
    3  name.j
>1
 ATOM, TERM, Z in FORMAT(A,A,F) :
>Li,2D,3.
 Give energy of core state (a.u.):
>-7.23641520
  ...
   Outer region reduced by   93points in CSOLVE.
   End of range:     16.060899430366 au.
          kdc      0.0070780       0.7874451   0.7874451 c  2.70D-01

          ITERATION NUMBER  1
          ----------------

          CONVERGENCE CRITERIA =  1.0D-06
          EL      ED/DELTA        AZ           NORM      DPM
          kdc     0.0166828       0.8331957  1.0581001 c  5.49D-02

          ITERATION NUMBER  2
          ----------------

          CONVERGENCE CRITERIA =  1.3D-06
          EL      ED/DELTA        AZ           NORM      DPM
          kdc     0.0162580       0.8316465  0.9981406 c  1.86D-03
  ...
AUTO.DAT

          Continuum State:
           1s( 2) kdc( 1)
               1S0     2D1     2D0
          Continuum electron: k^2/2=   2.00863 au.
          Discrete State, main component:
       1       1s( 1)  2p( 2)
               2S1     1D2     2D0
          Auto-ionization Data
                      Interaction, V        =  -8.2674E-03 au
                      Half-Line Width       =   2.1473E-04 au
                                                5.8427E-03 ev
                                                4.7127E+01 cm-1
                      Auto-ionization rate =    1.7755E+13 s-1
                      Half-Life            =    5.6324E-14 s
```

Table 10.6. Gencl and CI part of the calculation for the autoionization of the $1s2p^2$ $^4P_{5/2,3/2}$ for Li.

```
>Gencl
              Header  ?
>
      Closed Shells  ?
>
      Reference Set  ?
>1s(1)2p(2)
                  2  ?
>1s(1)2p(1)3p(1)
                  3  ?
>1s(1)3p(2)
                  4  ?
>1s(1)2s(1)3d3(1)
                  5  ?
>1s(1)3d2(2)
                  6  ?
>1s(1)2s(2)
                  7  ?
>
         Active Set  ?
>
       Replacements  ?
>
         Final Terms  ?
>
           ...
>Ci
  Name of State
>Li-even
  Is this a relativistic calculation ? (Y/N) :
>y
  Is mass-polarization to be included ? (Y/N) :
>y
     Gradient or Slater integral form ? (G/S) :
>g
  Default Rydberg constant (Y or N) ?
>y
  The size of the matrix is     28
  Enter the approximate number of eigenvalues required
>3
  Maximum and minimum values of 2*J ?
>5,1
  Do you want the matrix printed? (Y or N)
>n
```

Table 10.7. Analysis of results using LEVELS and COMP.

```
>Levels
 Enter name and type (.l or .j) of file
>LiBP.j
```

    ENERGY LEVELS
    Z =   3   3 electrons

Configuration	Term	J	Total Energy (a.u.)	Energy Level (cm-1)
1s.2s(2)	2S	0.5	-5.3966853	0.00
1s.2p(2)3P	4P	2.5	-5.2406540	34242.23
		0.5	-5.2406515	34242.78
		1.5	-5.2406435	34244.53
1s.2p(2)1D	2D	2.5	-5.2302250	36530.96
		1.5	-5.2301980	36536.88
1s.2p(2)3P	2P	0.5	-5.2080392	41399.80
		1.5	-5.2080227	41403.42

```
>Comp
 Enter name of .c file
>LiBP
  What tolerance? :
>.0001
 Compositions from:
      1   name.c
      2   name.l
      3   name.j
 Enter selection
>3
         ...
 1s.2p(2)3P2  4P    2.5   -5.24065400
    0.9916747  1s.2p(2)3P2  4P
    0.0804625  1s.3d2(2)3P2   4P
   -0.0773913  1s.3p(2)3P2   4P
   -0.0641634  1s.2p_3P.3p   4P
   -0.0008108  1s.2p(2)1D2  2D
   -0.0003068  1s.2s_3S.3d3  2D
   -0.0001124  1s.2s_1S.3d1  2D
```

Table 10.8. Breit–Pauli autoionization calculations for $1s2p^2$ $^4P_{5/2,3/2}$ levels of Li with target energy of -7.23694760 au.

```
>Auto
  Name of State
>Li-BP
 Select input file for discrete state(s):
    1  name.c
    2  name.l
    3  name.j
>3
 ATOM, TERM, Z in FORMAT(A,A,F) :
>Li,4P,3.
 There are   8 orbitals as follows:
    1s  2p  3p  2s  3d1  3d3  3d2  kdc
    Is continuum function to be computed? (y/n)
>y
   *** CALCULATIONS FOR k^2 =    3.99259
   Outer region reduced by   93points in CSOLVE.
   End of range:     16.060899430366
    kdc      0.0070227      0.7830293    0.7830293  c  2.77D-01


   ITERATION NUMBER   1
   ----------------
   CONVERGENCE CRITERIA =   1.0D-06
   EL      ED/DELTA       AZ          NORM       DPM
   kdc       0.0165765      0.8285622   1.0581496  c  5.50D-02

   ITERATION NUMBER   2
   ----------------
   CONVERGENCE CRITERIA =   1.3D-06
   EL      ED/DELTA       AZ          NORM       DPM
   kdc       0.0161535      0.8270186   0.9981370  c  1.87D-03
   . . .
```

1s(1) 2s(1) 3d3(1) should be 3d1 whenever the resultant coupling for 1s(1) 2s(1) is 1S0, as in table 10.5. (There is no simple way of achieving this with GENCL.) In the present case, there are no overlap integrals and BREIT is able to generate a correct integral list. CI needs to be run in order to obtain the eigenvector expansions. This is shown in table 10.6. Since Li is a light atom, the mass-polarization correction was included in generating the interaction matrix and approximately three eigenvalues/eigenvectors were requested. The results of this calculation are shown in table 10.7. Note the inversion of the order

Table 10.9. The auto.dat file for $1s2p^2\ {}^4P_{5/2,3/2}$.

```
        Energy of Target:     -7.23694760
        Continuum State:
          1s( 2) kdc( 1)
              1S0      2D1      2D0


        Continuum electron: k^2/2=   1.99629 au.
        Discrete State, main component:
  4         1s( 1)  2p( 2)                        2*J =   5
              2S1       3P2       4P0
        Auto-ionization Data
                  Interaction, V        =    7.8286E-06 au
                  Half-Line Width       =    1.9254E-10 au
                                             5.2390E-09 ev
                                             4.2257E-05 cm-1
                  Auto-ionization rate =    1.5920E+07 s-1
                  Half-Life             =    6.2815E-08 s


        Continuum electron: k^2/2=   2.00672 au.
        Discrete State, main component:
  2         1s( 1)  2p( 2)                        2*J =   5
              2S1       1D2       2D0
        Auto-ionization Data
                  Interaction, V        =   -8.1517E-03 au
                  Half-Line Width       =    2.0876E-04 au
                                             5.6804E-03 ev
                                             4.5818E+01 cm-1
                  Auto-ionization rate =    1.7261E+13 s-1
                  Half-Life             =    5.7933E-14 s
                  . . .
```

---

of the J values as shown by the LEVELS program. The composition of the eigenvector (with small components omitted) can be obtained with the use of the COMP program. Note, also, the rather small admixture of a ^2D component. To obtain the autoionization rates, the energy of the target needs to be added to the header of the file containing the wave function expansion (which should now be a relativistic value), the target with its continuum added to the expansion with a coefficient of unity (as shown in table 10.5), the NONH program run and the cfg.inp moved to LiBP.c. The file LiBP.w would be a copy of Li2D.w.

Table 10.8 shows part of the execution of the AUTO program and table 10.9 some of the output. The autoionization rate for $^2D_{5/2}$ is much greater than for $^4P_{5/2}$ from which it follows that the lifetime of the latter is much longer. The energy of the continuum electron for $^2D_{5/2}$ is now a bit lower, yielding a position of 59.99 meV, compared with the experimental value of 61.062 ± 0.006 meV (McIlrath and Mannervik 1977) mentioned earlier, but the width has decreased to 11.36 meV, a correction in the right direction.

10.5   Computational aspects

The numerical solution of the Hartree–Fock or close-coupling equation for the continuum orbital is closely related to that of a bound state orbital, with two differences:

(i)   the binding energy of the continuum electron is known
(ii)   there is no boundary condition at large r, other than amplitude.

As a result, it is possible to integrate outward for a solution having the proper behaviour near the origin.   Since the integro-differential equation is homogeneous, normalization is arbitrary; however, ultimately a properly normalized wave function is required and so the SCF procedure normalizes the orbital at each iteration. Consequently, as the iterations converge, the resulting computed solution should be normalized. This NORM is used as an indication of accuracy and convergence.

The present code was designed around the logarithmic grid of the MCHF program. This restricts its application to studies such as photoionization near threshold where the continuum orbital is not oscillating too rapidly for the underlying grid. The grid is scaled by $1/Z$ which, for a given r, will result in closer grid points. To help alleviate the problem associated with the logarithmic grid, outward integration proceeds in the usual fashion up to the classical turning point r_{NJ}, where

$$k^2 \approx -\frac{2}{r_{NJ}} \left[Z - Y_{kl}(r_{NJ}) + \frac{l(l+1)}{r_{NJ}^2} \right] \qquad (10.27)$$

after which the step-size is reduced by a factor of two.   This improves the accuracy of outward integration and normalization, but only values at the regular grid-points are saved for the computation of exchange and interaction with the perturber. The choice of outer cut-off r_M is dictated by the grid so that

$$k\Delta r < 1 \quad \text{if} \quad r < r_M, \qquad (10.28)$$

where Δr is the distance between two grid points. This assures that reasonable accuracy is maintained. For normalization, we also need to determine the onset of the Coulomb region. Two requirements need to be met, namely a constant effective nuclear charge and an exchange function that is sufficiently small. The point r_{MP} is such that

$$\frac{\Delta Y_{kl}(r)}{r} < 10^{-4} \quad \text{if} \quad r > r_{MP}, \qquad (10.29)$$

where $\Delta Y_{kl}(r)$ is the difference in the value of $Y_{kl}(r)$ in two consecutive grid points. The point r_{MJ} will be defined as the first point greater than r_{MP} where the exchange function is sufficiently small;

$$\frac{X_{kl}^{CH}(r)}{Y_{kl}(r)} < 0.0025 \quad \text{if} \quad r > r_{MJ}. \tag{10.30}$$

Clearly, in order to normalize our solution correctly, we require $r_{MJ} < r_M$. The parameters in these tests are somewhat arbitrary and depend on the accuracy to which a solution is required.

The normalization constant and phase shift can be determined simply from a few points near the end of the range. In the present atomic structure package, we used the Maple symbol manipulation package (Char *et al* 1990) and determine the iterates analytically up to four iterations. With a sufficiently large r value, the iterations can converge to an error of less than 10^{-9}. But because of the logarithmic grid, leading to constraints on the range, the accuracy in the present calculations is reduced in some cases to three or four significant digits.

Having $\zeta(r)$ determined at a given large r value, we can determine the phase function $\phi(r)$ from equation (10.7)), which gives

$$\tan \phi(r) = \zeta \left/ \left(\frac{P'}{P} + \frac{\zeta'}{2\zeta} \right) \right. \tag{10.31}$$

where P' and ζ' are the derivatives of $P(kl; r)$ and $\zeta(r)$, respectively, at the given r value. Then the energy-normalized regular and irregular Coulomb functions can be written as

$$F(r) = \sqrt{\frac{2}{\pi \zeta}} \sin \phi(r) \tag{10.32}$$

$$G(r) = \sqrt{\frac{2}{\pi \zeta}} \cos \phi(r). \tag{10.33}$$

Now let

$$P(kl; r) = aF(r) + bG(r). \tag{10.34}$$

The unknown coefficients (a and b) can be determined by matching either $P(kl; r)$ and $P'(kl; r)$ at a given point to the known Coulomb functions, or $P(kl; r)$ at two adjacent points. Let $\tan \delta_l = b/a$ and $c^2 = a^2 + b^2$. Then equation (10.34) becomes

$$P(kl; r) = \sqrt{\frac{2}{\pi \zeta}} c \, (\cos \delta_l \sin \phi(r) + \sin \delta_l \cos \phi(r))$$

$$= c \sqrt{\frac{2}{\pi \zeta}} \sin(\phi(r) + \delta_l). \tag{10.35}$$

Thus both the amplitude c and the phase shift δ_l of the computed solution have been determined. Note that the sign of the solution is still arbitrary and the phase shift may differ by $\pm\pi$. In CMCHF the phase factor is chosen so that the radial function is positive near the origin.

10.6 Exercises

(i)   Determine the oscillator strength distribution for excitation from the helium $1s^2$ ground state.

(ii) Determine the position and the width of the autoionizing $2s2p\ ^3P$ state of He. Begin with the simplest possible approximation and then systematically improve the description of the autoionizing state.

(iii) Perform the CMCHF calculation for the expansion of table 10.3 at an energy of $k^2 = 1.0$. Consider the diagonal energy parameters. Why are the parameters for 9p2, 9s2, 9d2, 9p3, 9f1 all the same? Can you explain the diagonal energy parameters for all the bound channels?

Appendix A

Angular Momentum Theory

In this appendix some of the angular momentum theory needed for the construction of configuration state functions is summarized. Also, spherical tensor operators and reduced matrix elements are discussed briefly. For a more comprehensive treatment we refer the reader to Cowan (1981). Not discussed here are the use of the creation operator for constructing antisymmetrized states, or the expression of sums of operators f_i acting on the single electron i, or sums of operators g_{ij} operating on the pair of electrons i and j, in terms of creation and annihilation operators, concepts that are part of second quantization. Together with the notion of quasispin, this approach has been shown to lead to faster methods for the evaluation of matrix elements (Gaigalas and Rudzikas 1996). For a review of second quantization, see Judd (1996).

A.1 Angular momentum operators

A general angular momentum operator, j, is defined as a set of three operators, (j_x, j_y, j_z), that satisfy the cyclic commutation relations

$$[j_x, j_y] = \mathrm{i}j_z, \qquad [j_y, j_z] = \mathrm{i}j_x, \qquad [j_z, j_x] = \mathrm{i}j_y. \qquad (A.1)$$

From these relations it is easy to show that

$$[j^2, j_z] = 0 \qquad (A.2)$$

where $j^2 = j_x^2 + j_y^2 + j_z^2$, and thus it is possible to find a complete set of simultaneous eigenfunctions to j^2 and j_z. The eigenfunctions depend on the space on which the angular momentum operators are acting, and normally there are several linearly independent eigenfunctions of j^2 and j_z with the same eigenvalue. These eigenfunctions can be separated by introducing additional observables, Γ, so that the simultaneous eigenfunctions of Γ, j^2 and j_z are completely determined by the corresponding quantum numbers.

Using the commutation relations (A.1) it can be shown that the possible eigenvalues of j^2 are

$$j(j+1) \qquad j = 0, \tfrac{1}{2}, 1, \tfrac{3}{2}, 2, \ldots . \qquad (A.3)$$

For any given value of j, the possible eigenvalues of j_z are

$$m = -j, -j+1, \ldots, j. \qquad (A.4)$$

If the corresponding eigenfunctions are denoted $|\gamma jm\rangle$, where γ are the additional quantum numbers needed to completely specify the eigenfunctions, we have the relations

$$j^2|\gamma jm\rangle = j(j+1)|\gamma jm\rangle, \qquad j_z|\gamma jm\rangle = m|\gamma jm\rangle \qquad (A.5)$$

$$(j_x \pm ij_y)|\gamma jm\rangle = e^{i\delta}[(j \mp m)(j \pm m + 1)]^{1/2}|\gamma jm \pm 1\rangle \qquad (A.6)$$

where $e^{i\delta}$ is an arbitrary phase factor which normally is choosen to be 1 (phase convention of Condon and Shortley).

A.1.1   Orbital angular momentum

The orbital angular momentum of an electron is described by the operator

$$l = -ir \times \nabla. \qquad (A.7)$$

The simultaneous eigenfunctions of l^2 and l_z are given by the *spherical harmonics* $Y_{lm_l}(\theta, \varphi)$

$$Y_{lm_l}(\theta, \varphi) = (-1)^{m_l} \left[\frac{(2l+1)(l-m_l)!}{4\pi(l+m_l)!} \right]^{1/2} P_l^{m_l}(\cos\theta)\, e^{im_l\varphi}, \qquad m_l \geqslant 0$$
$$Y_{l-m_l}(\theta, \varphi) = (-1)^{m_l} Y_{lm_l}(\theta, \varphi)$$

$$(A.8)$$

where $P_l^{m_l}(\cos\theta)$ is the associated Legendre polynomial of the first kind. In order for the spherical harmonics to be unique, m_l, and thus also l, must be integers and we have

$$l^2 Y_{lm_l}(\theta, \varphi) = l(l+1)Y_{lm_l}(\theta, \varphi), \qquad l_z Y_{lm_l}(\theta, \varphi) = m_l Y_{lm_l}(\theta, \varphi) \qquad (A.9)$$

where $l = 0, 1, 2, \ldots$ and $m_l = -l, -l+1, \ldots, l$. The first spherical harmonics are given in table A.1.

Table A.1. Spherical harmonics.

$$Y_{00}(\theta, \varphi) = \frac{1}{\sqrt{4\pi}}$$

$$Y_{1\pm1}(\theta, \varphi) = \mp\sqrt{\frac{3}{8\pi}}\,\sin\theta\,e^{\pm i\varphi}$$

$$Y_{10}(\theta, \varphi) = \sqrt{\frac{3}{4\pi}}\,\cos\theta$$

$$Y_{2\pm2}(\theta, \varphi) = \sqrt{\frac{15}{32\pi}}\,\sin^2\theta\,e^{\pm i2\varphi} \qquad (A.10)$$

$$Y_{2\pm1}(\theta, \varphi) = \mp\sqrt{\frac{15}{8\pi}}\,\sin\theta\,\cos\theta\,e^{\pm i\varphi}$$

$$Y_{20}(\theta, \varphi) = \sqrt{\frac{5}{16\pi}}(2\cos^2\theta - \sin^2\theta)$$

A.1.2 Spin angular momentum

In addition to the orbital angular momentum, the electron also has a spin angular momentum s. The spin is an intrinsic angular momentum with quantum numbers $s = 1/2$ and $m_s = \pm1/2$, introduced in 1925 by Uhlenbeck and Goudsmit on empirical grounds. Later, Dirac showed that the electron spin is a direct consequence of a proper relativistic treatment of the electron. The existence of the spin means that the electron is not just a point charge characterized by the three space co-ordinates, but has one more degree of freedom. We choose for this a spin co-ordinate, σ, which can only assume the values $\pm1/2$. Introducing the *spin-function* $\chi_{m_s}(\sigma)$, which is a simultaneous eigenfunction of s^2 and s_z

$$s^2\chi_{m_s}(\sigma) = \tfrac{3}{4}\chi_{m_s}(\sigma), \qquad s_z\chi_{m_s}(\sigma) = m_s\chi_{m_s}(\sigma), \qquad (A.11)$$

with the additional property

$$\chi_{m_s}(\sigma) = \delta_{m_s\sigma} \qquad (A.12)$$

the general wavefunction of the electron can be written

$$\phi(r, \sigma) = \phi(q) = \phi_1(r)\chi_{1/2}(\sigma) + \phi_2(r)\chi_{-1/2}(\sigma) \qquad (A.13)$$

where the space co-ordinates are allowed to run through the entire space, while σ assumes only the values $\pm1/2$. In this representation

$$|\phi(r, 1/2)|^2\,dr = |\phi_1(r)|^2\,dr \qquad (A.14)$$

and

$$|\phi(r, -1/2)|^2 \, \mathrm{d}r = |\phi_2(r)|^2 \, \mathrm{d}r \tag{A.15}$$

give the probability of finding the electron in the volume element $\mathrm{d}r$ centred at r with the spin up and down, respectively, and thus we have the following normalization condition

$$\sum_{\sigma=-1/2}^{1/2} \int_r |\phi(r, \sigma)|^2 \, \mathrm{d}r = \int_q |\phi(q)|^2 \, \mathrm{d}q = 1. \tag{A.16}$$

A.2 Coupling of two angular momenta

Two angular momentum operators j_1 and j_2 acting on different parts of a system, for example two different electrons or, respectively, the space- and spin-part of the electron wave function, can be coupled to form an angular momentum operator

$$J = j_1 + j_2 \tag{A.17}$$

for the total system. The operators Γ, j_1^2, j_2^2, J^2 and J_z define a complete set of commuting observables. The simultaneous eigenfunctions $|\gamma_1\gamma_2 j_1 j_2 J M\rangle$ corresponding to this set can be obtained as a linear combination

$$|\gamma_1\gamma_2 j_1 j_2 J M\rangle = \sum_{\substack{m_1, m_2 \\ m_1+m_2=M}} \langle j_1 j_2 m_1 m_2 | j_1 j_2 J M\rangle |\gamma_1 j_1 m_1\rangle |\gamma_2 j_2 m_2\rangle \tag{A.18}$$

where the coefficients $\langle j_1 j_2 m_1 m_2 | j_1 j_2 J M\rangle$ are known as vector-coupling, or Clebsch–Gordan, coefficients. For given j_1 and j_2 the possible values of J are $J = j_1 + j_2, j_1 + j_2 - 1, \ldots, |j_1 - j_2|$. With the normal phase conventions the Clebsch–Gordan coefficients are real and define a unitary matrix. Thus we have the following orthogonality relations:

$$\sum_{m_1 m_2} \langle j_1 j_2 m_1 m_2 | j_1 j_2 J M\rangle \langle j_1 j_2 m_1 m_2 | j_1 j_2 J' M'\rangle = \delta_{JJ'}\delta_{MM'} \tag{A.19}$$

and

$$\sum_{JM} \langle j_1 j_2 m_1 m_2 | j_1 j_2 J M\rangle \langle j_1 j_2 m_1' m_2' | j_1 j_2 J M\rangle = \delta_{m_1 m_1'}\delta_{m_2 m_2'}. \tag{A.20}$$

In addition, the Clebsch–Gordan coefficients have the following important symmetry property:

$$\langle j_1 j_2 m_1 m_2 | j_1 j_2 J M\rangle = (-1)^{j_1+j_2-J} \langle j_2 j_1 m_2 m_1 | j_2 j_1 J M\rangle. \tag{A.21}$$

Often the Clebsch–Gordan coefficients are given in terms of the more symmetric 3-j symbols

$$\langle j_1 j_2 m_1 m_2 | J M \rangle = (-1)^{j_1-j_2+M}[J]^{1/2} \begin{pmatrix} j_1 & j_2 & J \\ m_1 & m_2 & -M \end{pmatrix} \tag{A.22}$$

where we have introduced the notation

$$[j_1, j_2, \ldots, j_n] = (2j_1 + 1)(2j_2 + 1)\ldots(2j_n + 1). \tag{A.23}$$

The general $3j$-symbol

$$\begin{pmatrix} j_1 & j_2 & j_3 \\ m_1 & m_2 & m_3 \end{pmatrix} \tag{A.24}$$

has the symmetry properties

$$\begin{pmatrix} j_2 & j_1 & j_3 \\ m_2 & m_1 & m_3 \end{pmatrix} = \begin{pmatrix} j_1 & j_3 & j_2 \\ m_1 & m_3 & m_2 \end{pmatrix}$$

$$= (-1)^{j_1+j_2+j_3} \begin{pmatrix} j_1 & j_2 & j_3 \\ m_1 & m_2 & m_3 \end{pmatrix} \tag{A.25}$$

and is non-zero only if

$$\begin{aligned} m_i &= -j_i, -j_i + 1, \ldots, j_i, \quad i = 1, 2, 3 \\ m_1 + m_2 + m_3 &= 0 \\ j_1 + j_2 + j_3 &\quad \text{integral} \\ j_1 &\leqslant j_2 + j_3 \\ j_2 &\leqslant j_1 + j_3 \\ j_3 &\leqslant j_1 + j_2. \end{aligned} \tag{A.26}$$

A.3   Coupling of three and four angular momenta

Three angular momenta can be coupled together in two different ways that are grahically represented in figure A.1. Corresponding to these couplings there are two sets of eigenfunctions

$$|\gamma((j_1 j_2) J_{12} j_3) J M\rangle$$

and

$$|\gamma(j_1 (j_2 j_3) J_{23}) J M\rangle$$

where J_{12} and J_{23} are intermediate quantum numbers. Both these sets are complete, and thus there exists a transformation

$$|\gamma(j_1 (j_2 j_3) J_{23}) J M\rangle$$
$$= \sum_{J_{12}} \langle ((j_1 j_2) J_{12} j_3) J | (j_1 (j_2 j_3) J_{23}) J \rangle | \gamma((j_1 j_2) J_{12} j_3) J M \rangle.$$

$$\tag{A.27}$$

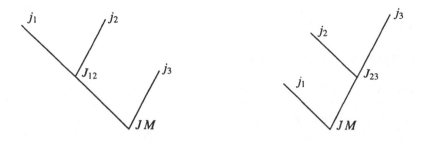

Figure A.1. Coupling of three angular momenta.

Here the summation need only to be done over J_{12}, since all other quantum numbers are common to the two sets of eigenfunctions. The coefficients are called recoupling coefficients and are often given in terms of *6-j symbols*

$$\langle((j_1 j_2) J_{12} j_3) J | (j_1 (j_2 j_3) J_{23}) J\rangle$$

$$= (-1)^{j_1+j_2+j_3+J} [J_{12}, J_{23}]^{1/2} \left\{ \begin{array}{ccc} j_1 & J_{12} & j_2 \\ j_3 & J_{23} & J \end{array} \right\}.$$

$$(A.28)$$

When dealing with four angular momenta, there is a whole set of different coupling schemes. Two possibilities are defined by the following sets of eigenfunctions

$$|\gamma((j_1 j_2) J_{12}(j_3 j_4) J_{34}) J M\rangle$$

and

$$|\gamma((j_1 j_3) J_{13}(j_2 j_4) J_{24}) J M\rangle$$

The transformation between these two sets is given by

$$|\gamma((j_1 j_3) J_{13}(j_2 j_4) J_{24}) J M\rangle$$

$$= \sum_{J_{12} J_{34}} \langle((j_1 j_2) J_{12} (j_3 j_4) J_{34}) J | ((j_1 j_3) J_{13} (j_2 j_4) J_{24}) J\rangle$$

$$\times |\gamma((j_1 j_2) J_{12}(j_3 j_4) J_{34}) J M\rangle. \qquad (A.29)$$

In this case the coefficients are given in terms of *9-j symbols*

$$\langle((j_1 j_2) J_{12} (j_3 j_4) J_{34}) J | ((j_1 j_3) J_{13} (j_2 j_4) J_{24}) J\rangle$$

$$= [J_{12}, J_{34}, J_{13}, J_{24}]^{1/2} \left\{ \begin{array}{ccc} j_1 & j_2 & J_{12} \\ j_3 & j_4 & J_{34} \\ J_{13} & J_{24} & J \end{array} \right\}.$$

$$(A.30)$$

The 3-j, 6-j and 9-j symbols are available from a number of tables, for example Rotenberg *et al* (1959).

A.4   Spherical tensor operators

Angular momentum operators can be used to classify not only wavefunctions, but also operators acting on these functions. If T denotes a general operator acting on the space spanned by the eigenfunctions of the angular momentum operator j, then a *spherical tensor operator* of rank k $\mathbf{T}^{(k)}$ can be defined as a set of $2k+1$ operators, $\{T_q^{(k)}; q = -k, -k+1, \ldots, k\}$, that fulfills the following commutation relations:

$$[j_z, T_q^{(k)}] = q T_q^{(k)}$$
$$[j_\pm, T_q^{(k)}] = [k(k+1) - q(q \pm 1)]^{1/2} T_{q\pm 1}^{(k)} \tag{A.31}$$

where

$$j_\pm = j_x \pm \mathrm{i} j_y. \tag{A.32}$$

As can easily be verified, the renormalized spherical harmonics

$$C_q^{(k)}(\theta, \varphi) = \sqrt{\frac{4\pi}{2k+1}} Y_{kq}(\theta, \varphi) \tag{A.33}$$

satisfy the commutation relations above.

Another important example is given by the vector operator $\mathbf{V}$. According to definition, the Cartesian components of a vector operator transform as the spatial co-ordinates x, y and z during a rotation. From these transformation properties it can be shown that the spherical components

$$V_1^{(1)} = -\frac{1}{\sqrt{2}}(V_x + \mathrm{i}V_y)$$
$$V_0^{(1)} = V_z \tag{A.34}$$
$$V_{-1}^{(1)} = \frac{1}{\sqrt{2}}(V_x - \mathrm{i}V_y)$$

define a spherical tensor operator $\mathbf{V}^{(1)}$ of rank 1. Specifically, the angular momentum operators themselves may by written as spherical tensor operators $\mathbf{j}^{(1)}$ of rank 1.

A.4.1   Scalar operators

According to the definition, an operator T that commutes with all components of the angular momentum operator is a scalar, or rank zero, operator. From the relations (A.31) it is easy to show that the matrix elements of T are diagonal in j and m

$$\langle \gamma j m | T | \gamma' j' m' \rangle = \delta_{jj'} \delta_{mm'} \langle \gamma j m | T | \gamma' j m \rangle. \tag{A.35}$$

In addition the matrix elements are independent of the m quantum number.

A.4.2 Matrix elements of scalar operators between coupled functions

Consider the coupled function

$$|\gamma_1\gamma_2 j_1 j_2 JM\rangle = \sum_{m_1 m_2}\langle j_1 j_2 m_1 m_2|j_1 j_2 JM\rangle|\gamma_1 j_1 m_1\rangle|\gamma_2 j_2 m_2\rangle. \quad (A.36)$$

If T is a scalar operator that operates only on the space spanned by $|\gamma_1 j_1 m_1\rangle$ then

$$\langle\gamma_1\gamma_2 j_1 j_2 JM|T|\gamma_1'\gamma_2' j_1' j_2' J'M'\rangle$$
$$= \sum_{m_1 m_2}\sum_{m_1' m_2'}\langle j_1 j_2 m_1 m_2|j_1 j_2 JM\rangle\langle j_1' j_2' m_1' m_2'|j_1' j_2' J'M'\rangle$$
$$\times \delta_{\gamma_2 j_2 m_2, \gamma_2' j_2' m_2'}\langle\gamma_1 j_1 m_1|T|\gamma_1' j_1' m_1'\rangle$$
$$= \delta_{\gamma_2 j_2, \gamma_2' j_2'}\langle\gamma_1 j_1 m_1|T|\gamma_1' j_1' m_1\rangle$$
$$\times \sum_{m_1 m_2}\langle j_1 j_2 m_1 m_2|j_1 j_2 JM\rangle\langle j_1' j_2' m_1 m_2|j_1' j_2' J'M'\rangle$$
$$= \delta_{\gamma_2 j_2, \gamma_2' j_2'}\delta_{j_1 j_1'}\delta_{JJ'}\delta_{MM'}\langle\gamma_1 j_1 m_1|T|\gamma_1' j_1 m_1\rangle$$

$$(A.37)$$

where $\langle\gamma_1 j_1 m_1|T|\gamma_1' j_1 m_1\rangle$ is independent of m_1. Similarly, if T operates only on the space spanned by $|\gamma_2 j_2 m_2\rangle$ then

$$\langle\gamma_1\gamma_2 j_1 j_2 JM|T|\gamma_1'\gamma_2' j_1' j_2' J'M'\rangle = \delta_{\gamma_1 j_1, \gamma_1' j_1'}\delta_{j_2 j_2'}\delta_{JJ'}\delta_{MM'}\langle\gamma_2 j_2 m_2|T|\gamma_2' j_2 m_2\rangle.$$

$$(A.38)$$

Thus *passive* or *spectator* angular momenta may be eliminated even though they are coupled to the *active* angular momentum function on which the operator acts.

A.4.3 Coupling of tensor operators

Two spherical tensor operators $\mathbf{T}^{(k_1)}$ and $\mathbf{U}^{(k_2)}$ acting on the space spanned by the eigenfunctions of j may be coupled to a spherical tensor operator $\mathbf{X}^{(K)}$ of rank K using the Clebsch–Gordan expansion

$$X_Q^{(K)} = \left(\mathbf{T}^{(k_1)} \times \mathbf{U}^{(k_2)}\right)_Q^{(K)} = \sum_{q_1 q_2}\langle k_1 k_2 q_1 q_2|k_1 k_2 KQ\rangle T_{q_1}^{(k_1)} U_{q_2}^{(k_2)}. \quad (A.39)$$

If the two tensor operators $\mathbf{T}^{(k_1)}$ and $\mathbf{U}^{(k_2)}$ have the same rank $k_1 = k_2 = k$ they can be combined to form a scalar tensor, i.e. a tensor of rank zero

$$X_0^{(0)} = \left(\mathbf{T}^{(k)} \times \mathbf{U}^{(k)}\right)_0^{(0)}$$
$$= \sum_{q_1 q_2}\langle kkq_1 q_2|kk00\rangle T_{q_1}^{(k)} U_{q_2}^{(k)}$$
$$= (-1)^k\frac{1}{[k]^{1/2}}\sum_q (-1)^q T_q^{(k)} U_{-q}^{(k)}. \quad (A.40)$$

Traditionally, the scalar product of two tensor operators is defined by

$$\mathbf{T}^{(k)} \cdot \mathbf{U}^{(k)} = \sum_q (-1)^q T_q^{(k)} U_{-q}^{(k)} \tag{A.41}$$

and thus

$$\mathbf{T}^{(k)} \cdot \mathbf{U}^{(k)} = (-1)^k [k]^{1/2} \left(\mathbf{T}^{(k)} \times \mathbf{U}^{(k)} \right)_0^{(0)}. \tag{A.42}$$

When $k = 1$ the tensor scalar product reduces to the ordinary Cartesian scalar product

$$\begin{aligned}
\mathbf{T}^{(1)} \cdot \mathbf{U}^{(1)} &= - T_1^{(1)} U_{-1}^{(1)} + T_0^{(1)} U_0^{(1)} - T_{-1}^{(1)} U_1^{(1)} \\
&= T_x U_x + T_y U_y + T_z U_z = \mathbf{T} \cdot \mathbf{U}.
\end{aligned} \tag{A.43}$$

A.5   The Wigner–Eckart theorem

From the commutation relations (A.31) it can be shown that the matrix element $\langle \gamma j m | T_q^{(k)} | \gamma' j' m' \rangle$ is proportional to a $3j$-symbol

$$\langle \gamma j m | T_q^{(k)} | \gamma' j' m' \rangle = (-1)^{j-m} \begin{pmatrix} j & k & j' \\ -m & q & m' \end{pmatrix} \langle \gamma j \| \mathbf{T}^{(k)} \| \gamma' j' \rangle. \tag{A.44}$$

This relation is known as the Wigner–Eckart theorem. The coefficient $\langle \gamma j \| \mathbf{T}^{(k)} \| \gamma' j' \rangle$ is called the *reduced matrix element* of $\mathbf{T}^{(k)}$ and is independent of the magnetic quantum numbers m, m' and q.

From the properties of the $3j$-symbol we obtain the following selection rules for the quantum numbers that have to be fulfilled for the matrix element $\langle \gamma j m | T_q^{(k)} | \gamma' j' m' \rangle$ to be non-zero.

$$m = q + m' \tag{A.45}$$

and

$$|j - j'| \leqslant q \leqslant j + j'. \tag{A.46}$$

As an example we consider the matrix elements of the renormalized spherical harmonic $\mathbf{C}^{(k)}$. From the Wigner–Eckart theorem we obtain

$$\langle l m | C_q^{(k)} | l' m' \rangle = (-1)^{l-m} \begin{pmatrix} l & k & l' \\ -m & q & m' \end{pmatrix} \langle l \| \mathbf{C}^{(k)} \| l' \rangle. \tag{A.47}$$

Noting that

$$\begin{aligned}
\langle l m | C_q^{(k)} | l' m' \rangle &= \sqrt{\frac{4\pi}{2k+1}} \int_0^\pi \int_0^{2\pi} Y_{lm}^* Y_{kq} Y_{l'm'} \sin\theta \, \mathrm{d}\theta \, \mathrm{d}\varphi \\
&= (-1)^{-m} [l, l']^{1/2} \begin{pmatrix} l & k & l' \\ 0 & 0 & 0 \end{pmatrix} \begin{pmatrix} l & k & l' \\ -m & q & m' \end{pmatrix}
\end{aligned}$$

$$\tag{A.48}$$

it can be seen that

$$\langle l||\mathbf{C}^{(k)}||l'\rangle = (-1)^l [l, l']^{1/2} \begin{pmatrix} l & k & l' \\ 0 & 0 & 0 \end{pmatrix}. \tag{A.49}$$

Another important example is the matrix elements of the angular momentum operator j. For the $j_0^{(1)} = j_z$ component we have

$$\langle \gamma j m | j_0^{(1)} | \gamma' j' m' \rangle = \langle \gamma j m | j_z | \gamma' j' m' \rangle = m \delta_{\gamma j m, \gamma' j' m'}. \tag{A.50}$$

Thus,

$$\langle \gamma j m | j_0^{(1)} | \gamma' j' m' \rangle = (-1)^{j-m} \begin{pmatrix} j & 1 & j' \\ -m & 0 & m' \end{pmatrix} \langle \gamma j || \mathbf{j}^{(1)} || \gamma' j' \rangle$$

$$= m \delta_{\gamma j m, \gamma' j' m'} \tag{A.51}$$

and

$$\langle \gamma j || \mathbf{j}^{(1)} || \gamma' j' \rangle = \delta_{\gamma j, \gamma' j'} [j(j+1)(2j+1)]^{1/2}. \tag{A.52}$$

A.6 Matrix elements of tensor operators between coupled functions

We will now evaluate matrix elements of the coupled tensor operator $\mathbf{X}^{(K)}$ (A.39). In most cases the tensor operators $\mathbf{T}^{(k_1)}$ and $\mathbf{U}^{(k_2)}$ act on different spaces, spanned by $|\gamma_1 j_1 m_1\rangle$ and $|\gamma_2 j_2 m_2\rangle$, respectively, and so we need to consider matrix elements between coupled functions

$$|\gamma_1 \gamma_2 j_1 j_2 J M\rangle = \sum_{m_1 m_2} \langle j_1 j_2 m_1 m_2 | j_1 j_2 J M\rangle |\gamma_1 j_1 m_1\rangle |\gamma_2 j_2 m_2\rangle. \tag{A.53}$$

In this case

$$\langle \gamma_1 \gamma_2 j_1 j_2 J || \left(\mathbf{T}^{(k_1)} \times \mathbf{U}^{(k_2)}\right)^{(K)} || \gamma_1' \gamma_2' j_1' j_2' J'\rangle$$

$$= [J, K, J']^{1/2} \begin{Bmatrix} j_1 & j_2 & J \\ j_1' & j_2' & J' \\ k_1 & k_2 & K \end{Bmatrix} \langle \gamma_1 j_1 || \mathbf{T}^{(k_1)} || \gamma_1' j_1' \rangle \langle \gamma_2 j_2 || \mathbf{U}^{(k_2)} || \gamma_2' j_2' \rangle. \tag{A.54}$$

From this relation several special cases can be obtained. Setting $K = 0$ and $k_1 = k_2 = k$ and using the properties of the $3j$- and $9j$-symbols it can be shown that

$$\langle \gamma_1 \gamma_2 j_1 j_2 J || \mathbf{T}^{(k)} \cdot \mathbf{U}^{(k)} || \gamma_1' \gamma_2' j_1' j_2' J'\rangle$$

$$= \delta_{JJ'} (-1)^{j_1' + j_2 + J} [J]^{1/2} \begin{Bmatrix} j_1 & j_2 & J \\ j_2' & j_1' & k \end{Bmatrix} \langle \gamma_1 j_1 || \mathbf{T}^{(k)} || \gamma_1' j_1' \rangle \langle \gamma_2 j_2 || \mathbf{U}^{(k)} || \gamma_2' j_2' \rangle. \tag{A.55}$$

Other special cases of equation (A.54) are obtained by setting $k_2 = 0$ and $k_1 = 0$, respectively and using the properties of the 9-j symbol

$$\langle \gamma_1 \gamma_2 j_1 j_2 J || \mathbf{T}^{(k)} || \gamma_1' \gamma_2' j_1' j_2' J' \rangle$$

$$= \delta_{\gamma_2 \gamma_2'} \delta_{j_2 j_2'} (-1)^{j_1 + j_2 + J' + k} [J, J']^{1/2} \left\{ \begin{matrix} j_1 & j_2 & J \\ J' & k & j_1' \end{matrix} \right\} \langle \gamma_1 j_1 || \mathbf{T}^{(k)} || \gamma_1' j_1' \rangle$$

$$(\text{A.56})$$

$$\langle \gamma_1 \gamma_2 j_1 j_2 J || \mathbf{U}^{(k)} || \gamma_1' \gamma_2' j_1' j_2' J' \rangle$$

$$= \delta_{\gamma_1 \gamma_1'} \delta_{j_1 j_1'} (-1)^{j_1 + j_2' + J + k} [J, J']^{1/2} \left\{ \begin{matrix} j_1 & j_2 & J \\ k & J' & j_2' \end{matrix} \right\} \langle \gamma_2 j_2 || \mathbf{U}^{(k)} || \gamma_2' j_2' \rangle .$$

$$(\text{A.57})$$

Setting $k = 0$, we obtain the reduced matrix element form of the uncoupling formulae (A.37) and (A.38) for the scalar operator.

Appendix B

The Dirac and Breit–Pauli Theory

B.1 Introduction

In this appendix we will outline the most important parts, for our purposes, of relativistic atomic physics. For a more detailed account of this theory we refer the reader to chapters 2–9 of Greiner (1990) and chapter 20 of Messiah (1965).

B.2 Dirac theory of one-electron systems

At first glance, the method for finding the relativistic counterpart to the Schrödinger equation seems straightforward. We know, from classical theory of special relativity, that the total energy of an electron in an electromagnetic field defined by the scalar and vector potentials $\phi(r)$ and $A(r)$ is given by

$$\{E + e\phi(r)\}^2 = c^2\{p + eA(r)\}^2 + m^2c^4. \tag{B.1}$$

Interpreting this relation quantum mechanically, we obtain the equation†

$$\{E + e\phi(r)\}^2\psi(r, t) = (c^2\{p + eA(r)\}^2 + m^2c^4)\psi(r, t) \tag{B.2}$$

where, as usual, $p = -i\hbar\nabla$ and $E = i\hbar\frac{\partial}{\partial t}$. This wave equation, known as the Klein–Gordon equation, is, however, unsatisfactory for a number of reasons. It does not incorporate the electron spin, and, more fundamentally, the time derivative enters this equation at second order. Therefore the solutions will not obey the superposition principle, and the wave function at the time $t = 0$ will not completely determine the system for all time. The square root of this equation is not Lorentz invariant, and time and spatial co-ordinates will not enter on an equal footing.

---

† For clarity, in the following sections we will explicitly write out e, $4\pi\epsilon_0$, m, $\hbar$ and c in all expressions. To obtain the expressions in atomic units the following substitutions should be made: $e = 4\pi\epsilon_0 = m = \hbar = 1$ and $c = \alpha^{-1}$.

Dirac (1928) showed the way out of this dilemma by postulating that the fundamental equation should have the more general form

$$\mathcal{H}_D \psi(r) = E \psi(r) \tag{B.3}$$

where

$$\mathcal{H}_D = c\alpha \cdot \{p + eA(r)\} - e\phi(r) + \beta mc^2 \tag{B.4}$$

is the one-electron Dirac Hamiltonian. As in the non-relativistic case the Hamiltonian is required to be Hermitian and so

$$\alpha = \alpha^\dagger, \quad \beta = \beta^\dagger. \tag{B.5}$$

In addition the solutions to the Dirac equation should, in accordance with the relation (B.1), also be solutions to the Klein–Gordon equation (the converse need not be true), leading to the commutation relations

$$\alpha_x^2 = \alpha_y^2 = \alpha_z^2 = \beta^2 = 1$$
$$[\alpha_x, \alpha_y]_+ = [\alpha_y, \alpha_z]_+ = [\alpha_z, \alpha_x]_+ = 0 \tag{B.6}$$
$$[\alpha_x, \beta]_+ = [\alpha_y, \beta]_+ = [\alpha_z, \beta]_+ = 0$$

where $[A, B]_+$ denotes the anticommutator

$$[A, B]_+ = AB + BA. \tag{B.7}$$

It can be shown that the conditions (B.5) and (B.6) may be satisfied if α and β are taken as (4×4) matrices. Within this representation there is an infinite set of solutions to (B.5) and (B.6) all leading to the same physical result. A particularly useful set of matrices for studying the non-relativistic limit is given by

$$\alpha = \begin{pmatrix} 0 & \sigma \\ \sigma & 0 \end{pmatrix}, \quad \beta = \begin{pmatrix} I & 0 \\ 0 & -I \end{pmatrix} \tag{B.8}$$

where the σ are the (2×2) Pauli spin matrices, and I and 0 the (2×2) unit and zero matrices, respectively. Since α and β are represented by (4×4) matrices, the relativistic wave function must be a (4×1) column matrix, or four-component spinor. In many cases it is convenient to split up the four-component spinor into two two-component spinors $\xi(r)$ and $\eta(r)$:

$$\psi(r) = \begin{pmatrix} \xi(r) \\ \eta(r) \end{pmatrix} \tag{B.9}$$

where

$$\xi(r) = \begin{pmatrix} \psi_1(r) \\ \psi_2(r) \end{pmatrix}, \quad \eta(r) = \begin{pmatrix} \psi_3(r) \\ \psi_4(r) \end{pmatrix} \tag{B.10}$$

in which case the Dirac equation takes the form

$$(E - mc^2)\xi(r) = c\sigma \cdot \{p + eA(r)\}\eta(r) - e\phi(r)\,\xi(r)$$
$$(E + mc^2)\eta(r) = c\sigma \cdot \{p + eA(r)\}\xi(r) - e\phi(r)\,\eta(r). \tag{B.11}$$

B.2.1   Commutation relations for the Dirac Hamiltonian

We now consider the commutation relations for the central-field Hamiltonian

$$\mathcal{H}_D = c\boldsymbol{\alpha} \cdot \boldsymbol{p} + V(r) + \beta mc^2, \tag{B.12}$$

which is the Hamiltonian relevant for describing an electron moving in the Coulomb field of an infinitely heavy nucleus. In contrast to the non-relativistic theory, the central-field Hamiltonian does not commute with the orbital angular momentum operator. Instead we have

$$[\mathcal{H}_D, \boldsymbol{l}] = -i\hbar c\boldsymbol{\alpha} \times \boldsymbol{p}. \tag{B.13}$$

Defining the four-component operator, $\boldsymbol{s}$, in terms of the Pauli spin matrices

$$\boldsymbol{s} = \frac{\hbar}{2}\begin{pmatrix} \boldsymbol{\sigma} & 0 \\ 0 & \boldsymbol{\sigma} \end{pmatrix}, \tag{B.14}$$

it follows from the commutation relations of the latter that

$$[s_x, s_y] = i\hbar s_z \tag{B.15}$$

and

$$s^2\psi(\boldsymbol{r}) = \frac{3}{4}\hbar^2\begin{pmatrix} I & 0 \\ 0 & I \end{pmatrix}\psi(\boldsymbol{r}). \tag{B.16}$$

Thus $\boldsymbol{s}$ can be interpreted as an angular momentum operator with quantum numbers $s = 1/2$ and $m_s = \pm 1/2$. We will refer to this operator as the spin angular momentum operator. Using the commutation relations of the Pauli matrices, it is easy to show that

$$[\mathcal{H}_D, \boldsymbol{s}] = +i\hbar c\boldsymbol{\alpha} \times \boldsymbol{p}. \tag{B.17}$$

If we now take the total angular momentum of the electron as the sum of the orbital and spin angular momenta $\boldsymbol{j} = \boldsymbol{l} + \boldsymbol{s}$, we see that $\boldsymbol{j}$ commutes with the Dirac Hamiltonian, i.e.

$$[\mathcal{H}_D, \boldsymbol{j}] = 0. \tag{B.18}$$

The interpretation of these commutation relations is that there is now an interaction, in accord with experiment, between the orbital and spin angular momenta of the electron.

In addition to the angular symmetry, the Dirac Hamiltonian has an inversion symmetry and we need to investigate the commutation relations for the parity operator. Since this Hamiltonian is linear in the momentum operator $\boldsymbol{p}$ it does not commute with the ordinary non-relativistic parity operator Π. From the relations $[\boldsymbol{\alpha}, \beta]_+ = 0$ it can, however, be seen that the desired commutation relation is restored for a parity operator of form $\beta\Pi$ and we have

$$[\mathcal{H}_D, \beta\Pi] = 0. \tag{B.19}$$

In addition to the commutation relation with the Hamiltonian, the parity operator commutes with the total angular momentum operator.

B.2.2 Solutions of the Dirac equation

The commutation relations established in the previous section imply that $\mathcal{H}_D, j^2, j_z$ and $\beta\Pi$ define a set of mutually commuting operators. The simultaneous eigenfunctions of these operators can be written

$$\psi(\mathbf{r}) = \begin{pmatrix} \xi(\mathbf{r}) \\ \eta(\mathbf{r}) \end{pmatrix} = \frac{1}{r} \begin{pmatrix} P(r)\Omega_{lsjm_j}(\theta, \varphi) \\ iQ(r)\Omega_{l'sjm_j}(\theta, \varphi) \end{pmatrix} \tag{B.20}$$

where $\Omega_{lsjm_j}(\theta, \varphi)$ are spherical two-spinors

$$\Omega_{lsjm_j}(\theta, \varphi) = \sum_{m_l m_s} \langle l\tfrac{1}{2} m_l m_s | l\tfrac{1}{2} jm \rangle Y_{lm_l}(\theta, \varphi) \chi_{\frac{1}{2}m_s}. \tag{B.21}$$

The spherical two-spinors are obtained by coupling the spherical harmonics $Y_{lm_l}(\theta, \varphi)$ and spin functions

$$\chi_{\frac{1}{2}\frac{1}{2}} = \begin{pmatrix} 1 \\ 0 \end{pmatrix}, \quad \chi_{\frac{1}{2}-\frac{1}{2}} = \begin{pmatrix} 0 \\ 1 \end{pmatrix}, \tag{B.22}$$

and thus $|l - 1/2| \leqslant j \leqslant l + 1/2$ and $m_j = -j, -j + 1, \ldots, j - 1, j$. In order for $\psi(\mathbf{r})$ to be an eigenfunction of the parity operator l' must be related to l according to

$$l' = \begin{cases} l + 1 & \text{for} \quad j = l + 1/2 \\ l - 1 & \text{for} \quad j = l - 1/2. \end{cases} \tag{B.23}$$

In relativistic theory it is convenient to define a quantum number κ:

$$\kappa = \pm \left(j + \tfrac{1}{2} \right) \quad \text{for} \quad l = j \pm \tfrac{1}{2} \tag{B.24}$$

and to express the eigenfunctions in terms of this quantum number, in which case we have

$$\psi(\mathbf{r}) = \frac{1}{r} \begin{pmatrix} P(r)\Omega_{\kappa m}(\theta, \varphi) \\ iQ(r)\Omega_{-\kappa m}(\theta, \varphi) \end{pmatrix}. \tag{B.25}$$

Inserting this expression into the Dirac equation and using the properties of the spherical two-spinors, we obtain a system of equations for the radial functions

$$\hbar \left(\frac{d}{dr} + \frac{\kappa}{r} \right) P(r) - \frac{1}{c} \left(E + mc^2 - V(r) \right) Q(r) = 0$$
$$\hbar \left(\frac{d}{dr} - \frac{\kappa}{r} \right) Q(r) + \frac{1}{c} \left(E - mc^2 - V(r) \right) P(r) = 0. \tag{B.26}$$

Since $\psi(\mathbf{r})$ must be finite everywhere, $P(r)$ and $Q(r)$ should satisfy the boundary conditions $P(0) = Q(0) = 0$. For $\psi(\mathbf{r})$ to be square integrable, it is necessary that both $P(r)$ and $Q(r)$ go to zero as r goes to infinity. Solutions to the system of radial equations satisfying the boundary conditions at the origin and at the infinity exist only for certain discrete energy values. Since the electron rest mass energy mc^2 is included in the Dirac Hamiltonian, the discrete energy values are shifted by roughly $E = mc^2$ compared with the non-relativistic case. For $E > mc^2$ the electron is no longer bound to the nucleus and it behaves, for large r, essentially like a free particle. In this case there is a solution to the radial equations for every E, and a continuous spectrum of eigenvalues adjoins the discrete levels. For wave functions belonging to the discrete and the positive continous spectrum it can be shown that $P(r)$ is of the order c larger than $Q(r)$. These components of the radial functions are therefore referred to as the *large* and *small* components.

In addition to the positive continuous eigenvalue spectrum there is a corresponding negative one in the range $-\infty < E < -mc^2$. The wave functions corresponding to the negative energies describe positron states, that is, states of a particle with the same mass as the electron but with the opposite charge. We will not be concerned with the negative energy solutions, but refer the reader to chapter 12 of Greiner (1990) for a comprehensive treatement.

B.2.3   Pauli approximation for one-electron systems

We will now investigate the non-relativistic limit of the one-electron Dirac equation and derive a wave equation for a two-component function similar to the Schrödinger equation that contains relativistic corrections to the order $(1/c)^2$. The present treatment closely follows the treatment of Beresteskii *et al* (1971). Before starting, we note that the relativistic electron energy includes rest mass energy mc^2. In the non-relativistic limit the latter energy should be left out. Therefore we transform to a new energy E'

$$E = E' + mc^2 \tag{B.27}$$

in which case the Dirac equation can be written

$$E'\xi(r) = c(-i\hbar\sigma \cdot \nabla)\eta(r) + V(r)\xi(r) \tag{B.28}$$

$$(E' + 2mc^2)\eta(r) = c(-i\hbar\sigma \cdot \nabla)\xi(r) + V(r)\eta(r). \tag{B.29}$$

From these equations it is seen that $\eta(r)$ is of the order $(1/c)$ times $\xi(r)$ and thus it seems natural to eliminate $\eta(r)$ from the second equation and insert it into the first to obtain an equation that only contains the two-component function $\xi(r)$. This should then be the wave equation we are looking for. The above proceedure is, however, not correct since, to the order of $(1/c)^2$, it gives a probability density

$$\rho(r) = |\xi(r)|^2 + |\eta(r)|^2 = |\xi(r)|^2 + \frac{\hbar^2}{4m^2c^2}|\sigma \cdot \nabla\xi(r)|^2. \tag{B.30}$$

This expression differs from the one we expect from a solution $\xi(r)$ of a Schrödinger-type equation. To find the correct wave equation, instead of $\xi(r)$ we must introduce a new two-component function $\psi(r)$ such that

$$\rho(r) = |\psi(r)|^2. \tag{B.31}$$

To find the transformation between $\xi(r)$ and $\psi(r)$ we look at the normalization integral

$$\int \psi^* \psi \, dr = \int \left[\xi^* \xi + \frac{\hbar^2}{4m^2c^2} (\nabla \xi^* \cdot \sigma)(\sigma \cdot \nabla \xi) \right] dr. \tag{B.32}$$

A partial integration gives (the same holds with ξ^* and ξ exchanged)

$$\int (\nabla \xi^* \cdot \sigma)(\sigma \cdot \nabla \xi) \, dr = -\int \xi^* (\sigma \cdot \nabla)(\sigma \cdot \nabla) \xi \, dr = -\int \xi^* \nabla^2 \xi \, dr, \tag{B.33}$$

and thus we have

$$\int \psi^* \psi \, dr = \int \left[\xi^* \xi - \frac{\hbar^2}{8m^2c^2} \left(\xi^* \nabla^2 \xi + \xi \nabla^2 \xi^* \right) \right] dr \tag{B.34}$$

leading to the relations

$$\psi = \left(1 - \frac{\hbar^2}{8m^2c^2} \nabla^2 \right) \xi, \qquad \xi = \left(1 + \frac{\hbar^2}{8m^2c^2} \nabla^2 \right) \psi. \tag{B.35}$$

Solving equation (B.29) for η we have

$$\eta = \frac{1}{E' + 2mc^2 - V} c(-i\hbar \sigma \cdot \nabla) \xi. \tag{B.36}$$

Expanding the denominator in orders of $1/c$ and substituting into (B.28) gives

$$E'\xi = \left[-\frac{\hbar^2}{2m} \left\{ 1 - \frac{E' - V}{2mc^2} \right\} (\sigma \cdot \nabla)^2 - \frac{\hbar^2}{4m^2c^2} (\sigma \cdot \nabla V)(\sigma \cdot \nabla) + V \right] \xi. \tag{B.37}$$

We now substitute (B.35) into the equation above and use the relation, correct to order $(1/c)^2$,

$$\left(1 - \frac{\hbar^2}{8m^2c^2} \nabla^2 \right)^{-1} = 1 + \frac{\hbar^2}{8m^2c^2} \nabla^2 \tag{B.38}$$

to obtain

$$E'\psi = \left[-\frac{\hbar^2}{2m} \left\{ 1 - \frac{E' - V}{2mc^2} \right\} (\sigma \cdot \nabla)^2 + V \right.$$
$$\left. - \frac{\hbar^2}{4m^2c^2} \left\{ (\sigma \cdot \nabla V)(\sigma \cdot \nabla) - \frac{1}{2}(\nabla^2 V + V \nabla^2) \right\} \right] \psi. \tag{B.39}$$

Simplifying and using the approximation

$$E' - V(r) \approx \frac{p^2}{2m} = -\frac{\hbar^2}{2m}\nabla^2 \tag{B.40}$$

and the relation

$$i\boldsymbol{\sigma} \cdot (\nabla V \times \nabla \psi) = -\frac{2}{\hbar^2}\frac{1}{r}\frac{dV}{dr}\boldsymbol{l}\cdot\boldsymbol{s} \tag{B.41}$$

we obtain the desired equation

$$\mathcal{H}_P\psi(\boldsymbol{r}) = E'\psi(\boldsymbol{r}) \tag{B.42}$$

where

$$\mathcal{H}_P = -\frac{\hbar^2}{2m}\nabla^2 + V - \frac{\hbar^4}{8m^3c^2}\nabla^4 + \frac{1}{2m^2c^2}\frac{1}{r}\frac{dV}{dr}\boldsymbol{l}\cdot\boldsymbol{s} + \frac{\hbar^2}{8m^2c^2}\nabla^2 V \tag{B.43}$$

is the so-called Pauli Hamiltonian†. The first two terms of the Hamiltonian are, respectively, the kinetic and potential terms of the non-relativistic Schrödinger equation. The third term is the classical relativistic mass correction which can be obtained from expanding $(c^2p^2 + m^2c^4)^{1/2} - mc^2$ into powers of $1/c$. The fourth term is the spin–orbit interaction, which was first included by Pauli on empirical grounds, but is a direct consequence of a proper relativistic treatment. The last term is the so-called one-body Darwin term and has no simple classical interpretation. In the case of a Coulomb potential $V(r) = -Ze^2/4\pi\epsilon_0 r$, the Darwin term is given by

$$\frac{\hbar^2}{8m^2c^2}\nabla^2 V(r) = \frac{\pi\hbar^2}{2m^2c^2}\left(\frac{Ze^2}{4\pi\epsilon_0}\right)\delta(\boldsymbol{r}), \tag{B.44}$$

where $\delta(\boldsymbol{r})$ is the three-dimensional delta function.

In all the formulae in this apppendix we have used SI units. For actual calculations, however, it is more convenient to use atomic units, and from now on we will assume that $e = m = \hbar = 4\pi\epsilon_0 = 1$ and $c = \alpha^{-1}$.

B.2.4 Energy corrections to one-electron systems

To investigate the effect of the relativistic correction terms we now treat them in first-order perturbation theory using spin-orbitals, which are solutions to the ordinary non-relativistic equation, as zero-order functions. The spin-orbitals are not eigenfunctions of j^2 and j_z, as the correct eigenfunctions of the Pauli Hamiltonian should be. However, by applying the Clebsch–Gordan expansion to

† In the Pauli Hamiltonian $\boldsymbol{s} = \boldsymbol{\sigma}/2$ is the ordinary two-component spin operator.

the product of spherical harmonics and the spin functions we obtain zero-order functions with the correct angular symmetry:

$$\psi(nlsjm_j; r) = \sum_{m_l, m_s} \langle lm_l \tfrac{1}{2} m_s | jm_j \rangle \frac{P(nl; r)}{r} Y_{lm_l}(\theta, \varphi) \chi_{\frac{1}{2} m_s}. \qquad \text{(B.45)}$$

For hydrogen-like ions with $V(r) = -Z/r$ the expectation value of the mass velocity term can be straightforwardly evaluated to yield an energy shift

$$E_v = -E_n \frac{(Z\alpha)^2}{n^2} \left[\frac{3}{4} - \frac{n}{l+1/2} \right] \qquad \text{(B.46)}$$

where $E_n = -Z^2/2n^2$ is the non-relativistic energy. To evaluate the value of the spin–orbit energy we note that

$$l \cdot s = \tfrac{1}{2}(j^2 - l^2 - s^2) \qquad \text{(B.47)}$$

leading to the expression

$$E_{so} = \zeta_{so}(nl) \left[j(j+1) - l(l+1) - \tfrac{3}{4} \right] \qquad \text{(B.48)}$$

where

$$\zeta_{so}(nl) = \frac{\alpha^2 Z}{2} \int_0^\infty r^{-3} P^2(nl; r) \, dr \qquad \text{(B.49)}$$

is the spin–orbit radial integral. Evaluating the radial integral we obtain for $l \neq 0$

$$E_{so} = -E_n \frac{(Z\alpha)^2}{2nl(l+1/2)(l+1)} \times \begin{cases} l & \text{for} \quad j = l + 1/2 \\ -l - 1 & \text{for} \quad j = l - 1/2. \end{cases} \qquad \text{(B.50)}$$

For $l = 0$ the spin–orbit interaction vanishes and therefore $E_{so} = 0$ in that case. From the expression above it is seen that a level characterized by the quantum numbers l and s is split into two so-called fine-structure levels corresponding to, respectively, the quantum numbers $j = l + 1/2$ and $j = l - 1/2$. Finally, the expectation value of the Darwin term can be evaluated for the case $l = 0$ to yield

$$E_{D1} = -E_n \frac{(Z\alpha)^2}{n}. \qquad \text{(B.51)}$$

Adding the three contributions to the non-relativistic energy we obtain

$$E_{rel} = E_n \left[1 + \frac{(Z\alpha)^2}{n^2} \left(\frac{n}{j+1/2} - \frac{3}{4} \right) \right]. \qquad \text{(B.52)}$$

This energy agrees to order $(Z\alpha)^2$ with the exact energy

$$E_{Dirac} = mc^2 \left[\left\{ 1 + \left(\frac{\alpha Z}{n - |\kappa| + \sqrt{\kappa^2 - \alpha^2 Z^2}} \right)^2 \right\}^{-1/2} - 1 \right] \qquad \text{(B.53)}$$

for the Dirac equation in the Coulomb potential $V(r) = -Z/r$.

B.3   The relativistic wave equation for many-electron systems

In the non-relativistic treatment it was straightforward to include the interaction between the electrons and find a wave equation for a many-electron system. In the relativistic case this has to be done with care, since the instantaneous Coulomb interaction $1/r_{ij}$ is not sufficient to describe the electron–electron interaction. First of all, it is not Lorentz invariant, and therefore not appropriate for a relativistic theory. It also neglects the magnetic properties of the electron, introduced via the orbital and spin motions. Finally, the speed of light is finite in a relativistic model, and retardation effects should be expected to appear.

In spite of these shortcomings, it is still a common approach to combine the instantaneous Coulomb interaction with the one-electron operators of the Dirac theory to yield the *Dirac–Coulomb* Hamiltonian

$$\mathcal{H}_{DC} = \sum_{i=1}^{N} \mathcal{H}_D(i) + \sum_{i<j}^{N} \frac{1}{r_{ij}}. \tag{B.54}$$

To account for magnetic interactions and retardation effects to order $1/c^2$, Breit (1929, 1930) showed that the operator (known as the Breit operator)

$$B_{ij} = -\frac{1}{2r_{ij}} \left[(\boldsymbol{\alpha}_i \cdot \boldsymbol{\alpha}_j) + \frac{(\boldsymbol{\alpha}_i \cdot \boldsymbol{r}_{ij})(\boldsymbol{\alpha}_j \cdot \boldsymbol{r}_{ij})}{r_{ij}^2} \right]. \tag{B.55}$$

should be added to the instantaneous Coulomb operator, and we arrive at the *Dirac–Coulomb–Breit* Hamiltonian

$$\mathcal{H}_{DCB} = \sum_{i=1}^{N} \mathcal{H}_D(i) + \sum_{i<j}^{N} \left(\frac{1}{r_{ij}} + B_{ij} \right). \tag{B.56}$$

When QED effects are unimportant this Hamiltonian, although not Lorentz invariant, has shown itself to give results in good agreement with experiment.

Appendix C

Fundamental Constants

In table C.1 some of the fundamental physical constants are shown. In addition, the permittivity ϵ_0 and permeability μ_0 of free space must be fixed. These two constants are related according to

$$\epsilon_0 \mu_0 c^2 = \kappa^2. \tag{C.1}$$

In the rationalized MKSA (SI) system $\kappa = 1$ and $\mu_0 = 4\pi \times 10^{-7}$ H m^{-1}.

From the above constants two important quantities, the fine-structure constant $\alpha = \mu_0 c e^2 / 2h$ and the Rydberg constant $R_\infty = m_e c \alpha^2 / 2h$, can be formed. The values of these two constants are shown at the foot of table C.1.

Table C.1. The fundamental physical constants.

Constant	Symbol	Recommended value
Velocity of light in free space	c	299 792 458 m s^{-1}
Rest mass of the electron	m_e or m	9.109 389 7(54) $\times$ 10^{-31} kg
Rest mass of the proton	m_p	1.672 623 1(10) $\times$ 10^{-27} kg
Electron/proton mass ratio	m_e/m_p	5.446 170 13(11) $\times$ 10^{-4}
Elementary charge	e	1.602 177 33(49) $\times$ 10^{-19} C
Planck's constant	h	6.626 075 5(40) $\times$ 10^{-34} J s
Fine-structure constant	α	0.007 297 353 08(33)
Rydberg constant	R_∞	10 973 731.568 34(24) m^{-1}

C.1 Atomic units

In this book we have used the atomic units, that is the unit system in which the numerical values of the basic units in table C.2 are unity. Other units in

Table C.2. Basic quantities for the atomic unit system.

Constant	Symbol
Rest mass of the electron	m_e
Elementary charge	e
Planck's constant divided by 2π	$\hbar = h/2\pi$
4π times the permittivity of free space	$4\pi\epsilon_0$

Table C.3. Quantities for the atomic unit system.

Constant	Symbol	Recommended value
Length, Bohr	$a_0 = 4\pi\epsilon_0\hbar^2/m_e e^2$	$5.291\ 772\ 49(24) \times 10^{-11}$ m
Velocity	$v_B = \alpha c$	$2.187\ 691\ 42(10) \times 10^6$ m s^{-1}
Energy, Hartree	$E_h = \hbar^2/m_e a_0^2$	$4.359\ 748\ 2(26) \times 10^{-18}$ J
Time	$\tau_0 = \hbar/E_h$	$2.418\ 884\ 326\ 555(53) \times 10^{-17}$ s
Magnetic dipole moment	$\mu_B = e\hbar/2m_e$	$9.274\ 015\ 4(31) \times 10^{-24}$ J T^{-1}
Electric dipole moment	$d_0 = ea_0$	$8.478\ 357\ 9(26) \times 10^{-30}$ C m

the system are obtained by combining these four basic quantities and are shown in table C.3. From the expression for the velocity it is seen that the velocity of light has the value $\alpha^{-1} = 137.035\ 989\ 5(61)$ in atomic units.

C.2 Additional units

In atomic spectroscopy the photon wavelength λ of a transition is often measured in ångströms, $1\ \text{Å} = 10^{-10}$ m. In addition the photon wavenumber $\sigma = 1/\lambda$ is given in units of cm^{-1} instead of in units of m^{-1}. Sometimes the unit Kayser is used for the wavenumber where $1\ \text{K} = \text{cm}^{-1}$. The wavenumber $2R_\infty = 2 \times 109\ 737.315\ 34(13)$ cm^{-1} corresponds to an energy separation of $1\ E_h$. Ionization potentials and electron affinities are often expressed in eV, where $1\ E_h = 27.211\ 396\ 1(81)$ eV.

For hyperfine structures and isotope shifts the unit MHz is frequently used. This unit can be related to an energy and wavenumber through the relations

$$E = h\nu \tag{C.2}$$

and

$$c = \lambda\nu. \tag{C.3}$$

1 MHz corresponds to a wavenumber of $3.335\ 641\ 0 \times 10^{-5}$ cm^{-1} and an energy separation of $1.519\ 829 \times 10^{-10}\ E_h$.

A more detailed discussion of units and their values may be found on page 1 of the *Atomic, Molecular & Optical Physics Handbook* (Baylis and Drake 1996).

Appendix D

Program Input Parameters

The examples in the book have, with a few exceptions, illustrated the more straightforward use of the codes where many default values of parameters were set. Here we concentrate on the more non-standard features of the programs that may not have been mentioned in the book. Also, for each program, the complete reference to a published long write-up of the program and its organization is included. Complete examples of calculations too complex to consider in this book are included with the code.

HF

Input files:	wfn.inp, optional
Output files:	wfn.out
Reference:	Froese Fischer C 1987 *Comput. Phys. Commun.* **43** 355, Gaigalas G and Froese Fischer C 1996 *Comput. Phys. Commun.* **98** 255

This program has some built-in help features. By entering H (or h) a brief summary of possible responses is provided.

ATOM	1–6 character label for the calculation.
TERM	LS term value or 'AV' or 'av' (for average energy).
Z	Atomic number, real; non-integers are allowed.
S	Screening parameter for the orbital.
IND	0—use screened hydrogenic functions as initial; estimates; 1—leave radial function unchanged (already in memory); −1—search for function in wfn.inp; otherwise same as 0.
ACC	Accelerating parameter (see section 3.4).
NO	Maximum number of points in the range of the function ($\leqslant 220$).
STRONG	If true, orthogonalize after each orbital update (see section 3.4). Enter t for true, f for false.

PRINT   If t (true) radial functions are printed.

SCFTOL   Initial value of the self-consistency criterion.

NSCF   The maximum number of SCF cycles.

IC   Number of orbitals to be updated using the least-self-consistent criteria.

TRACE   If t (true), detailed information about the SCF energy adjustment process is printed.

A feature of the HF program is that the occupation numbers need not be integer. In order to study the $2s$–$2p$ transition in Be, for example, orbitals could be computed for the configuration 2s(1.5)2p(0.5) in which case average energy calculations will be performed for

$$(1/2)\left[E_{av}(2s^2) + E_{av}(2s2p)\right].$$

Also, various expectation values may be printed at the end of a calculation. A sample input data line is given for each case and, ideally, the input should be aligned with the sample. However, the format for the input is also provided. Here it is helpful to know the following format rules for a line of input:

nX   Skip n positions on the line (i.e. enter n blanks).

An   The next n positions on the line will be interpreted as characters.

In   The next n positions on the line will be interpreted as an integer.

More details can be found in any FORTRAN text.

It should also be remembered that HF is a program for simple cases: if there are two or more open shells, an LS calculation may not be possible using HF. Configurations of the form $nl^w n's$ generally are allowed, $l \leqslant 3$; configurations $np^w n'l$, $l \leqslant 2$ may request information about the parent term for np^w; CSFs of the form $nd^w n'l$ are not allowed except for $n'l = n's$; open f-shells may have any occupation, but any other open shell may only be a single s electron. For more complex situations NONH and MCHF should be used, though these codes have not yet been extended to open f-shells.

GENCL

Input files:	None
Output files:	cfg.inp
Reference:	Froese Fischer C and Liu B 1991 *Comput. Phys. Commun.* **64** 406

This program comes with a series of examples which are displayed upon entering h to the initial prompt. One feature of the program is that b may be entered at most stages in which case the menu is turned *back* to allow a change or correction of the input data.

Many examples of the use of GENCL are included in the book, thus the meaning of different queries will not be explained. However, it should be mentioned that the program maintains an order for the orbitals, determined from the order in which an orbital is first encountered. Thus, with a reference set of 4s(2)3d(1) and an active set of 3d,4s,4p the file cfg.inp will not have orbitals in a consistent ordering. The ordering of orbitals in the reference set, the active set and the virtual set should always be the same in the sense that one orbital will always appear *before* or *after* another. Sometimes this is referred to as the *after relation*.

NONH

Input files:	cfg.inp
Output files:	int.lst
Reference:	Hibbert A and Froese Fischer C 1991 *Comput. Phys. Commun.* **64** 417

The default input for NONH is very simple, but the program also has a feature for generating the interaction matrix for a first-order correction to a zero-order wave function in which all interactions are computed within the zero-order set and all interactions between the zero-order and first-order correction, but only the diagonal interactions for the latter. For large calculations this greatly reduces the amount of data generated and the time for an MC-SCF calculation. The response y to the query ALL INTERACTIONS ? (Y/N) is the normal situation, but n allows for a possible reduction of computational effort.

NEW If only a portion of the interaction matrix is to be computed, NEW specifies the number of CSFs at the end that are new. This feature is not supported in other codes, but could allow for the expansion of a matrix by adding a few rows or columns.

NZERO The size of the expansion of the zero-order wave function, consisting of the first NZERO CSFs.

MCHF

Input files:	wfn.inp (optional); cfg.inp and int.lst
Output files:	wfn.out and cfg.out
Reference:	Froese Fischer C 1991 *Comput. Phys. Commun.* **64** 431

The program has many default parameters. These parameters can be redefined by responding n to queries about the default values. The FORMAT of the response is given in each case, as to the data type. Commas will separate all entries.

ATOM 1–6 character label; it does not affect the calculation.

TERM 1–6 character label; does not affect the calculation.

Z	Atomic number, real.
NIT	Number of orbitals to be varied, counted from the end of the list.
S	Screening parameter for the orbital.
IND	0—use screened hydrogenic function as initial estimate; 1—leave radial function unchanged (already in memory); −1—read from wfn.inp; if not present, same as 0.
METH	One of three methods: 0, 1, 3 (see section 3.11.4).
ACC	Accelerating parameter (see section 3.4).
NO	Maximum number of points in the range of the function, ⩽220.
REL	Logical variable; if true, relativistic shift corrections will be included in the computation of the interaction matrix.
STRONG	If true, orthogonalize orbitals after each update (see section 3.4).
PRINT	If t (true) radial information is printed.
CFGTOL	Convergence criteria for the total energy; the default is 1.D-10.
SCFTOL	Initial value of the self-consistency criterion; the default is 1.D-7.
NSCF	Maximum number of SCF cycles; the default is 12 cycles.
IC	Number of orbitals to be updated using the least-self-consistent criteria. The default is zero for an MCHF calculation.
ACFG	Accelerating parameter for the expansion coefficients.
LD	Logical variable; if false, mixing coefficients will not be changed.
TRACE	If t (true), SCF energy adjustment information is printed.

In atomic structure calculations, once the wfn.out file is sufficiently converged, it is recommended that if be saved as name.w, where name is a convenient name for the calculation. Often cfg.out is also saved as name.c.

BREIT

Input files:	cfg.inp
Output files:	int.lst
Reference:	Hibbert A, Glass R and Froese Fischer C 1991 *Comput. Phys. Commun.* **64** 455

In the default mode, all interactions are derived for the selected operators. But this program can generate a lot of information, not all of which will contibute significantly. Different Breit–Pauli operators may be selected (see section 7.5) but like NONH, some of the interactions also can be omitted through the control of parameters.

NEW	If only a portion of the interaction matrix is to be computed, NEW specifies the number of CSFs at the end of the list that are new. This feature is not supported in other codes, but could allow for the expansion of a matrix by adding a few rows or columns.
NZERO	The size of the CSF list (from the beginning) that will contain the zero-order wave function. If the type of the operator selected was either 1 or 2 and NZERO is not equal to all CSFs, then

- A subset of the first NZERO may be selected to determine the zero-order set.
- The diagonal relativistic corrections outside the zero-order set may be omitted.

Restricted 2-body	If y, the two-body Breit–Pauli operators are to be evaluated *only when the spin–orbit operator contributes to the interaction.*

Once the BREIT calculation is completed, cfg.inp often needs to be saved as name.c.

CI

Input files:	name.c, name.w where name is specified by input data, and an int.lst for the CSFs of name.c
Output files:	name.l and name.j for a relativistic calculation
Reference	Froese Fischer C 1991 *Comput. Phys. Commun.* **64** 473

The CI program sets up an interaction matrix and determines a few selected eigenvalues and eigenvectors. It assumes an int.lst has already been derived. Since the finite mass and relativistic shift effects are derived from the int.lst from a NONH run, these effects can be computed without using a BREIT run to produce the int.lst. If a mass correction is requested, the user must indicate whether the gradient or Slater integral form is to be used. Generally the gradient form is recommended.

Selected eigenvalues are found using EISPACK (Smith *et al* 1974) routines that find all eigenvalues in a specified range, (RLB, RUB), where RLB is a lower bound and RUB an upper bound. The program sorts the diagonal elements and sets RLB $= 1.5 * \text{LOWEST}$ (most negative); if n eigenvalues are requested, then RUB is the average of the nth and $(n + 1)$th diagonal element unless $n = \text{NCFG}$, in which case RUB $= (2/3) * \text{HIGHEST}$ diagonal. The upper and lower bounds are determined from the non-fine-structure matrix, so that for certain high J values, only a few eigenvectors are found, whereas for J values with many possible states, more are found. In general, the number of eigenvalues should be the largest number for any single *J* value. Often, when correlation is strong, this algorithm will yield a few more eigenvalues than expected because correlation lowers the energy levels of the states.

MLTPOL

Input files:	name.c, where name is specified on input
Output files:	mltpol.lst
Reference:	Froese Fischer C and Godefroid M R 1991 *Comput. Phys. Commun.* **64** 486

The calculation is defined by two prompts:

name   Valid name for a file (not including the extension).
type   Type of transition E1, E2, ... or M1, M2, ... or *.

A calculation may produce angular integrals for more than one type of transition, with * terminating the calculation.

LSTR and LSJTR

Input files:	name.c, name.w, and mltpol.lst. For LSJTR also name.l or name.j
Output files:	tr.lsj
Reference:	Froese Fischer C and Godefroid M R 1991 *Comput. Phys. Commun.* **64** 501

Only a few prompts are required to specify the calculation. Besides the names for the initial and final state, the program asks whether intermediate results are to be printed and allows the user to cut down on the information generated by specifying a tolerance for printing: this tolerance is applied to the matrix element in the length form where matrix elements smaller in absolute value than the tolerance will not be printed. Finally, the program asks whether the default Rydberg constant is to be used: if not, the user will be asked to specify the atomic mass. The same cutoff tolerance is used to determine whether the transition data should be written to the tr.lsj file, though here the criterion applies to the *gf*-value. Since the latter depends on the *square* of the transition matrix elements, it may happen that intermediate data are displayed for a transition without any information being appended to tr.lsj. A value of TOL = 0 will include *all* information in the printout.

CMCHF

Input Files:	cfg.inp, wfn.inp, and int.lst
Output Files:	cfg.out, wfn.out, phase.dat

Reference: Froese Fischer C (unpublished)

Input parameters are similar to those of MCHF with the addition of a request for the range of values for k^2, in the form of MIN, DELTA, MAX. Examples are 0.1, 0.1, 1.0 but, if convergence problems arise, it is also possible to use 1.0, -0.1, 0.1, in which case the calculations proceed in the reverse direction with results from one energy used as initial esimates for the next. cfg.out and wfn.out will then contain results for a series of calculations and phase.dat will contain information about perturbers and the phase shift. Upon completion, the output files should be saved as name.c and name.w, where name is an appropriately chosen name for the calculation.

PHOTO

Input files: name.c, name.w and mltpol.lst
Output files: photo.out
Reference: Froese Fischer C (unpublished)

This program is essentially LSTR, which computes an allowed transition, modified for a series of calculations of photodetachment or photoionization cross section.

AUTO

Input Files: name.c, name.w, and int.lst; name.l or name.j (for calculation of type 2 or 3)
Output Files: auto.dat and auto.log
Reference: Froese Fischer C and Brage T 1993 *Comput. Phys. Commun.* **74** 381

Two energies are needed for an autoionization calculation (see section 10.4). Either the energy E_b of the perturber or the energy E_t of the target is always taken from the file providing the expansion information for the perturber which depends on the type of calculation as described below.

1. The expansion for the perturber is obtained from an MCHF calculation in which case E_b is taken from the header of name.c and the energy of the target must be provided as input.

2. The expansion for the perturber is obtained from a name.l file which includes E_b and E_t is taken from the header of the name.c file.

3. The expansion for the perturber is obtained from a name.j file which includes E_b and E_t is taken from the header of the name.c file. Since the name.j file may contain expansions for a series of J values, a series of autoionization rates will be computed.

HFS

Input files:	name.c, name.w and name.l or name.j, if the configuration weights are obtained from a CI calculation
Output files:	name.h
Reference:	Jönsson P, Wahlström C-G and Froese Fischer C 1993 *Comput. Phys. Commun.* **74** 399

The user may select one of three calc *B* factors larger than a specified tolerance for printing are printed out for every pair of configurations.

Expansion coefficients are obtained from name.c if an MCHF calculation is indicated; for a CI calculation, they are obtained from name.l for a non-fine-structure result, and name.j for *J*-dependent data. In the last two cases, the file may contain eigenvectors for a number of states. The user may specify the 'order number' of the state to be used.

ISO

Input files:	name.c, name.w, and int.lst name.l or name.j may also be selected
Output files:	None
Reference:	Froese Fischer C, Smentek-Mielczarek L, Vaeck N and Miecznik G 1993 *Comput. Phys. Commun.* **74** 415

After selecting the source for the wave function expansion from a menu, the type of calculation is selected—mass shift, field shift or both. Detailed information about contributions can be obtained, with printing limited by a tolerance option.

UTILITIES

In addition to the programs that were described separately, a number of short utility programs are available that assist in the processing and the managing of the data.

1. COMP

    Print dominant components of a wave function expansion in decreasing order of magnitude, for expansion coefficients obtained either from a name.c, name.l, or name.j file.

2. CONDENS

Read the name.c and optionally also name.l or name.j file to create a new configuration list in cfg.out containing only those configurations with a mixing coefficient greater or equal to a specified tolerance. When input is from name.c the output will retain the expansion coefficients of the input file, but for input from name.l or name.j, where the file may contain a number of eigenvectors, the tolerance criterion is applied to the maximum absolute value of the expansion coefficient for all eigenvectors in the file, and the expasion coefficient in cfg.out is this maximum.

3. LEVELS

Program to sort the energy levels in a name.l or name.j file and print energy levels in atomic units and cm^{-1}, relative to the lowest.

4. LINES

Print transition data in a tr.lsj file (produced by the LSJTR program) in sorted (increasing) order, according to a number of different criteria. A maximum of 1000 lines may be processed. The tr.lsj files from a number of runs may be concatenated before processing.

5. PLOTW

This program produces a file plot.dat which, for each radial function not deleted by entering d or D in repsonse to the electron label, contains a header line followed by pairs of numbers corresponding to $\sqrt{r}$ and $P(nl; r)$. This file contains data suitable for plotting. By tabulating the radial function versus $\sqrt{r}$, more detail is provided near the origin and extended exponential tails are compressed. The program can easily be changed.

6. PRINTW

This program selectively prints the data in wfn.inp: the electron label is displayed and unless d or D is entered, the radial function is printed.

7. RELABEL

Because the electron labels must match exactly (UPPERCASE and lowercase as well as blank fields) it may be necessary to change the electron label. This routine lists the labels of all electrons found in wfn.inp and selectively allows the user to change the labels before the radial function is written to a file. After each label is printed, the user may respond in three ways:

    d   Radial function is not copied (deleted).
    (Blank) No change in the label, radial function copied to wfn.out.
    AAA   Change label to AAA, then copy to wfn.out.

This routine may also be used to eliminate some radial functions from the file. All .w files can always be concatenated.

8. TERMS

A routine to print the terms allowed for the different groups of equivalent electrons. See table 2.1.

References

Aboussaïd A, Godefroid M R, Jönsson P and Froese Fischer C 1995 *Phys. Rev.* A **51** 2031

Abramowitz M and Stegun I A 1965 *Handbook of Mathematical Functions* (New York: Dover)

Accad Y, Pekeris C L and Schiff B 1971 *Phys. Rev.* A **4** 516

Arimondo E, Inguscio M and Violino P 1977 *Rev. Mod. Phys.* **49** 31

Armstrong L Jr 1983 *Atomic Physics* vol 8, ed I Lindgren, A Rosén and S Svanberg (New York: Plenum) p 129

Aufmuth P, Heilig K and Steudel A 1987 *At. Data Nucl. Data Tables* **37** 455

Barnett A R, Feng D H, Steed J W and Goldfarb L J B 1974 *Comput. Phys. Commun.* **8** 377

Bastin T, Baudinet-Robinet Y, Garnir H P and Dumont P D 1992 *Z. Phys.* D **24** 343

Baylis W E and Drake G W F 1996 *Atomic, Molecular & Optical Physics Handbook* ed G W F Drake (Woodbury, NY: American Institute of Physics) p 1

Beckmann A, Böklen K D and Elke D 1974 *Z. Phys.* **270** 173

Bell K L, Ramsbottom C A and Hibbert A 1992 *J. Phys. B: At. Mol. Opt. Phys.* **25** 1735

Berestestkii V B, Lifschitz E M and Pitaevskii 1971 *Relativistic Quantum Theory* (Reading, MA: Addison-Wesley)

Berry H G, Desesquelles J and Dufay M 1972 *Phys. Rev.* A **6** 600

Bethe H A and Salpeter E E 1957 *Quantum Mechanics of One- and Two-Electron Atoms* (Berlin: Springer)

Beverini N, Maccioni E, Pereira D, Strumia F, Vissani G and Wang Y 1990 *Optics Communications* **77** 299

Botch B H, Dunning T H and Harrison J F 1981 *J. Chem Phys.* **75** 3466

Blume M and Watson R E *Proc. R. Soc.* A **270** 127

Bohr N 1922 *Z. Phys.* **9** 1

Brage T and Froese Fischer C 1994 *J. Phys. B: At. Mol. Opt. Phys.* **27** 5467

Brage T and Hibbert A 1989 *J. Phys B: At. Mol. Opt. Phys.* **21** 2563

Brage T, Froese Fischer C and Vaeck N 1993 *J. Phys. B: At. Mol. Opt. Phys.* **26** 621

Breit G 1929 *Phys. Rev.* **34** 553

——1930 *Phys. Rev.* **36** 383

——1932 *Phys. Rev.* **39** 616

Brillouin L 1926 *C. R. Acad. Sci. Paris* **183** 1926

——1932 *J. Physique* **3** 373

——1934 *Act. Sci. Ind.* No 159 (Paris: Hermann)

Brink D M and Satchler G R 1968 *Angular Momentum (Oxford Library of the Physical Sciences)* (Oxford: Clarendon) app 6

Burgess A 1963 *Proc. Phys. Soc.* **81** 442

Burke P G 1996 *Atomic, Molecular & Optical Physics Handbook* ed G W F Drake (Woodbury, NY: American Institute of Physics) p 536

Burke P and Seaton M J 1971 *Methods Comput. Phys.* **10** 1

Carlsson J 1988 *Z. Phys.* D **9** 147

Carlsson J, Jönsson P and Froese Fischer C 1992a *Phys. Rev.* A **46** 2420

Carlsson J, Jönsson P, Sturesson L and Froese Fischer C 1992b *Phys. Scr.* **46** 394

Cederquist H and Mannervik S 1985 *Phys. Rev.* A **31** 171

Char B W, Geddes K O, Gonnet G H, Monagan M B and Watt S M 1990 *A Tutorial Introduction to Maple* (Waterloo: Watcom)

Cowan R D 1981 *The Theory of Atomic Structure and Spectra* (Berkeley, CA: University of California Press)

Davidson E R 1975 *J. Comp. Phys.* **17** 87

——1989 *Comput. Phys. Commun.* **53** 49

deShalit A and Feshbach H 1974 *Theoretical Nuclear Physics* vol 1 (New York: Wiley)

Dirac P A M 1928 *Proc. R. Soc.* A **117** 610

——1930 *The Principles of Quantum Mechanics* (Oxford: Oxford University Press)

Edlén B 1964 *Handbuch der Physik* ed S Flügge (Berlin: Springer) p 80

Einstein A 1917 *Physik. Zeit.* **18** 121

Elton L R B 1968 *Nuclear Sizes (Oxford Library of the Physical Sciences)* (Oxford: Clarendon)

Engström L, Denne B, Huldt S, Ekberg J O, Curtis J L, Veje E and Martinson I 1979 *Phys. Scr.* **20** 88

Fano U 1961 *Phys. Rev.* **124** 1866

——1965 *Phys. Rev.* A **140** 67

Fano U and Cooper J W 1968 *Rev. Mod. Phys.* **40** 441

Fleming J, Brage T, Bell K T, Vaeck N, Hibbert A, Godefroid M R and Froese Fischer C 1995 *Ap. J.* **455** 758

Fletcher R 1987 *Practical Methods of Optimization* (New York: Wiley)

Fock V 1930 *Z. Phys.* **61** 126; **62** 795

Fritsche S and Grant I P 1994 *Phys. Scr.* **50** 473

Froese C 1966 *Phys. Rev.* **150** 1

——1971 *Int. J. Quantum Chem.* **4** 95

Froese Fischer C 1971 *Can. J. Phys.* **49** 1205

——1980 *Phys. Rev.* **22** 551

——1986 *Comput. Phys. Rep.* **3** 273

Froese Fischer C, Gaigalas G and Godefroid M 1997 *J. Phys B: At. Mol. Opt. Phys.* submitted

Gaigalas G A and Rudzikas Z B 1996 *J. Phys. B: At. Mol. Opt. Phys.* **29** 3303

Gaunt J A 1929 *Proc. R. Soc.* A **285** 581

Gaupp A, Kuske P and Anrä H J 1982 *Phys. Rev.* A **26** 3351

Gelebart F, Tweed R J and Peresse J 1976 *J. Phys. B: At. Mol. Phys.* **9** 1739

Glass R and Hibbert A 1978 *Comput. Phys. Commun.* **16** 19

Godefroid M R, Froese Fischer C and Jönsson P 1996 *Phys. Scr.* T **65** 70

Grant I P 1961 *Proc. R. Soc.* A **262** 555

——1988 *Meth. Comp. Chem.* **2** 1

——1996 *Atomic, Molecular & Optical Physics Handbook* ed G W F Drake (Woodbury, NY: American Institute of Physics) p 258

Greiner W 1990 *Relativistic Quantum Mechanics* (Berlin: Springer)

Hartree D R 1927 *Proc. Camb. Phil. Soc.* **24** 89 and 111

——1957 *The Calculation of Atomic Structures* (New York: Wiley)

Hibbert A 1975 *Rep. Prog. Phys.* **38** 1217

——1979 *J. Phys. B: At. Mol. Phys.* **12** 22

Hibbert A, Froese Fischer C and Godefroid M 1988 *Comput. Phys. Commun.* **51** 285

Hicks P J and Comer J 1975 *J. Phys. B: At. Mol. Phys.* **8** 1866

Holmström J-E and Johansson L 1969 *Ark. Fys.* **40** 113

Hughes D S and Eckart C 1930 *Phys. Rev.* **36** 694

Hylleraas E and Undheim B 1930 *Z. Phys.* **65** 759

Jackson J D 1962 *Classical Electrodynamics* (New York: Wiley)

Johansson L 1962 *Ark. Fys.* **23** 119

Johnson W R 1983 *Atomic Physics* vol 8, ed I Lindgren, A Rosén and S Svanberg (New York: Plenum) p 149

Johnson W R, Kolb D and Huang K-N 1982 *At. Data Nucl. Data Tables* **28** 333

——1983 *At. Data Nucl. Data Tables* **28** 333

Jönsson P, Johansson S G and Froese Fischer C 1994 *Ap. J.* **429** L45

Jönsson P, Ynnerman A, Froese Fischer C, Godefroid M R and Olsen J 1996 *Phys. Rev. A* **53** 4021

Judd B R 1996 *Atomic, Molecular & Optical Physics Handbook* ed G W F Drake (Woodbury, NY: American Institute of Physics) p 88

King W H 1984 *Isotope Shifts in Atomic Spectra* (New York: Plenum)

Klein H 1985 *J. Phys. B: At. Mol. Phys.* **18** 1483

Koopmans T A 1933 *Physica* **1** 104

Kramers H A 1926 *Z. Phys.* **39** 828

Kuhn W 1925 *Z. Phys.* **33** 408

Kwong V H S, Fang Z, Gibbons T T, Parkinson W H and Smith P J 1993 *Ap. J.* **441** 431

Larsson S 1970 *Phys. Rev. A* **2** 1248

Landau L D and Lifschitz E 1965 *Quantum Mechanics* (Reading, MA: Addison-Wesley)

Layzer D 1959 *Ann. Phys.* **8** 271

Layzer D, Horak Z, Lewis M N and Thompson D P 1964 *Ann. Phys* **29** 101

Lindgren I and Rosén A 1974 *Case Studies in Atomic Physics* **4** 151

Löwdin P-O 1955 *Phys. Rev.* **97** 1509

MacDonald J K L 1933 *Phys. Rev.* **43** 830

McAlexander W I, Abraham E R I and Hulet R G 1996 *Phys. Rev. A* **54** R5

McIlrath T J and Lucatorto T B 1977 *Phys. Rev. Lett.* **38** 1390

Martin W C 1987 *Phys. Rev. A* **36** 3575

Martin W C and Wiese W L 1996 *Atomic, Molecular & Optical Physics Handbook* ed G W F Drake (Woodbury, NY: American Institute of Physics) p 135

Messiah A 1965 *Quantum Mechanics* vols I and II (Amsterdam: North-Holland)

Moore C E 1949 *Atomic Energy Levels (Circl. National Bureau of Standards 467)* (Washington, DC: US Government Printing Office)

Nielson C W and Koster G F 1963 *Spectroscopic Coefficients for p^n, d^n and f^n Configurations* (Cambridge, MA: MIT Press)

Norcross D W and Seaton M J 1976 *J. Phys. B: At. Mol. Phys.* **9** 2983

Oates C, Vogel K and Hall J L 1996 *Phys. Rev. Lett.* **76** 2866

Olme A 1970 *Phys. Scr.* **1** 256

Orth H, Aackermann H and Otten E W 1975 *Z. Phys.* A **273** 221

Pauli W 1924 *Naturwissenschaften* **12** 74

——1925 *Z. Phys.* **31** 765

Racah G 1943 *Phys. Rev.* **63** 367

Raghavan P 1989 *At. Data Nucl. Data Tables* **42** 189

Reistad N, Brage T, Ekberg J O and Engström L 1984 *Phys. Scr.* **30** 249

Robb W D 1973 *Comput. Phys. Commun.* **6** 132

Rotenberg M, Bivins R, Metropolis N and Wooten J K Jr 1959 *The 3-j and 6-j symbols* (Cambridge, MA: MIT Press)

Sakurai J J 1967 *Advanced Quantum Mechanics* (Reading, MA: Addison-Wesley)

Schiff L I 1968 *Quantum Mechanics* McGraw-Hill international student edn (Tokyo: McGraw-Hill–Kogakusha)

Schüler H 1927 *Z. Phys.* **42** 487

Schwartz C 1963 *Methods in Computational Physics* vol 2, ed B Alder, S Fernbach and M Rotenburg (New York: Academic) p 241

Slater J C 1960 *Quantum Theory of Atomic Structure* (New York: McGraw-Hill)

Smith B T, Boyle J M, Garbow B S, Ikebe Y, Klema V C and Moler C B 1974 *Matrix Eigensystem Routines—EISPACK Guide* (Berlin: Springer)

Sobelman I I 1979 *Atomic Spectra and Radiative Transitions (Springer Series in Chemical Physics 1)* (Berlin: Springer)

Sugar J and Musgrove A 1993 *J. Phys. Chem. Ref. Data* **22** 1213

Sundholm D and Olsen J 1992 *Phys. Rev. Lett.* **68** 927

Thomas W and Reiche F 1925 *Naturwissenschaften* **13** 627

Uhlenbeck G E and S Goudsmit 1925 *Naturwissenshaften* **13** 953

Vinti J P 1940 *Phys. Rev.* **56** 1120

Volz U, Majerus M, Liebel H, Schmitt A and Schmoranzer H 1996 *Phys. Rev. Lett.* **76** 2862

Wapstra A H and Audi G 1985 *Nucl Phys.* A **432** 1

Wentzel G 1926 *Z. Phys.* **38** 518

Worsley B H 1958 *Can. J. Phys.* **36** 289

Zatsarinny O 1996 *Comput. Phys. Commun.* **98** 235

Index

active set, AS, 71
allowed transitions, 179, 189, 191
antisymmetry, 2
autoionization, 218
  decay rate, 230
  autoionizing width, 230
average energy of configuration, 31

Blume–Watson spin–orbit, 139
branching ratio, 181, 210
BREIT, 136, 138
Breit–Pauli, 129, 130
  radial integrals, 135
Brillouin's theorem, 53

central field, 8
CI, 137
close-coupling equations, 220
Coulomb functions, 219
coefficients of fractional parentage, 23
complete active space, CAS, 73
complex, 69
component strength, 180
configuration, 10
  interaction, CI, 16
  state functions, 12, 24
continuum
  functions, 6, 218
  HF, 220
  oscillator strength, 225
  processes, 217
correlation, 67

core–core, 71
core–valence, 71
core-polarization, 141

Darwin term
  one-body, 130
  two-body, 130

energy parameters, 52
  diagonal, 84
exchange potential, 38
exchange terms, 17

Fermi contact, 166
field shift, 157
fine-structure, 130
  levels, 132
  operators, 130
forbidden transitions, 179, 209

generalized Brillouin theorem, GBT, 85
GENCL, 27, 28, 72, 142
gf-value, 181
golden rule, 230

Hartree equations, 17
Hartree–Fock method, 17
Hartree–Fock, 35
hyperfine
  structure, 165
  electric quadrupole, 166
  polarization effects, 171

277